CONSTRUCTION LAW FOR MANAGERS, ARCHITECTS, AND ENGINEERS

Construction Law for Managers, Architects, and Engineers

Nancy J. White

THOMSON

DELMAR LEARNING

Australia Canada Mexico Singapore Spain United Kingdom United States

Construction Law for Managers, Architects, and Engineers

Nancy J. White

Vice President, Technology Professional Business Unit:
Gregory L. Clayton

Product Development Manager:
Ed Francis

Product Manager:
Stephanie Kelly

Editorial Assistant:
Nobina Chakraborti

Director of Marketing:
Beth A. Lutz

Executive Marketing Manager:
Taryn Zlatin

Marketing Specialist:
Marissa Maiella

Director of Production:
Patty Stephan

Production Manager:
Andrew Crouth

Content Project Manager:
Kara A. DiCaterino

Art Director:
Marissa Falco

Library of Congress Cataloging-in-Publication Data:
White, Nancy J.
  Construction law for managers, architects, and engineers / Nancy J. White.
      p. cm.
   Includes index.
   ISBN-13: 978-1-4180-4847-1
   ISBN-10: 1-4180-4847-X
1. Construction contracts--United States. 2. Construction industry--Law and legislation--United States. I. Title.

KF902.W46 2008
343.73'078624--dc22

2007045732

ISBN-10: 1-4180-4847-X
ISBN-13: 978-1-4180-4847-1

NOTICE TO THE READER

Dedication

Dedicated to my parents, "Whitey" (Floyd) and Shirley White, who wondered why a girl wanted to go to college but were proud of me when I did.

TABLE OF CONTENTS

About the Author

Nancy J. White, JD, is an Associate Professor in the Department of Finance and Law at Central Michigan University. She has been teaching the basics of construction law for over 10 years. She received her juris doctorate from Loyola Law School, Los Angeles, California and practiced business law for over 10 years before turning to teaching.

To the Reader

None of the information in this book can be used as a substitute for professional legal consultation with an attorney with knowledge of the law of the reader's state or the state where the project will be built. All information in this book is general in nature, and the laws of any particular state may vary greatly from the information here.

All forms are for informational purposes only. Do not use any forms without the advice of a lawyer with knowledge of the jurisdiction in which those forms will be used.

This book is intended as a reference, and it is expected that the reader will only read short portions of the book for specific information. For this reason, material and examples may be repeated. This is done so that the reader need not refer back to prior pages in the book. If extensive material, previously discussed, is necessary to understand the current topic or is closely related to the current topic, it is referenced.

CHAPTER 1

Introduction to Government and Dispute Resolution Processes in the United States

■ Overview

This chapter contains a basic overview of the legal systems and dispute resolution processes in use in the United States. As such, it, or parts of it, may be too basic for many and is included here for reference and background. Sometimes it is difficult to understand why a case is handled in a certain manner without understanding the difference between, for example, how an issue of law as compared to an issue of fact is handled by the court system. This chapter includes a description of the various government entities, the types of law, classification of the law, and dispute resolution processes, including a basic overview of the litigation process.

■ Federal, State, and Tribal Governments

In order to understand the law applicable to any construction project, it is necessary to have a basic understanding of the structure and interrelationships between the various governmental entities in operation in the United States: the federal government, the 50 state governments, and the tribal governments. Essentially each government has its own job and operates independently of the other governments. Some overlap does exist.

The state and tribal governments operate within a particular geographic area or jurisdiction. The term **jurisdiction** is often used in the law to refer to a particular geographic area over which a particular government has power. Each state or tribe is a different and separate jurisdiction, as is the federal government (Figure 1-1).

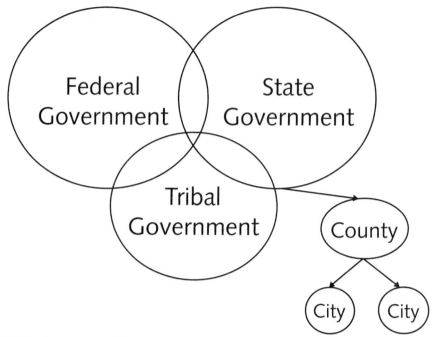

FIGURE 1-1 Governments in Operation in the United States

Branches of Government

Federal and state governments are divided into three parts or branches: legislative, executive, and judicial. Administrative agencies have increased in importance in recent decades such that they are depicted in Figure 1-2 as being a separate part of the government. Actually all administrative agencies are created by one of the other branches of the government to perform some particular purpose such as collecting federal income taxes, a function of the Internal Revenue Service, or monitoring the environment, a function of the Environmental Protection Agency.

Each branch of the government and each administrative agency has a specific job to do. The legislative branch makes most of the laws, called statutes, effecting the population. The executive branch enforces the laws made by the legislative branch and administers many of the laws, usually through administrative agencies. The judicial branch interprets and clarifies the law and also provides dispute resolution services. Administrative agencies are charged with some specific duty (Figure 1-2). Some tribal governments combine the executive and legislative branches into one tribal council headed by the chief.

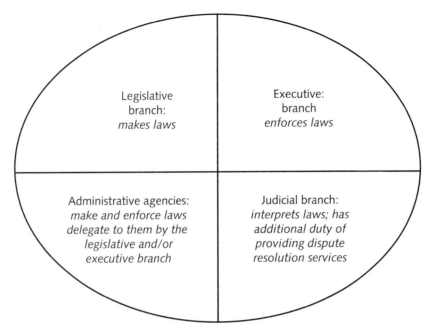

FIGURE 1-2 Four Branches of a Typical Government

Federal Government

The federal or central government of the United States was formed by the original 13 states in the late 1700s and went into operation in 1790. According to the United States Constitution, the federal government has limited powers, and most governmental power was to be kept at the state level. For various historical and cultural reasons, governmental power has shifted away from the state governments and to the federal government. However, state governments still control many aspects of the construction industry. Major federal laws that impact the construction industry and that are discussed in this book include labor and employment law, environmental law, and bankruptcy law. State law controls contract and tort law. In addition, subdivisions of a state, such as a county or city, will have building codes applicable to those areas.

State Government

Every state has its own laws and governs within its geographical boundaries. State governments establish counties, towns, cities, townships, and other similar entitles, and they delegate certain powers to those entities. Of particular importance to the construction industry are county governments, which often set construction standards and/or building codes. Because of the problems with several different and sometimes conflicting county building codes, many states have adopted or are in the process of adopting uniform, statewide building codes.

Tribal Governments

In recent decades, tribal governments have grown in importance and power. Unlike state governments, which have independent existence apart from the federal government, tribal governments exist at the sufferance of the federal government and are ultimately under the control of the federal government. However, for general application in the construction industry, a tribal government is a separate government, and tribal laws are applicable within the reservation. Many tribes have not waived their sovereign immunity, meaning that no entity can sue the tribe without its permission. Sovereign immunity is discussed below.

Individual tribal members may enter into construction contracts on Indian lands with non-Indian entities. If the non-Indian entity wishes to sue the tribal member, it must do so in the tribal court and the matter is subject to tribal law.[1]

If the tribal member wishes to sue the non-Indian, the law is more complex. In some tribes, tribal law states that tribal courts only have jurisdiction, or the power to hear a case, over non-Indians if the non-Indian agrees. Assuming the non-Indian does not agree to being sued in the tribal court, the tribal member must sue in state court.

If the tribal law does not limit the tribal court's jurisdiction to voluntary non-Indian defendants only, the tribal member may sue the non-Indian in either tribal or state court.[2]

Sovereign Immunity

Sovereign immunity is an old, common-law doctrine that originally held that no one could sue the king or sovereign. This doctrine has been interpreted to mean that no one can sue a government without the government's permission.[3]

Most states and the federal government have waived sovereign immunity in whole or in part. Few tribal governments have done so.

MANAGEMENT TIP

Before entering into a contract with a government entity, a business should know the extent to which the government entity has waived sovereign immunity. If possible, negotiate a waiver of sovereign immunity in the contract.

■ Sources of Law and Hierarchy of Law

The fundamental source of law in a federal, state, or tribal government is the constitution. A **constitution** establishes a government and outlines what it can and cannot do. Other sources of law in a government are the branches of the government and the administrative agencies. Table 1-1 lists the branches of a typical government that are the source of law, the name of the law made by that branch, and the general purpose of that law. The law in Table 1-1 is also listed in hierarchical order. That is, if laws conflict, the law higher on the list controls. In addition, a law higher on the list can modify a law lower on the list.

TABLE 1-1 Sources of Law in Hierarchical Order

Source of law	Name of law coming from this source	Purpose(s) of this law	Example
Constitution and cases interpreting a constitution; federal, state, and many tribal governments are formed by a constitution.	Constitutional law	Outline the form and operation of the government; establish fundamental legal principles; protect fundamental rights.	"Congress shall make no law respecting an establishment of religion..." First Amendment to the U.S. Constitution "Equality under the law shall not be denied or abridged because of sex, race, color, creed, or national origin." Texas Constitution, Art. 1, Sec. 3a "The constitution is either a superior, paramount law, unchangeable by ordinary means, or it is on a level with ordinary legislative acts, and, like other acts, is alterable when the legislature shall please to alter it... Thus, the particular phraseology of the constitution of the United States confirms and strengthens the principle, supposed to be essential to all written constitutions, that a law repugnant to the constitution is void. [And no court need uphold that law]..." *Marbury v. Madison*, U.S. Supreme Court, 1803.
Executive branch	Executive orders	Various orders to administer government operations and projects under the control of the executive branch	Executive order of President Clinton, later rescinded by President Bush. In certain federal government contracts, if a new contractor replaces old, new contractor must hire old contractor's employees, if employees want to stay. Example: cleaning contract
Legislative branch	Statute, legislation	Day-to-day running of the government	Americans with Disabilities Act
Administrative Agency	Regulations	Enforce a task given to the agency by the legislature	Each employer (1) shall furnish to each of his employees employment and a place of employment which are free from recognized hazards that are causing or are likely to cause death or serious physical harm to his employees... (OSHA Sec. 5).
Judicial branch a.k.a. courts	Case law a.k.a. judge-made law a.k.a. common law	Clarify any of the above laws if requested to do so by a party filing a lawsuit	Architects have no duty to contractors or subcontractors and therefore cannot be sued by them for negligence. *Amazon v. British Am. Dev. Corp.*, 216 A.D.2d 702, 628 N.Y. S.2d 204 (1995).

EXAMPLE

The court in a certain state held that architects have no legally recognized duty to contractors and therefore cannot be sued by contractors for negligence. The state legislature passed a statute saying architects can be sued for negligence by contractors. The statute controls the case law, because the statute is higher on the list.

EXAMPLE

The U.S. Congress passes a statute making it illegal to burn the American flag. The U.S. Supreme Court nullifies the statute because it violates the U.S. Constitution's First Amendment granting the right to freedom of speech. The case controls over the statute because the matter is one of constitutional law.

Statutes

Statutes are laws passed by a legislative body. An **act** is the name given to several related statutes passed at one time, such as the Social Security Act. A **code** is a collection of related statutes.

EXAMPLE

The following are some titles of federal codes. Each code contains statutes related to the title of that code:

Title 11: Bankruptcy
Title 15: Commerce and Trade
Title 16: Conservation
Title 17: Copyrights
Title 23: Highways

EXAMPLE

The following are some of the titles of California codes. Each code contains the statutes related to the topic of that code:

    Business and Professions Code
    Civil Code
    Code of Civil Procedure
    Commercial Code
    Corporations Code
    Evidence Code
    Insurance Code
    Public Contract Code

Regulations

Regulations are laws enacted by administrative agencies such the Environmental Protection Agency (EPA), an agency of the federal government, and the Department of Transportation (DOT), an agency of a state government. Many regulations impact the construction industry. The Occupational Health and Safety Administration (OSHA) has enacted thousands of regulations relating to safety on job sites. The EPA has enacted regulations to protect the environment.

Judge-Made Law, Case Law, Common Law

In the event that parties disagree about the meaning of a constitution, statute, or regulation, a lawsuit can be filed asking a judge to clarify the meaning of that constitution, statute, or regulation. Many jurisdictions also require that some actual injury have occurred; that is, no lawsuit can be filed merely to clarify a constitution, statute, or regulation. When a judge clarifies a constitution, statute, or regulation, the judge makes a law and the name of that law made by the judge has several names: **judge-made law, case law, or common law.** Some jurisdictions have adopted other processes to clarify constitutions, statutes, and regulations such as an attorney general's opinions or an administrative agency's letters. These interpretations are not binding on judges, however, should the law under review by the attorney general or agency come to the court.

EXAMPLE

Texas Civil Practice Code §16.009 states: "A person must bring suit for damages for a claim...against a person who constructs or repairs an improvement to real property not later than 10 years after the substantial completion of the improvement..." This is a statute of repose. In a series of cases, the Texas courts wrestled with the issue of whether or not this statute prevented lawsuits against *manufacturers* of items such as elevators and heaters after 10 years.

Eventually the Texas State Supreme Court resolved the issue in the case of *Sonnier v. Chisholm-Ryder Co., Inc.*, 909 S.W.2d 475 (Tex. 1995) and held that §16.009 only prevents lawsuits against contractors and does not prevent lawsuits against manufacturers. All of the courts were clarifying the meaning of the statute. The Texas Supreme Court made the final clarification.

Maxims of Law and Basic Premises of Contract Law

All law in the United States can be summarized in the following three premises:

- Do not take property not belonging to you.
- Uphold contracts entered into.
- Be reasonable.

The maxims of law and the basic premises of contract law are fundamental legal principles upon which much of the judge-made law is based. Some states have adopted the maxims in statute; some recognize them only in the case law.[4]

The maxims of law[5] most applicable in the construction industry are as follows:

- *Mistakes should be fixed, not taken advantage of.* Many disputes in the construction industry could be avoided if the parties understood this maxim.
- *The law ultimately controls, not the contract.*
- *The law never requires impossibilities.* This very important maxim is applicable to the construction industry. See the section on "Objective Impossibility."
- *Parties must come to court with clean hands.* Known as the "clean hands doctrine," this maxim states it is unlikely that a party who has treated the other party unfairly or dishonestly will get a favorable ruling from the court.
- *The law does not require useless or idle acts.*
- *The law abhors waste.*
- *Liability follows responsibility.*
- *For every wrong, there is a remedy.*
- *No one should suffer by the act of another.*
- *Substance, not form, controls.* In other words, fairness and justice are important, not technicalities.
- *The one who takes the benefit must bear the burden.*

The Basic Premises of Contract Law are as follows:

- *A party must honor its contract or respond in reasonable damages.* American law does not require any contract to be performed, and it is not illegal to fail to perform a contract. It is only illegal to fail to pay reasonable damages.
- *Parties are presumed to know the contents of their contracts.*

■ Classifications of Law

Law is generally organized by topic to aid in finding and understanding it. Law can be divided into two broad categories: criminal law and civil law. Civil law is further divided into tort and contract law.

Criminal Law

Criminal laws are adopted by societies to protect the society as a whole rather than any particular individual. People who commit crimes are considered to have harmed society as a whole and not just an individual member of that society. For this reason, all criminal matters are under the control of the government, usually through an agency called the district attorney or similar name. The victim of a crime cannot bring a criminal action, the victim can only make a criminal complaint. The district attorney decides whether or not to prosecute the matter. The injured party or victim of a crime is a witness in the government's case against the alleged criminal. Punishments for crimes can include jail time, prison time, fines, and/or community service.

It has become more common in recent decades for criminal courts to award some damages to a victim of a crime; however, these are usually much less than the amount awarded in a civil matter, as discussed below. Of course, it is usually difficult, if not impossible, to collect damages from a person in jail, and the victim of a crime usually has no way of obtaining any compensation for her injuries.

The crimes of embezzlement and theft are most likely to occur on a construction project. **Embezzlement** is generally defined as the misappropriation of the property of a principal or employer. **Theft or larceny** is the wrongful taking of the property of another.

Civil Law

Civil law is the broad area of law dealing with all noncriminal disputes. Civil law is frequently divided into the two broad categories of tort and contract law. **Tort law** is accident or injury law, and the most common tort is negligence. **Contract law** is the law that governs the enforceability of contracts.

Tort (Injury) Law

The law imposes certain duties or obligations upon entities dealing with each other in a society. A violation of one of these legally imposed duties is a tort and the law can require the **tortfeasor,** the person who violated the duty, to compensate the injured party. The most common tort is negligence, which is the duty required by law, to act reasonably in interactions with others. An unreasonable act causing harm to another may result in liability. An unreasonable act is anything a jury, or a judge if a jury is waived, says is unreasonable given the circumstances of the case. Torts are discussed in Chapter 12.

EXAMPLE

Don, the roofing subcontractor, places a tarp over a hole in the floor on the second story of a construction project. Tammy, an electrical subcontractor, walks across the tarp, falls through the hole, and is injured. Tammy cannot sue Don for breach of contract, because there is no contract between them. She should sue Don for negligence, a tort.

Contract Law

The other broad category of civil law is contract law. Although the law upholds contracts, a contract duty does not enjoy the same importance under the law as a tort duty. A tort duty is, after all, a duty established by the law, and the law puts great importance upon the fulfillment of that duty. A contract duty, on the other hand, is merely a duty established by two people or businesses; it is not established by the law, and therefore the law may or may not uphold the contract or any part of the contract. In other words, the law does not enforce a contract duty with the same vigor and zeal as a tort duty. An understanding of this principle will avoid many contract disputes, because the law may not place as much importance upon the terms of a contract as a party to the contract thinks it should.

EXAMPLE

A contract between the parties for construction of a small strip mall imposes $10 million per day in liquidated damages. The law will not enforce this clause. See Chapter 9.

EXAMPLE

A contract between the parties states the contractor cannot rescind its bid for any reason. The law will allow the contractor to rescind its bid if it contains a major mistake and work has not yet begun. See Chapter 2.

EXAMPLE

A contract between the parties states all change orders must be in writing. The law will enforce verbal change orders in some circumstances. See Chapter 8.

■ Dispute Resolution

Even though relatively few disputes in the construction industry are resolved in a court, the power of the law as applied by the strong judicial branches of governments greatly influences how entities elect to handle their disputes in the United States. Because the judicial branch is powerful, respected, and relatively accessible, entities may use it rather than other informal means to resolve disputes. Although not universally true, Americans tend to believe that the law, rather than the wealth or influence of any entity, should control the outcomes of a dispute; this is known as the **rule of law.** The government structures in the United States are based on the rule of law.

Cultural aspects of American society also support the use of the judicial system rather than informal methods of dispute resolution. Americans tend to desire to signal strength to an opponent rather than signal cooperation. Willingness to cooperate is often viewed by Americans as weakness. One method of signaling strength to an opponent in a dispute is to hire lawyers to handle the dispute. Because lawyers are trained to use the judicial system to resolve disputes, and because lawyers make the most money when the judicial system is used to resolve disputes, the judicial system is more often the starting point for the resolution of disputes in the United States than in other countries.[6]

This section discusses various methods of dispute resolution. The construction industry has been a leader in using alternative forms of dispute resolution such as arbitration and mediation. The methods discussed here are negotiation, when no lawyers are involved, negotiation, when lawyers are involved, contractually required arbitration, contractually required mediation, court-ordered arbitration, court-ordered mediation, and litigation.

Negotiation—No Lawyers Involved

Negotiation, when no lawyers are involved, is the attempt to resolve a dispute without outside or professional help; the parties discuss the dispute and come to a resolution. The resolution usually requires each party to give up some of what she desires. The advantages of negotiation are flexibility and low cost. One party to the negotiation may believe she has been taken advantage of because of her lack of power, knowledge, and[or assertiveness, leading to unwillingness to engage in further business or negotiation with the other party. That is, the relationship of the parties may be destroyed.

EXAMPLE

A dispute arises between the architect and the contractor about which of them will pay for cabinets that do not fit. The architect and the contractor meet to discuss the matter. If the parties desire to maintain a good relationship with each other, they are likely to cooperate and reach a mutually agreeable solution. If one or both of the parties is very assertive in the pursuit of her own goals, the relationship may break down, causing problems with the project.

Negotiation—Lawyers Involved

When an entity decides to hire an attorney to aid in the negation of a dispute, the entity should realize the potential ramifications this could have on the dispute. The entity should be clear on what its goals are in hiring the attorney. Generally two goals exist: to enforce the legal rights of the hiring entity and/or to resolve a dispute.

Both of these goals can be achieved through the legal system; however, it is usually extremely expensive for the hiring entity to achieve both goals. Unfortunately, many inexperienced entities have the misunderstanding that enforcing rights should be like breathing: free. This is not the case. Enforcing rights is an extremely expensive goal. Be sure you are willing to pay for that goal before pursuing it. It is often necessary to hire an attorney to achieve the goal of enforcing rights, and attorneys are trained to attain that goal for their clients. Attorneys also charge high fees to attain that goal. If an entity wants its legal rights enforced, it should realize the potential costs of achieving that goal. The costs are not limited to the huge attorney fees that can accrue but also include such items as time spent by employees in the litigation, damage to any ongoing relationship, and damage to the reputation of the businesses.

If the primary goal is to resolve a dispute or maintain an ongoing relationship, then care must be taken to hire an attorney who is capable of and comfortable with the goal of dispute resolution. Some attorneys are not trained in dispute resolution or do not have the personality to resolve disputes without litigation.

Many entities hire a law firm to resolve disputes. This has the advantage of allowing two or more attorneys to be assigned the matter. One attorney is the "good cop" who attempts to resolve the dispute, and the other is the "bad cop" who threatens to litigate the matter "all the way up to the Supreme Court." This avoids the appearance of weakness that a single, cooperative attorney may give to an opponent.

EXAMPLE

Ace entered into a contract with the U.S. Army Corps of Engineers to construct a runway and related improvements at an Army airfield. The parties disagree on which of two different testing methodologies can be used to determine if the finished concrete is sufficiently smooth and level. The Army wants the contractor to use the profilograph testing method, which is more expensive. The contractor wants to use the straightedge method.[7]

Attorneys will tend to approach the resolution of this dispute by looking at the contract and the law. If the goal is to resolve the dispute quickly, but attorneys are hired to aid in the resolution of the dispute, it is important to hire attorneys who are comfortable with a primary goal of dispute resolution rather than enforcement of rights.

Depending on the specific wording of the contract, which is not given here, the law relating to this issue is found in Chapter 8.

Arbitration—Contractually Required

Arbitration is a nongovernment dispute resolution process through which the parties present their dispute to a person or panel specifically hired by the parties to resolve the dispute. The arbitration process resembles litigation but is less formal. The arbitrator(s) hears the dispute and decides who wins. Parties often hire attorneys to prepare and present their cases to the arbitrators. The decision of an arbitrator is called an **award** and can be enforced by filing a copy of the award in a court and following the government-provided enforcement of judgment procedures. Enforcement of judgment is discussed below.

Many construction contracts contain mandatory binding arbitration clauses. This clause closes the doors of the court to the parties agreeing to arbitration.

The advantages of arbitration are as follows:

- Less costly both in terms of time and money. Arbitrations are usually faster than a trial, because they are less formal and no jury is involved.
- No jury is involved. This lessens the possibility that a decision will be made on an emotional response to the evidence.
- Arbitration awards cannot be appealed. The arbitrator's decision is final, and the losing party has no further recourse.
- The process is private. Arbitrations are not open to the public.
- Arbitrators with knowledge of the construction industry can be hired to resolve the dispute. For example, an architect, engineer, and[or contractor can be an arbitrator.

Drawbacks to arbitration include the following:

- Arbitrators may not be familiar with construction law or any law. This could result in a legally incorrect decision.
- Courts seldom overturn arbitration decisions. Arbitration decisions are not reviewed for accuracy by courts.

Mediation—Contractually Required

In **mediation,** a person, called a mediator, is hired to *help* the parties resolve a dispute. The mediator does not decide who wins and does not apply the law to the matter. The mediator employs different techniques designed to help the parties come to a mutually agreeable settlement. Mediation is not to be confused with arbitration. See Table 1-2 for a comparison between the two processes.

TABLE 1-2 Comparison of Arbitration and Mediation

Arbitration	Mediation
Arbitrator decides who wins the dispute	Mediator helps the parties settle the dispute; seldom does either party win the dispute
Parties are told how the dispute will be settled	Parties settle the dispute
Legal and factual issues are important	Legal and factual issues of little importance; settlement of the claim of primary importance
Private	Private

TABLE 1-2 Comparison of Arbitration and Mediation (continued)

Arbitration	Mediation
Arbitrator issues an award that may be enforced through the government's enforcement of judgment process	If the parties settle the dispute, a contract is normally drawn up at that time; the contract can be enforced through a breach-of-contract lawsuit
No appeal; the decision is final	If parties do not settle, they may proceed with arbitration or litigation
Similar to litigation	Similar to negotiation

Parties to any dispute can seek the help of a mediator at any time. Many counties provide free or low-cost mediation services. Some contracts are now requiring mediation before the filing of arbitration or a lawsuit.

Mediators meet with all of the parties and their attorneys, if any. Each side outlines the claim as they see it. The attorneys play less of a role in mediation than other forms of dispute resolution. The mediator requires the parties to listen to the opposing side or argument without interrupting. Often, this is the first time each side has looked at the matter from the viewpoint of the other.

It is not uncommon for the mediators to separate the parties at some point and speak to them individually. This process of separating the parties is called a **caucus** and is never used during litigation or arbitration. The mediator encourages the parties to generate novel ways of resolving the dispute and to compromise. Sometimes solutions suggested by a mediator are more palatable to the parties, and the mediator may go back and forth between the parties with offers and counteroffers.

If a settlement is reached, the mediator will strongly encourage the parties to prepare a settlement agreement immediately and sign it. If the settlement agreement is not prepared and signed immediately, it is common for the parties or their attorneys to add new and/or different terms, which may destroy the settlement.

The settlement agreement is a valid contract. If either party fails to uphold the settlement agreement, it can be enforced as any other contract and taken to court. Mediated settlement agreements are seldom litigated as the parties have mutually agreed to the terms.

If the parties fail to settle a matter at mediation, they may schedule additional mediation sessions, engage in negotiation, proceed to arbitration, or proceed to litigation. Contract provisions may control subsequent procedures.

Court-Ordered Arbitration and Mediation

Because private arbitration and mediation have proved so effective, many courts have incorporated these into the traditional litigation process in an attempt to settle cases early and keep costs down. Today, many courts will require the parties to engage in some type of arbitration or mediation before trial. Because the arbitration or mediation is court-ordered rather than voluntary, parties may not engage in the process in good faith. Parties may see it merely as a waste of time or as a means of obtaining information for use in a trial.

If the parties do not resolve their disputes through the court-ordered arbitration or mediation, the case will proceed through the litigation process. Judges cannot require the parties to settle the matter at arbitration or mediation because of the due process requirements of the U.S. Constitution.

EXAMPLE

An owner and a contractor have entered into a contract that does not contain a mandatory arbitration or mediation clause. A dispute arises as to the cost of a change ordered by the owner. The parties cannot come to a resolution, and the contractor files a lawsuit against the owner. The judge reviews the case and orders the parties to court-ordered mediation. The parties do not reach a settlement, and the matter proceeds to trial.

Litigation

Litigation is the government-provided dispute resolution system and as such is complex, time-consuming, and costly, both to the parties involved and the taxpayers who pay for the process. Most parties in the construction industry who engage in litigation use lawyers, although this is not required. The litigation process is generally open to the public. All hearings, trials, and even the documents filed in the case are subject to public scrutiny.

In the United States, the court system operates under the **adversary system of justice**. Under this system, the goal of each side is to win within the bounds of the law and legal ethics. The theory behind the adversary system of justice is that through this battle to win, truth and justice will emerge, and the judge and[or jury can make the right decisions. Under this system, the lawyers have a great deal of control over the litigation process. The judge and jury are, for the most part, observers and make decisions based upon the evidence and legal arguments made by the lawyers.

Federal, State, and Tribal Court Organization

The judicial branch of most governments is divided into three levels or stages: trial courts, appeal courts, and a supreme court (Figure 1-3). All lawsuits begin by filing in a trial court. The parties are entitled to a review of the trial court's decision in an appeal court. The parties may request a further review in the supreme court, but the supreme court usually has discretion on whether or not to grant an appeal to that level. Each government has a different name for its trial, appeal, and possibly supreme court. Some of the names of trial courts used in different states are as follows:

District court

County court

Superior court

Supreme court

States usually organize the trial court level into several different parts with different parts taking certain types of cases. For example, cases valued at less than $25,000 may be filed and heard in municipal court, and cases valued at more than $25,000 may be filed and heard in a superior court. Other cases may be heard in a small claims court, family court, or probate court. Each of these is a trial court.

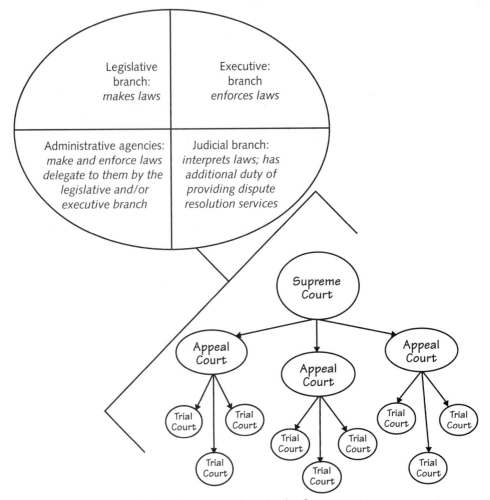

FIGURE 1-3 Typical Organization of a Judicial Branch of a Government

Federal, State, and Tribal Court Jurisdiction (Power)

Jurisdiction is the power or ability of a court to hear a case. All courts in the United States have limits on the types of cases that may be brought before them. Most construction-related litigation will be heard in state courts unless the U.S. government is a party to the contract. In that event, the litigation will be heard in the Court of Federal Claims. It is unlikely, although possible, for a construction case to be heard in a federal court.

State Court Jurisdiction

State courts have personal jurisdiction or power over any defendant who is a resident of or doing business in that state. States have limited power over residents or businesses of other states but can hear such cases if the out-of-state defendant is doing business in the state or owns property in the state.

EXAMPLE

A designer is incorporated in New York and has its principle place of business there. The designer is licensed in New York and New Jersey. The designer enters into a contract in New York to be performed in New Jersey. The owner, who is a resident of Connecticut, claims that the designer breached the contract. The owner can sue in either New York or New Jersey.[8]

The domicile of the plaintiff is irrelevant; state courts are, for the most part open to any who wish to file a lawsuit. Some exceptions applicable to the construction industry exist. If the state requires licensing, such as an engineering license or general contractor's license, most state courts will not allow unlicensed persons or businesses to file a lawsuit. Certain types of cases, such as bankruptcy and patent cases, must be filed in federal courts.

EXAMPLE

An unlicensed contractor builds a room addition onto a house, and the owner does not pay the contractor. The contractor files a lawsuit against the owner to collect the amount owed. The judge dismisses the lawsuit, because the contractor is not licensed.

Some states will allow the contractor to continue a suit that has not yet been dismissed if the contractor obtains a license before the case is dismissed. However, some states require that the license be in the contractor's possession at the time the work is completed and will not allow the contractor to file suit even if a license is later obtained.

Federal Court of Claims   If the contractor has a contract with the federal government, any disputes must be filed in the Federal Court of Claims. If subcontractors are involved, those lawsuits must be filed in the state court. This can cause the litigation to become extremely complex.

EXAMPLE

A contractor licensed and doing business in Colorado enters into a contract with the federal government to build a federal building in Colorado. The contractor enters into subcontracts with Colorado subcontractors. A dispute arises between the owner, the contractor, and a subcontractor. The dispute between the contractor and the federal government must be filed in the Court of Federal Claims. The dispute between the contractor and the subcontractor must be filed in the Colorado state court.

Jurisdiction of Federal District Courts   The federal district courts can hear the following types of cases:

- Federal subject matter jurisdiction: Any matter involving federal law such as the Americans with Disabilities Act, federal environmental laws, or federal labor laws.

EXAMPLE

An employee of a designer claims that Design Corporation discriminated against employee in violation of the federal Age Discrimination Act. This lawsuit could be filed in the federal court, and where the parties live or are doing business is irrelevant.

- Diversity jurisdiction: Cases involving state law, such as contracts or torts, but only if the plaintiff and the defendant are from different states *and* the amount in controversy is over $75,000.

EXAMPLE

While flying from his home in Arizona to Rhode Island, Tomas makes a stop at the Chicago airport. He is injured when a walkway collapses at the airport. Tomas' medical bills related to the collapse are $150,000. The walkway was designed by Chicago Design, Inc., a corporation incorporated in Illinois with its principle place of business in Illinois. The walkway was constructed by Airport Walkways, Inc., a corporation also incorporated in Illinois and with its principle place of business in Illinois. Even though this is a state law matter involving alleged negligence, Tomas can file the case in a federal court. Tomas could file in the Arizona federal court, but the matter would probably be removed to an Illinois federal court under the doctrine of *forum non conveniens,* which allows the federal court to remove a case to the most convenient federal court. If Tomas's injuries were less than $75,000, he would have to file his lawsuit in an Illinois state court.[9]

Typical Path Taken by a Lawsuit

The typical stages of litigation are as follows: filing of pleadings, discovery, motion for summary judgment, trial, appeal, and enforcement of judgment.

Filing of Pleadings   The **plaintiff,** the party starting the lawsuit, files the complaint to commence the litigation. The defendant files the answer. Other common pleadings include counterclaims, or the claims defendants may raise against the plaintiff, cross complaints, or claims the defendants may raise against other defendants or to bring other parties not sued by the plaintiff into the lawsuit.

EXAMPLE

An owner enters into contracts with a designer and a builder. A dispute arises, and the owner files a complaint against the contractor. The contractor files an answer in response to the owner's complaint. The contractor also files a cross complaint against the architect, claiming that the owner's damages were actually caused by faulty designs. This brings the designer into the litigation. The designer files a counter claim against the contractor, claiming the damages were caused by faulty construction.

Discovery   After the pleadings are filed, the discovery stage begins. During this stage of the litigation, the parties attempt to discover all of the evidence available. The most common discovery tools are interrogatories, requests for documents, and depositions.

Interrogatories are written questions from one party to another. They cannot be sent to third parties not involved in the lawsuit.

EXAMPLE

What is the name, address, and telephone number of all persons on the construction site on May 10, 2007?

A Request for Documents or Things is sent to another party and requires that party to produce documents or other things for review or copying.

EXAMPLE

Produce the Contract(s), General Conditions, Supplemental Conditions for the construction project for review and copying on _____ (fill in date) at _____ . (Fill in location).

A **deposition** is a face-to-face meeting between the attorneys and a potential witness, whether party or not, in the presence of a court reporter. The court reporter swears in the witness, takes the testimony as elicited by the attorneys, records the testimony, and makes it available in a booklet for review by the attorneys, parties, and **deponent,** the person giving the deposition. The deposition can be used in court if the witness should die or be otherwise unavailable. For this reason, it is common to take the depositions of all potential witnesses.

Motion for Summary Judgment, Summary Adjudication, or Motion to Dismiss   After discovery is completed, it is common, though not mandatory, for parties to file a motion requesting the judge to review the law and evidence and determine if the matter must go to trial. This motion goes by various names in the different jurisdictions including **summary judgment, summary adjudication,** or **motion to dismiss.** These motions allow the judge to review the claims and make decisions. If a factual issue exists, the matter cannot be resolved at this point but must go to the jury. However, it is very common for parties to negotiate a settlement if this motion fails because of the uncertainty associated with jury trials.

Trial   At trial, the parties present the judge with a legal brief summarizing the law applicable to the case. If the parties disagree on what law is applicable to the case or disagree on what the law means, a legal issue is raised. Legal issues are discussed in more detail below.

The parties then present their version of the facts to the jury, or to a judge if a jury trial is waived, through witness testimony, documents, and other forms of physical evidence. The jury decides any factual issues and the amount of damages (see "Relief or Remedy" below).

The judge or winning party prepares a **judgment,** a piece of paper signed by the judge outlining who has won the case and what they win.

Appeal   If either of the parties is of the opinion that the trial judge has made an error, she may appeal the trial judge's ruling(s). The appeal court, consisting of a panel of judges, will review the trial court's rulings for errors. If the error is significant, the matter can be sent back to the trial court for retrial. The federal government and most states have another level of appeal, frequently called the supreme court. The supreme court reviews the trial court and]or appeal court judges' rulings for errors.

Enforcement of Judgment   Once the lawsuit and the appeals, if any, are over, the winning party must enforce the judgment. Neither the judge nor the court system helps the winning party obtain any money owed. If the losing party does not voluntarily pay the winning party, the winning party must take the judgment to the county marshal or other police agency assigned this duty and inform the marshal of the location of the losing party's assets. The marshal will then confiscate the assets, sell them, and turn over the proceeds to the winning party. The law in each state limits the types of assets the winning party can seize in satisfaction of the judgment. For example, the tools of a person's trade or pension plans may be protected.

Issues of Law and Issues of Fact

Every lawsuit contains two categories of issues: issues of fact and issues of law. In deciding whether or not to pursue litigation, it is important to know whether or not the issues in the matter are issues of law or issues of fact. Issues of fact contain more risk.

Issues of Fact   An issue of fact is an issue where the parties disagree about who, what, when, where, how, and[or why. Issues of fact are the most common types of issues raised by the parties and the most problematic. Unless waived by the parties, factual issues are decided by juries and therefore less predictable. Because of the increased risk associated with factual issues, it may be in a party's best interest to settle prior to a trial. In addition to the unpredictability of answers to factual issues, in the U.S. legal system, the decision of jury is seldom reviewed on appeal. Except in rare cases, judges take as true whatever the jury decides. Issues of fact always involve weighing the credibility of the evidence, particularly testimony, which can be particularly self-serving. In the legal system in the United States, it is the job of the jury, not the judge, to do this. A jury can be waived, however, and in that event the judge decides the factual issues.

EXAMPLE

Meshaw, an owner, and Tenbusch, a contractor, entered into an oral contract for repainting of certain rooms and installing wallpaper and borders for $10,000. The owner claimed the contract was for the four rooms: the living room, dining room, kitchen, connecting hallways, and the master bedroom. The contractor claimed that the contract was only for the living room, dining room, kitchen, and connecting hallway. A factual issue exists. Each side's version of the oral contract is presented to the jury, together with any supporting evidence, and the jury decides what the contract says.

EXAMPLE

Beneco entered into a contract with the National Park Service for maintenance work to be done at the Grand Canyon. Beneco entered into a subcontract with EPC to perform excavation and work related to the installation of utility lines. The price of the subcontract was about $3.8 million.

EPC filed a proof of claim against Beneco, seeking $ 2,791,617.65 for work performed by EPC. This amount included $1,323,020.98 for work performed under the subcontract and $1,468,596.67 for additional or changed work not contained in the subcontract.

Beneco claimed that EPC was not entitled to the sums due for changed work because EPC did not follow the contract requirements for changed work. The contract contained the following clauses:

Beneco may order changes in the Work. No alteration, addition, omission, or change shall be made in the Work or the method or manner of performance of the Work except upon the *written change order* of Beneco....

The contract may not be changed in any way...and no term or provision hereof may be waived by Beneco except in writing signed by its duly authorized officer or agent.

The factual issue raised is this: Did the parties orally modify the contracts requirements regarding changed work during the course of their relationship?

The law in the matter is not at issue. Utah has long recognized that "parties to a written agreement may not only enter into separate, subsequent agreements, but they may also modify a written agreement through verbal negotiations subsequent to entering into the initial written agreement, even if the agreement being modified unambiguously indicates that any modifications must be in writing."[10]

continued

At trial, EPC puts its project manager on the stand, and the project manager testifies to oral conversations he had with Beneco's project manager, who orally ordered changes to the contract and told EPC that it would be paid for the changes. In addition, documentary evidence shows that, on numerous occasions, EPC was paid for changed work that was not subject to written change orders. Beneco's project manager disputes EPC's version of the events and says he never told EPC that it would be paid for changed work and in fact told EPC to make sure it got the written change orders in.

The jury believes EPC's version of the events. It determines that the parties orally modified the contract and that EPC is entitled to the $1,468,596.67 for additional or changed work not contained in the subcontract. Beneco appeals the jury's decision. The appeal court will not overturn the jury's decision because evidence exists to support the decision.

Issues of Law   Issues of law are disputes about what the law says, means or what law applies. Issues of law are decided by trial court judges, but they are subject to appellate review by both the appeal and supreme courts of the jurisdiction.

EXAMPLE

The state statute of repose states, "Constructors of improvements to real property cannot be sued 10 years after substantial completion of the project." An owner was injured when the elevator failed 11 years after substantial completion. The owner sued the elevator manufacturer under the theories of product liability and negligence. The elevator manufacturer moved for a dismissal of the action, claiming that the above statute prevented it from being sued because more than 10 years had elapsed.

The legal issue in the case is the meaning of the statute. Does the statute prevent lawsuits against manufacturers or just contractors? After several cases worked their way through the system, the state supreme court determined that the statue only protects contractors and not manufacturers.[11]

Judges also make decisions during the trial regarding what evidence the jury can hear. When attorneys object to evidence presented by the other party, the judge must decide, based on the law of evidence, whether or not the evidence can be presented to the jury. These decisions are subject to appeal at both the appeal court and supreme court level.

Jury Trials and Bench Trials   In a jury trial, a jury determines the factual issues and the judge determines the legal issues. A **bench** trial is a trial without a jury; the judge determines all legal and factual issues. Jury trials tend to be much longer and more formal than bench trials, and the type of evidence that can be presented to a jury is carefully controlled.

Statutes of Limitation

Statutes of limitation specify how long after an event a lawsuit can be filed. Each state has its own set of statutes of limitation. The statute of limitation for breach-of-contract cases is usually between four and ten years. The statute of limitation for tort cases usually between one and four years.

The time starts to run when the event giving rise to the lawsuit occurs or when the injured party should reasonably have known an event giving rise to a lawsuit has occurred. If the injury is hidden, such as a latent defect in construction, the statute of limitations in many states does not begin to run until the homeowner knows or should have known of the defect. In theory, the event giving rise to a lawsuit could occur decades after the completion of the project. and the time does not begin to run until the event giving rise to the lawsuit arises. Some states may still follow the older rule holding that a cause of action based on negligent design and construction accrues and that the statute of limitation begins to run at the time the project is substantially completed.

EXAMPLE

Bergey Construction builds an apartment complex for Damisis, the owner. The complex is designed by Chauvin. Assume the statute of limitations for breach of contract in the jurisdiction is four years. After construction of the project, the roof immediately shows signs of minor leakage, but Damisis makes no claim against either the contractor or the designer because the leakage is minor. Six years after completion of the house, the roof collapses and injures Lacross, the tenant.

In all jurisdictions, Damisis cannot maintain a lawsuit against the designer or the contractor for breach of contract, because the four-year statute of limitations has run. The statute began to run on the date the leaks become apparent. Lacross, the tenant, can sue the designer, the contractor, and the owner for negligence at any time within two years of the roof collapse, because the tort statute of limitation does not begin to run until the collapse. In order to recover for negligence, Lacross would have to prove the elements of negligence. Negligence is discussed in Chapter 12.

In the previous example, if the leaks were not apparent and the owner had no reason to know of a defect in the roof, the statute of limitations in many states would not begin to run until the roof actually collapsed. The owner would still have to prove negligence on the part of the contractor, the designer, or both. That may be extremely difficult after six years, but the owner's claim will not be dismissed because the statute of limitations has run.

In cases not involving defective construction, the statute of limitations does not begin to run until the events giving rise to the lawsuit occur.

EXAMPLE

A contractor sued a designer for negligent misrepresentation related to designs of a highway to be reconstructed. After work had commenced, it was determined that the designs were flawed. The contractor sued the designer for losses related to the flawed designs. No contract existed between the contractor and the designer. The statute of limitation was four years. The designer moved to dismiss the case, because it was filed more than four years after the alleged misrepresentations had been made. The court held that the statute did not begin to run until the contractor was actually damaged by the alleged misrepresentations.[12]

Statutes of Repose

Statutes of repose are different from statutes of limitations in that the time allowed for filing a lawsuit begins to run upon substantial completion of a project and not upon the occurrence of the injury. Under a statute of repose, the plaintiff is given a certain window of time during which it can sue the designer or contractor. The time begins to run upon substantial completion of a project and lasts between 5 and 20 years depending on the jurisdiction. Statute of repose generally protect only designers, contractors, and in some states, manufacturers. Owners are not protected.

EXAMPLE

Haggart designs and Keeley Construction builds a project for the owner, Leach, in a state with a 10-year statue of repose and a 2-year statute of limitations. At 13 years after substantial completion of the project, the roof collapses, injuring both the owner and a tenant. Neither the owner nor the tenant can sue the designer or contractor, because the statute of repose has run. The tenant can sue the owner for negligence within two years of the date of the injury.

States differ on how they handle a situation in which the injury occurs within the last years of statute of repose. In some jurisdictions, the time to sue is strictly limited to the time limit of the statute of repose.

EXAMPLE

Edick designs and Lowell Construction builds a project for Martin, the owner. The state has a ten-year statue of repose and a two-year statute of limitations. Nine years and eleven months after substantial completion, the roof collapses and injures Martin and a tenant. Martin and the tenant must file a claim against Edick and Lowell within the remaining month or be barred by the statue of limitations. The tenant can still pursue a claim against Martin within two years of the injury.

Some states extend the time limit for suing by the statute of limitations. As long as the injury occurs within the statute of repose time period, the injured party can sue within the applicable statute of limitations.

EXAMPLE

McGhee designs and Charles Construction builds a project for Antonides, the owner. The state has a 10-year statue of repose and a 2-year statute of limitations. Nine years after substantial completion of the project the roof collapses, and injures the owner and a tenant. The owner and the tenant can sue McGhee and Charles within two years of the date of injury or up to 11 years after substantial completion (Figure 1-4). The tenant can also sue the owner within the two-year period.

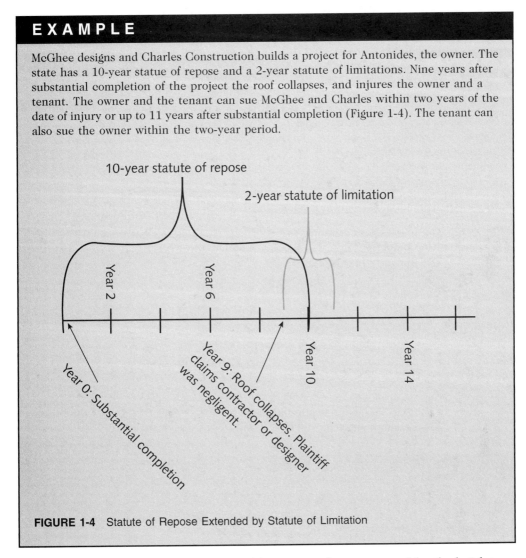

FIGURE 1-4 Statute of Repose Extended by Statute of Limitation

While not every state court has upheld a statute of repose passed by the legislature, most state courts have.[13] The trend in the law is to limit the statutes to patent or obvious defects and problems.[14] This means that if the defect is latent, the injured party can sue after the statute of repose.

Choice of Law

The term "choice of law" refers to the issue of which jurisdiction's law governs the lawsuit or claim. This question is often important in the construction industry, as state courts are open to all persons, and anyone can file a lawsuit in a state court against a resident of that state.

EXAMPLE

Chadwick Construction has its principle place of business in California and is a California corporation. It enters into a contract in New York to build a casino in Nevada. The owner is Rabaut and Richards, Inc., a New Jersey corporation. A dispute arises, and Chadwick files a lawsuit against the owner in New Jersey. The New Jersey court has jurisdiction. The issue then arises: Which state's law controls? Just because the lawsuit is filed in New Jersey does not mean New Jersey law applies.

The following choice of law rules are most applicable to the construction industry:

1. The parties may agree in the contract which state's law applies. The chosen state law must have a reasonable relationship to the transaction.

FORM

The law of Nevada will control this transaction.

2. If the parties have not stipulated which law applies, the law of the jurisdiction in which the contract was entered into usually applies.

EXAMPLE

A contract is entered into between an entity from Florida and one from Georgia for a project to be constructed in Louisiana. The contract is signed in Georgia. The parties do not stipulate which law applies. The law of Georgia applies.

Tort lawsuits are usually governed by the law where the tort occurred.

EXAMPLE

A contract is entered into between an entity from Florida and one from Georgia for a project to be construction in Louisiana. A subcontractor is injured on the job by the negligent act of another subcontractor. The law of Louisiana applies in the lawsuit between the injured subcontractor and the alleged tortfeasor.

Relief or Remedy

The terms **relief** and **remedy** generally apply to the list of things, such as attorney fees or punitive damages, that a court can award to a winning party. The most common type of relief is **damages** or money paid to the injured party. Another form of relief is an **injunction,** which is a court order to do or not to so something without involving a contract. The remedy of specific performance is available but rare in contract cases. **Specific performance** is a court order to perform a contract. The types of relief in contract and tort cases are different. See Chapter 7 for a discussion of the types of relief available in contract actions. See Chapter 12 for a discussion of the types of relief available in tort actions.

■ Appendix: Finding Law

The Web site for the Legal Information Institute at Cornell, *http://www.law.cornell.edu/*, contains links to state and federal law. State law is accessed by clicking the link at the left, <Law by Source or Jurisdiction>, and then clicking on the link at the right of the page, <Listing by Jurisdiction>, or by going directly to *http://www.law.cornell.edu/states/listing.html*.

■ Endnotes

1 *Williams v. Lee*, 358 U.S. 217 (1959).

2 William C. Canby, *American Indian Law* (4th ed., West, 2004).

3 *Kiowa Tribe v. Manufacturing Technologies*, 523 U.S. 751 (1998), holding "tribes enjoy immunity from suits on contracts, whether those contracts involve governmental or commercial activities and whether they were made on or off the reservation."

4 California Civil Code § 3509 (1999) states: "The maxims of jurisprudence hereinafter set forth are intended not to qualify any of the foregoing provisions of this Code, but to aid in their just application." See also North Dakota Cent. Code, § 31-11-05 (1999), *Maxims of Jurisprudence*, with a list of 34 maxims.

5 Many of the maxims listed here are based upon the author's familiarity with construction law.

6 Nancy J. White & Jun-Young Lee, *Dispute Resolution in the Korean and United States Markets: A Comparison*, Mid-American Journal of Business (2004).

7 Example based on *ACE Constructors, Inc. v. U.S.*, 70 Fed. Cl. 253 (Ct. Cl. 2006).

8 Under rare circumstances, the owner might be able to sue in Connecticut but that is unlikely.

9 The law would allow him to sue in Arizona if the designer and the contractor happened to be in Arizona and Tomas were able to serve them with the lawsuit in Arizona. This is unlikely to happen.

10 *U.S. v. Travelers Casualty & Surety Company of America*, 423 F. Supp.2d 1016 (D. Az. 2006) internally quoting R. T. Nielson Co. v. Cook, 2002 UT 11, 40 P.3d 1119, 1124 n.4 (Utah 2002).

11 *Sonnier v. Chisholm-Ryder Co., Inc.*, 909 S.W.2d 475 (Tex. 1995).

12 Example based on *Hardaway Co. v. Parsons, Brinckerhoff, Quade & Douglas, Inc.*, 267 Ga. 424, 479 S.E.2d 727 (Ga. 1997).

13 Jane Massey *Draper, Validity and Construction, as to Claim Alleging Design Defects, of Statute Imposing Time Limitations upon Action Against Architect or Engineer for Injury or Death Arising Out of Defective or Unsafe Condition of Improvement to Real Property*, 93 A.L.R.3d 1242 (2004).

14 *Tomko Woll Group Architects, Inc. v. Superior Court of L.A. County*, 46 Cal. App. 4th 1326, 54 Cal. Rptr. 2d 300 (Cal.App. 1996).

CHAPTER 2

Mistakes in Bids, Defective Construction, and Differing and Unforeseen Site Conditions

■ Mistakes in Bidding

Mistakes in bid law can be divided into three categories: obvious mistakes, mistakes made by the contractor, and mistakes made by the subcontractor. As a general statement, the law's attitude is that mistakes should be fixed if possible and no one should take advantage of the mistake of another. Figure 2-1 visually depicts this complex area of the law. Later in this chapter, Table 2-1 shows a summary of the basic rules applicable to contractor and subcontractor mistakes in bids.

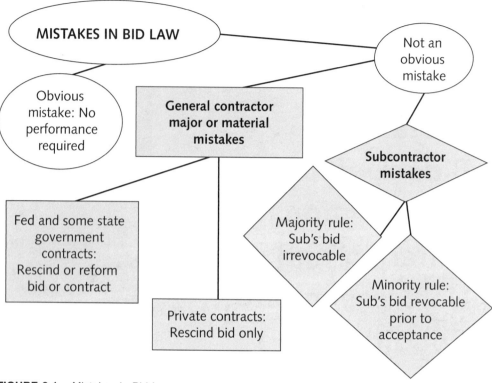

FIGURE 2-1 Mistakes in Bid Law

Obvious Mistake

A bid or a contract formed on a bid containing an obvious mistake is voidable by either party. The mistake must be obvious to an objective third party. **Voidable** means that a party has the option of performing the contract or not, without legal repercussions.

EXAMPLE

Mounger Construction orders 10 new refrigerators, at a cost of $1,000 each, from Southwell Supplier. Southwell mistakenly sends an invoice for $100 instead of $10,000. Mounger signs the invoice and faxes it back to Southwell with the words, "Offer accepted." Before shipping the refrigerators, Southwell notices the error and sends another invoice for $10,000. Mounger calls Southwell and says it has a contract for 10 refrigerators at a cost of $100 each and if the supplier does not supply them, the contractor will sue for breach of contract. The supplier need not ship the refrigerators, because the contract is voidable by the supplier.

TABLE 2-1 Summary of Mistakes in Bid Law

Mistakes in bids—Private contracts (all are governed by state law)		Summary of law
Obvious mistake: Any party		A bid or a contract formed on a bid containing an obvious mistake is voidable, that is, it can be rescinded.
Nonobvious mistake: Contractor		Contractor may rescind bid or contract formed on bid within a reasonable time. Note: A very few jurisdictions do not recognize this rule and require the contractor to perform or pay damages if the owner has accepted the bid.
Nonobvious mistake: Subcontractor	Majority rule	Subcontractor may not rescind its bid for a reasonable time after award of the prime contract even if the subcontractor has made a mistake.
	Minority rule	Subcontractor can rescind its bid at any time prior to contractor's acceptance of the bid.
Mistakes in bids—Federal government and some state government contracts		Summary of law
Obvious mistake: Any party		A bid or a contract formed on a bid containing an obvious mistake is voidable, that is, it can be rescinded.
Reformation: Contractor mistake		If the contracting officer has or should realize a mistake has been made, the contracting officer must verify the bid. Failure to do so allows the contractor to reform the bid or the contract formed on the bid to include the mistaken sum.

It is common to see language similar to that in the form below in bidding documents. This language is contrary to the principle that no contract can be formed upon an obvious mistake. The language in the following form, or language similar to it, is not legally enforceable in the event of an obvious mistake.

FORM

Bidder may not modify, withdraw, or cancel its bid. If the bidder refuses to sign the contract after award, the bid bond shall be paid to the owner as liquidated damages.

EXAMPLE

Bidding instructions contain the following language: "Bidder may not modify, withdraw, or cancel its bid. If the bidder refuses to sign the contract after award, the bid bond shall be paid to the owner as liquidated damages."

General contractor Arnold submitted a bid and a bid bond of $50,000 to the owner for the construction of a hospital wing. After opening the bids, the owner noticed that Arnold had not included any sum for plumbing. Arnold was therefore the lowest bidder, as all other bidders included an amount equal to approximately $4 million for plumbing.

The owner accepted Arnold's bid, and Arnold refused to sign the contract. The owner called the bonding company and demanded a check for $50,000 for the bid bond. The bid was voidable, Arnold did not need to perform, and the owner was not entitled to the bid bond. The owner's only recourse was to award the contract to the lowest bidder who did not make a mistake. The owner is not entitled to the amount of the bond.

To review an interesting case that appears to be contrary to the law, see *White Hat Management v. Ohio Farmers Ins. Co.,* 2006 Ohio 3280; 2006 Ohio App. LEXIS 3211 (Ohio App. 2006). In this case, the low bid for window replacement was about $87,000, and the next highest bid was about $167,000. This should have put the owner on notice of an obvious mistake and given the bidder relief. The bidder did inform the owner of the mistake shortly after being awarded the contract. However, the court allowed the owner to collect on the bond. The case did not discuss the law of mistaken bids or obvious mistake. Either the court ignored the law of obvious mistake or these issues were never raised. Because it did not discuss the law of mistaken bids or obvious mistake, the case was not in the line of authority on those topics. It was either an erroneous decision, based on attorney or judge error, or it may signal a change in the law in that jurisdiction due to a more conservative bench.

Nonobvious Contractor Mistakes in Bid

Contractor mistake in bid law varies depending on whether state or federal law controls. In addition, at the state level, it may vary depending on whether or not the state government is the project owner or not. The first rule discussed in the section entitled "Rescission for Major or Material Mistake" applies to all contracts, both private and government construction. The "Reformation for Mistake" section below discusses the rule applicable to federal government and some state government contracts. Selected states law applicable to private construction contracts is discussed in the section entitled "State Law in Selected Jurisdictions."

Rescission for Major or Material Mistake

In the event the contractor has made a nonobvious but major or material mistake, the contractor may void or rescind its bid or the contract formed upon the mistaken bid. **Rescind or rescission** means to take back or nullify and, in the context of a mistake in bid, the bid or contract with the major mistake is nullified at the contractor's option. If there is a bid bond, the owner cannot recover on it.

A more complete statement of this law is as follows: "A bid or contract can be rescinded for a contractor mistake if the contractor can prove each of the following elements:

1. The mistake is of so great a consequence that to enforce the contract as made would be unconscionable; and
2. The mistake relates to a major or material feature of the contract; and
3. The mistake was made regardless of the exercise of ordinary care; and
4. The parties can be placed in status quo [neither party will be prejudiced, except the owner has lost this bargain but can give it to the next bidder]."[1] This element is satisfied if work has not yet begun. The owner is expected to give the contract to the next-highest bidder.

EXAMPLE

The sewer district, a government agency, advertised for bids for the construction of improvements to two existing sewage treatment plants. The work was divided into four separate sections: general construction, plumbing/equipment, electrical, and heating.

Balaban submitted bids for each of the four separate sections but was the lowest bidder, at $2.25 million, for the general construction work section. Balaban's bid was $330,000 lower than the next-highest bidder. Balaban's bid for the plumbing and equipment was $375,000 higher than the only other bidder for that work.

The day after the bids were opened, Balaban sent a telegram to the engineer for the sewer district stating that in computing its bid for the general construction contract, it erroneously omitted certain items of equipment and included these items in computing its bid for the plumbing/equipment contract. Balaban also stated that it wished to withdraw its bid for the general construction work.

A few days later, the parties met to review the worksheets, and Balaban showed the engineer where it had erroneously included omitted items in the general construction contract and placed them in the plumbing/equipment contract. The sewer district refused to permit Balaban to withdraw its bid and awarded Balaban the general construction contract.

The court stated that Balaban had made a mistake but was not required to perform the contract. "The courts of this State have long recognized that one who makes a bid based on an honest and unintentional mistake can in the interest of equity be relieved from his contractual obligations....The fact that the mistake may have been due to the fault or negligence of the contractor does not bar relief....I find here that the mistake of the plaintiff [Balaban] was made in good faith and was not due to gross negligence. Nor was the defendant sewer district damaged by the mistake. There was no opportunity for it to suffer a change in position in reliance on plaintiff's bid, for the plaintiff gave notice of it the day after the bids were opened....Under these circumstances, the sewer district cannot retain the advantage of plaintiff's honest mistake; fair dealing requires that [Balaban] be relieved of its bid."[2]

The third element of the law, "the mistake was made regardless of the exercise of ordinary care," is often problematic for the contractor to prove. Usually if the contractor has informed the owner before the bid opening or very shortly after of the mistake, the law tends to favor the contractor and allow the contractor to rescind its bid. It is as if the bid were never made. The law prefers the owner to just pretend the mistaken bid was never submitted and give the contract to the next-highest bidder.

EXAMPLE

Cape, a road building contractor, submitted a bid and a $100,000 bid bond to the owner. Shortly before the bid opening, Cape attempted to withdraw its bid because of a major error. The owner awarded the contract to the next-higher bidder and attempted to collect on Cape's $100,000 bid bond. The owner claimed that Cape had not proved it "was free from carelessness, negligence, or inexcusable neglect" as required by statute. On appeal, the court went into the bidding process in detail and found that the mistake was not the result of contractor carelessness, negligence, or inexcusable neglect. The court said the following:

"The department solicited bids for construction of a highway interchange in Milwaukee. The bids were to be opened at 9:00 a.m. on October 10, 2000, as were bids on several other highway projects. As is apparently the custom, prospective bidders, including Cape, assembled personnel in rooms at a Madison hotel on the evening before the bid opening in order to compile and rework bids for submission just before the 9:00 a.m. deadline. Cape's bid team included some eight persons, equipped with several computers and printers, project folders, and telephones.

"The Cape bid team worked through the evening and into the morning hours, receiving proposals from prospective subcontractors, rejecting or accepting them, and incorporating the accepted proposals into Cape's final bids. Hard copies of the subcontractor proposals are customarily placed in appropriate folders and the figures are also entered into a computer so that comparisons can be made and final bid figures computed. One prospective subcontractor, Zenith Tech, submitted a proposal to Cape at about 6:00 a.m. Cape personnel identified Zenith Tech's bid as the low one for a component of the interchange project and entered the quotation into its bid calculations. At about 8:00 a.m., Zenith Tech discovered an error in the proposals it had submitted to Cape and two other contractors. Accordingly, Zenith Tech notified the three contractors by phone between 8:00 and 8:30 a.m. of an upward revision in its quoted price. Cape personnel compared the revised quotation with a proposal it had received for the item from a different subcontractor and concluded that Zenith Tech was still low for the item.

"Revisions from prospective subcontractors are apparently not uncommon during the final bid preparation process, and Cape personnel have several mechanisms available for incorporating last-minute revisions in its final bid, including making handwritten changes on the final bid sheet. The higher figure for Zenith Tech's work did not, however, become a part of the 'final schedule' Cape personnel printed out sometime between 8:35 a.m. and 9:00 a.m. Cape personnel reviewed the final schedule and made a handwritten change to another item on it, but no change was made to the Zenith Tech item, which continued to reflect Zenith Tech's original, lower quotation. As a result, Cape's bid was $450,450 lower than if the revised Zenith Tech figure had been inserted. The final bids of two other contractors who utilized Zenith Tech for the item in question reflected the revised, higher Zenith Tech quotation."

continued

Cape's bid was the lowest submitted for the interchange project, and it discovered the erroneous Zenith Tech figure in its bid soon after the bid was submitted. Cape notified the department, in writing, on the same day the bids were opened that its bid contained an error. Cape requested that either its bid be raised $450,450 or, if that were not possible, the bid be rescinded and no claim be made on the guarantee. The state rescinded the bid but made a demand on the guarantee.

This is a good example of the owner attempting to take advantage of a contractor's mistake it could easily have ignored. The owner was not allowed to collect on the bid bond.[3]

MANAGEMENT TIP

A wide range between the low bid and the other bids should put the owner on notice of a possible error.[4] If a wide range exists, the owner should inquire as to the accuracy of the bid before accepting it.[5]

An alternative legal theory that has been used to get a bid or contract rescinded when the mistake is major or material is the lack of a "meeting of the minds." Contract law requires a meeting of the minds in order for a contract to be formed. **Meeting of the minds** refers to requirement that both parties to a contract have the same understanding of what the contract means before a contract can be formed. If one of the parties has made a major or material mistake then there is no meeting of the minds. "Where there is a mistake that on its face is so palpable as to place a person of reasonable intelligence upon his guard, there is not a meeting of the minds of the parties and consequently there can be no contract."[6]

In *Kutsche v. Ford,*[7] the contractor had submitted a mistaken bid on a Michigan school building and attempted to withdraw the bid before contract award because of a mistake. The state Supreme Court said, "Where a mistake is of so fundamental a character that the minds of the parties have never, in fact, met, or where an unconscionable advantage has been gained by mere mistake or misapprehension, and there was no gross negligence on the part of the plaintiff, either in falling into the error or in not sooner claiming redress, and no intervening rights have accrued, and the parties may still be placed *in status quo,* equity will interfere in its discretion, to prevent intolerable injustice." This statement summarizes the attitude of the law in these situations: do not try and take advantage of another's mistake. States with similar attitudes include Nebraska,[8] New Jersey,[9] New York,[10] Ohio,[11] Pennsylvania,[12] Texas,[13] and Illinois.[14]

Using this rule, the contractor can only rescind by withdrawing its bid or the contract. The contractor cannot reform or revise its bid and remain in the running for the award. **Reform** means to change or revise. If the law allowed reformation of bids for mistakes, contractors would have a tendency to bid low and then reform the bid up to just below the next-lowest bidder. Reformation of a bid or contract is allowable in some circumstances in government contracts and is discussed below in the section entitled "Reformation of Mistake."

States with contrary law, that is where contractors do *not* get relief from a major or material mistake, include Alaska[15] and Kansas.[16] In Alaska, a state agency resolved a conflict between a dollar figure printed in numerals and words by following the state regulations, even though this produced a result obviously out of proportion with other bids. The state action was upheld because proper agency regulations had been followed. In Kansas, case law had held that a bid cannot be withdrawn for any reason after bid opening, only before, despite the fact that the bidder had attempted to withdraw its bid shortly after bid opening.

MANAGEMENT TIP

Contractors in states that do not allow rescission of the bid for a major or material mistake may need to bid slightly higher to cover the increased risk.

The reason for litigation on this topic is not that the owner wants to force the low, but mistaken, bidder to perform the contract. Few owners want to work with a contractor who is expecting to lose a lot of money on a project; the quality of the project will likely suffer. What the owners are trying to do is take advantage of the contractor's mistake and collect on the bid or performance bond. However, the law does not allow the owner to collect on the bid bond when there has been a major or material mistake and the owner can merely ignore the mistaken bid and award the contract to the next bidder.

Reformation of Mistake

Federal government and some state government contracts can be reformed in certain limited circumstances. The contractor can revise the bid or the contract based on the bid for a higher sum. If the contractor has made a computational mistake in a bid, *not* a mistake based on a judgment call, the federal government and some states allow the contractor to reform the bid or the contract formed on the bid *if* the government knew or should have known of the error.[17] This is not an easy theory to recover on. The contractor must prove that

1. the contracting officer had actual or constructive notice of a bidder's mistake; and
2. the contracting officer failed to verify the bid; and
3. the bid is accepted.

If the above can be proved, the law presumes that the contracting officer has acted in bad faith and taken advantage of the bidder, and reformation will be permitted.[18]

Contract officers routinely contact the contractor to verify a bid that appears to be low or mistaken. Once the contractor verifies the bid, the government has completed its obligation, and if the contractor later finds a mistake, the contractor is precluded from recovering. Even if the difference between the contractor's bid and the next-highest bid is large, as long as the contracting officer verifies the bid, the contractor is precluded from recovering.

EXAMPLE

The government requested Allsteel, the contractor, to submit a bid on a project for repair of latrine facilities. Allsteel's bid of $370,000 was $600 lower than the government's estimate for the job and 13% lower than the only other bid on the project. The contracting officer contacted Allsteel and asked it to verify its bid, which Allsteel did. Later Allsteel discovered a $42,000 mistake in the bid due to the failure to include a price for 6,700 square feet of ceramic tile. No recovery was possible for the contractor, as the mistake was a unilateral error in estimation of the necessary materials and the government had requested Allsteel to verify the bid.[19]

This rule does not allow the contractor to claim it made an in-house error in estimating a particular item and then lower that item so it is the low bidder. In order for the contractor to be entitled to reform the contract, the error must be obvious on the face of the bid or an obvious mistake.

EXAMPLE

The bids on a project were as follows:

	Contractor A's Bid	Contractor B's Bid
Part A	$1.2 million	$1.2 million
Part B	$2.2 million	$2.1 million
Part C	$1.2 million	$1.2 million
Total	$4.6 million	$4.5 million

Contractor A cannot reform its bid by claiming it made an error in estimating Part B and that Part B should have been $2.0 million.

If, after work has begun on the project, the contractor finds a mistake in the bid *and* the contractor can prove what the correct bid amount should have been, the contract can be reformed to either delete work or increase the amount payable to the contractor. The amount payable to the contractor can only be up to the amount of the next-lowest acceptable bid.[20]

EXAMPLE

The bids on a project were as follows:

	Contractor C's Bid	Contractor D's Bid
Part A	$118,000	$117,000
Part B	$2,200	$2,100
Part C	$6,000	$8,000
Total	$120,800	$127,100

Contractor C was awarded the job and began work. Later it was determined that the total for Contractor C's bid was incorrect and should have been $126,200. Contractor C's contract can be reformed to $126,200 since it is still the lowest bidder.

EXAMPLE

The bids on a project were as follows:

Contractor E's Bid	Contractor F's Bid
Part A $116,000	$117,000
Part B $2,500	$2,100
Part C $6,000	$5,000
Total $121,500	$124,100

Contractor E was awarded the job and began work. Later, it was determined that the total for Contractor E's bid was incorrect and should have been $124,500. Contractor E's contract can only be reformed to $124,099.99 because Contractor E's reformed contract must still be lower than Contractor F's bid.

State Law in Selected Jurisdictions

Some states follow the federal model and allow contractors to reform a state government contract. California follows the federal model and grants contractors the right to reform the bid if the government is the owner and the mistake is obvious on the face of the bid.[21] Florida grants the contractor this right to reform the bid if the mistake is made by an employee of the bidder[22] but not if actually made by a principal in the firm doing the bidding.[23] In Massachusetts, relief is governed by statute,[24] and the bid deposit can be returned in case of "death, disability, bona fide clerical or mechanical error of a substantial nature, or other similar unforeseen circumstances affecting the general bidder."

Nonobvious Subcontractor Mistakes in Bid

While most states allow a contractor that has made a major or material mistake to rescind its bid, most states do not allow a subcontractor to do so if the contractor has relied upon the subcontractor's bid in the contractor's bid to the owner. State law governs all contracts between a general contractor and a subcontractor; federal law never applies. Two different state approaches have emerged: the majority rule, also called the *Drennan* rule,[25] and the minority rule, also called the *James Baird* rule.[26] Both are discussed in more detail below. In addition, if the subcontractor's mistake is obvious, then the rule discussed in the section entitled "Obvious Mistake" applies.

Majority or Drennan Rule

The majority of jurisdictions follow the *Drennan*[27] rule, which states that *if* the contractor has reasonably relied upon the subcontractor's bid, the subcontractor cannot rescind or reform its bid for a reasonable time after award of the general contract *even if* the subcontractor has made a mistake. Because the subcontractor cannot rescind its bid, once the contractor accepts the bid, the contract is formed. Contract formation law is discussed in Chapter 7.

EXAMPLE

Tech, the general contractor, submitted a bid to Baltimore County for construction of the Winfield Police Athletic League Center project. Tech solicited proposals from potential subcontractors for the roofing portion of the project and received a bid of $39,000 from Citiroof. Citiroof had sent Tech an unsolicited bid based upon copies of plans it had received from another subcontractor. Citiroof later claimed the plans received from this other subcontractor were at half scale, and Citiroof did not realize it at the time it submitted its bid. The half-size plans were not admitted into evidence.

Tech also received a second bid of $62,000 from Jottan, Inc. Tech's president testified he realized "somebody has got to be right, and somebody has got to be wrong." Tech's president telephoned Citiroof, informed it the price was "rather low" and asked if "Citiroof was comfortable with the bid," because Tech intended to use Citiroof's bid. Citiroof confirmed its bid. After Tech was awarded the job, Citiroof faxed a letter to Tech rescinding its bid; it did not mention it had made an error. Later, when the lawsuit was filed, Citiroof revealed that it had bid on half-size plans. Citiroof's costs to perform the subcontract would be $50,000 without any profit.

Tech entered into a contract with Jottan for $59,000, sued Citiroof for $20,000, and won. The court said, "It's unfortunate for the original sub [Citiroof] that that error was made, but...it's initially [Citiroof's] error. And in order to pass it off to [Tech], you have to establish that they [Tech] were unreasonable in relying on it. And I think that under all of the facts of the case, that the Court cannot find that Tech was unreasonable in [its] reliance."[28]

EXAMPLE

AA, the general contractor, solicited bids for certain subcontracts on a project to build a Home Depot store. Grand State faxed a written but unsigned bid to AA in the amount of $115,000 for installation of the Exterior Insulation Finish System on the project. Grand State's bid stated: "Our price is good for 30 days." Because Grand State's bid was the lowest, AA included it in its bid on the general contract.

AA was awarded the job and forwarded a subcontract to Grand State within in the 30-day price-is-good period. Grand State refused to sign the subcontract, stating it had entered into four other contracts within the last few days and did not have the resources to complete AA's project.

AA entered into a contract with another subcontractor for $130,000 and sued Grand State for the difference of $15,000. Judgment was entered into in favor of AA and Grand State had to pay AA $15,000. AA reasonably relied upon Grand State's bid, and Grand State could not rescind its bid.[29]

The majority or *Drennan* rule is based on the doctrine of promissory estoppel, also called detrimental reliance. The Restatement (2nd) of Contracts (1981) §90 defines **promissory estoppel** as follows:

"A promise which the promisor [the party making the promise, that is, the subcontractor] should reasonably expect to induce action or forbearance on the part of the promisee [the party to whom the promise is made, that is, the contractor] or a third person and which does induce such action or forbearance is binding if injustice can be avoided only by enforcement of the promise."

MANAGEMENT TIP

The cost of litigation when a subcontractor refuses to perform will usually outweigh the benefits and also creates bad feelings. The contractor should carefully consider the ramifications before suing a subcontractor.

Bid Shopping

The majority or *Drennan* rule only obligates the subcontractor to the contractor; it does not obligate the contractor to the subcontractor, and the contractor can legally engage in bid shopping. **Bid shopping** occurs when a contractor attempts to get a subcontractor to lower its bid after the contractor has been awarded the project.

EXAMPLE

Pienkosz, the general contractor, obtains several bids for the electrical subcontract as follows:

Subcontractor A	$150,000
Subcontractor B	$142,000
Subcontractor C	$145,000
Subcontractor D	$135,000

Pienkosz uses Subcontractor D's bid of $135,000 for the electrical work as part of its bid to the owner. After Pienkosz is awarded the project, it gives the electrical subcontract to Subcontractor A, who agrees to do it for $130,000. Subcontractor D has no legal recourse against Pienkosz.

If the contractor has *not* relied upon the subcontractor's bid or the contractor cannot prove that it relied on the subcontractor's bid, the subcontractor may withdraw its bid at any time prior to acceptance of the bid. Some states may be amenable to an argument that evidence of bid shopping is proof the general contractor did not rely on the subcontractor's bid and therefore the subcontractor can withdraw its bid at any time prior to acceptance by the contractor. This rule is more fully discussed below in the section entitled "Minority Rule."

EXAMPLE

Thelen obtained several bids for a painting subcontract as follows:

Subcontractor A	$150,000
Subcontractor B	$142,000
Subcontractor C	$145,000
Subcontractor D	$135,000

Thelen bids a lump sum for the entire project and does not rely on any particular subcontractor's bid. After award of the contract, Thelen shops the painting subcontract but no subcontractor will perform for less than $142,000. Subcontractor D informs Thelen it made a mistake in its bid and cannot perform for less than $142,000.

Thelen cannot hold Subcontractor D to its bid of $135,000, because it did not rely on Subcontractor D's bid. In some jurisdictions, the evidence of bid shopping would also preclude Thelen from holding Subcontractor D to its bid.

Bid shopping is considered by many in the industry to be unethical, but it is not usually illegal. Some states limit the contractor's ability to bid shop on public projects. See California Gov. Code, §4100 *et seq.*

In discussing the unethical nature of bid shopping, one court said, "Allowing the listing of alternate subcontractors would conflict with the overall purpose behind section 11-35-3020(2)(b) [Southern Carolina state statute]...the underlying goals of the State Procurement Code are, *inter alia,* to ensure standards for the "fair and equitable treatment of all persons."...To these ends, a primary objective of the bid listing provisions, particularly regarding subcontractors, is to prevent bid shopping and peddling. Allowing the listing of alternate subcontractors would only serve to foster these unethical practices, because it would give contractors the opportunity to choose from among several prospective subcontractors, depending on who offered the lowest bid post-award."[30] The court held the contractor's bid to be nonresponsive.

MANAGEMENT TIP

Bid shopping is considered unethical in the industry. Economic theory supports the concept that contractors who engage in bid shopping are not as respected by the subcontracting community and therefore may not get the highest quality subcontractors to work for them. This results in lower-quality work. "Someone who is known not to perform his side of bargains will find it difficult to find anyone willing to make exchanges with him in the future, which is a costly penalty for taking advantage of the vulnerability of the other party....Thus the fundamental function of contract law (and recognized as such at least since Hobb's day [citing Thomas Hobbs, Leviathan, 1914, p. 1651]) is to deter people from behaving opportunistically toward their contracting parties..."[31]

Minority Rule

In the minority of jurisdictions, the subcontractor can rescind its bid at any time prior to acceptance of the bid by the contractor. Prior to acceptance by the contractor then, the subcontractor can rescind its bid for any reason or no reason. Courts in these jurisdictions do not apply the doctrine of promissory estoppel to protect the contractor. Once a bid has been accepted by the contractor, a contract is formed. See discussion of contract formation law in Chapter 7.

EXAMPLE

Whitaker, the general contractor, obtained several bids for the electrical subcontract as follows:

Subcontractor A	$150,000
Subcontractor B	$142,000
Subcontractor C	$145,000
Subcontractor D	$135,000

Whitaker used Subcontractor D's bid of $135,000 for the electrical work as part of its bid to the owner. Prior to acceptance of the bid by Whitaker, Subcontractor D informs Whitaker it cannot perform for $135,000 for whatever reason or even giving no reason. Whitaker cannot reform its bid to the owner, although it could possibly rescind its bid if the mistake is major or material under the law discussed in the section above entitled "Rescission for Major or Material Mistake." Whitaker is forced to hire Subcontractor B to do the job and has no recourse against Subcontractor D.

MANAGEMENT TIP

In jurisdictions using the minority rule, the general contractor has no assurance that subcontractors will perform at the price quoted. Contractors should include this risk in bidding contracts, and one would expect construction contracts in such jurisdictions to be slightly more costly than in jurisdictions applying the majority rule discussed above. "Risk aversion is not a universal phenomenon....But economists believe, with some evidence (notably the popularity of insurance), that most people are risk averse most of the time."[32]

The minority rule is based on the more general law of contract formation, specifically withdrawal of offers law. A **bid** is considered an offer under the law. **Withdrawal of an offer or bid** law states that an offer or bid can be withdrawn at any time prior to acceptance. A bid or offer cannot be withdrawn after acceptance, because at the moment of acceptance, a contract is formed. Failure to perform a contract is usually a breach of contract, and the party failing to perform must pay damages. A more complete discussion of these topics is in Chapter 7.

As with the majority rule, the minority rule allows the prime contractor to bid shop. Bid shopping is discussed above.

Defective Construction

Defective construction caused by faulty construction practices is a breach of contract, and the contractor is liable to the owner for damages. The damages due the owner are not always the cost of repair. For a more complete discussion of the amount of damages owed (see Chapter 7). This section addresses only the contractor's liability to pay the owner some amount of damages. The *amount* of damages to be paid is discussed in the above-referenced chapter.

Usually at substantial completion of the project, the contractor, the owner or the CM (construction manager), and the designer inspect the site and generate a punch list. A **punch list** is a list of minor defects or problems in the construction that the contractor needs to fix or remedy prior to receiving final payment or any retainage. **Retainage** is an amount due the contractor from a particular invoice but held by the owner and not paid until the end of the project. The retainage is a fund available to the owner should the contractor breach the contract. The normal practice is for the contractor to fix the items on the punch list or be in breach of contract.

Failure of the owner to list an item *visible during the inspection* on the punch list may be considered a waiver of the contractor's legal obligation to perform that particular part of the contract. "It is true that an owner is entitled to have a structure built in keeping with contract specifications which govern the work, and that a departure from those specifications—absent the consent of the owner or his authorized agent, or a waiver—will render the contractor liable for the necessary cost of bringing the structure into compliance with the specifications."[33] A **waiver** is the *knowing* relinquishment of a *known* right.

> ## EXAMPLE
>
> During the walk-through for the punch list, the owner notices that the contractor painted a certain room a color different from that required by the plans. However, the owner does not list it on the punch list. The owner cannot later require the contractor to return and repaint the room.

If a defect is *not* visible during the inspection of the project, the defect is not waived, and the contractor is liable to the owner. The owner may sue the contractor for damages at any time prior to the running of the statute of limitations.

> ## EXAMPLE
>
> Three months after the completion of the project and the punch list, the first major rainstorm of the season occurs. This is the first time leaking is visible. The leaking is caused by improper placement of the roofing system. The contractor is liable to the owner for damages. The statute of limitations begins to run when the leaking is first visible.

The contractor is liable to the owner for damages during the warranty period as long as the owner can prove causation. **Proving causation** requires the owner to prove the damage was caused by some mistake of the contractor's. Warranties may be set by the contract or established by law. Warranties are discussed in Chapter 12.

■ Differing and Unforeseen Site Conditions

Unexpected conditions can precipitate serious problems for the owner, the contractor, and the designer. This type of problem is fairly common in remodeling jobs and any job requiring excavation. Two different categories exist: differing site conditions, called Type 1 differing site conditions in federal law, and unforeseen site conditions, called Type 2 differing site conditions under federal law. The two conditions are treated exactly the opposite by the law. That is, the basic rule is the owner is liable for a differing (or Type 1) site condition, and the contractor is liable for an unforeseen (Type 2) site condition.

The law does not require a prebid site inspection, but most contracts will require one. When such a requirement exists, the contractor is required to include in its bid any costs associated with any visible condition at the site. In other words, if the contract contains a prebid site inspection requirement, a condition visible on the site can never be a differing or an unforeseen site condition. A contract clause requiring a prebid site inspection does not require the contractor to take boring logs, dig, or analyze the site, although the parties can certainly require such inspections in the bid documents or as part of the contract.

Differing Site Conditions

A **differing site condition** or a **Type 1 site condition** is defined as a condition at the site differing from what the plans, specifications, or other contract documents state or picture. Nonphysical conditions such as economic downturns, labor shortages, weather, and wars are never considered differing site conditions. A differing site condition is similar to a breach of the contract and has nothing to do with whatever anyone thought or assumed the site would be like. It derives strictly from the representations made in the contract documents. The owner warrants to the contractor that the plans and specifications are accurate, and if the contractor builds according to them, an acceptable product will result. See information on torts and warranties in Chapter 12. If the plans or specifications prove to be inaccurate, the cost to remedy the inaccuracy remains with the owner.

EXAMPLE

Smoot entered into a contract with an owner to build a prison facility. The contracts and the boring logs indicated that trench footings could be used. However, after construction began, the soil conditions for four of seven buildings required the use of form footings. A differing site condition existed, and the owner was responsible for the costs of the form footings.[34]

EXAMPLE

The contract indicated that only silty and blue clays and a small amount of shale would be encountered on the site. A large quantity of shale was actually encountered instead of the silty and clay material. The contractor had intended to use the excavated silty and blue clay for fill at another location on the site, thereby reducing the need to purchase fill. However, the excavated material would not compact well, and the contractor had to purchase fill from another source. This was a differing site condition, because the description of the site in the contract differed from what was actually found on the site. The owner was responsible for the cost of the fill.

EXAMPLE

The owner solicited bids for the renovation of a wing of university building. The contract included replacement of the drainage system around the foundation. The bid documents required bidders to conduct a prebid site inspection in order to familiarize themselves with the conditions in the field, and the contractor conducted a prebid site inspection. After construction had begun, the contractor encountered subsurface sewer lines that interfered with excavation for the drainage system. The subsurface sewer lines had not been indicated on the as-built drawings. The location of the sewer line was shown in records available at the municipal wastewater authority. An expert witness testified that subsurface sewer lines are normally indicated on as-built drawings. This was a differing site condition, because the contractor could rely upon the lack of subsurface sewer lines in the as-builts to mean there were none. The availability of the information at the municipal wastewater authority was irrelevant; the law did not require the contractor to conduct such a search as part of a prebid site inspection or for any other reason.[35]

If the contract documents are silent about the conditions at the site, no differing site condition can exist. By definition, a differing site condition exists *only* if the contract, plans, and/or specifications contain information about the site that later proves to be inaccurate.

EXAMPLE

The owner awards a contract to a contractor for construction of a small bridge over a river. The design calls for the bridge to be supported by concrete pilings resting on bedrock. No boring logs were included. The contractor bid the job assuming bedrock at 1–2 feet below the surface of the water. After construction begins, it is determined that at the place where the pilings are to be placed, the river contains pockets of silt, some

continued

as deep as 20 feet. This will greatly increase the cost of the project. This is *not* a differing site condition, because the contract says nothing about the conditions in the riverbed. Note that if the cost rises to the level of practical impossibility, the contractor would be entitled to relief. This theory is discussed below in the section entitled "Avenues of Possible, but Not Probable, Recovery for Contractors if No Unforeseen Site Condition Clause Exists."

EXAMPLE

The owner solicited bids for the renovation of a wing of university building. The contract included replacement of all windows and other renovations. The bid documents required bidders to conduct a prebid site inspection in order to familiarize themselves with the conditions in the field, and the contractor conducted a prebid site inspection. The owner also supplied as-built drawings to the contractor. The contractor, who submitted the low bid, found concealed metal framing around the doors when he had expected wooden framing. The door framing material was not included in the as-built drawings and increased the contractor's costs to perform the renovation. The contractor submitted a claim for increased costs. An expert witness testified that as-built drawings **normally do not contain** indications of specific materials used in the construction. This is not a differing site condition.

It is possible for the contract to be unclear or ambiguous concerning a condition at the site. In that event, the law must first decide what the contract says in order to determine if a differing site condition exists. See Chapter 9 for the rules used to determine the meaning of an ambiguous contract or clause. If after the contract is clarified, pursuant to the rules in that chapter, what the contract says differs from what is found at the site, a differing site condition exists.

Unforeseen Site Conditions

An **unforeseen site condition,** called a **Type 2 differing site condition** in federal law, is a condition at the site that was unforeseen or unexpected. When the contractor finds something unexpected at the site, it usually falls into this category.

Unforeseen site conditions might include rock,[36] subsurface concrete structures,[37] unusual soil conditions,[38] underground utilities,[39] a Conex® shipping container and a dredge stern swivel,[40] or anything that is unusual for the specific site. Nonphysical conditions such as economic downturns, labor shortages, weather, and wars are never considered unforeseen site conditions. If the parties do not agree that a particular condition is an unforeseen site condition, an issue of fact exists and must be determined by the trier of fact. See Chapter 1 in the section entitled "Issues of Law and Fact." Basically, the jury, or judge if a jury has been waived, or arbitrator will need to hear testimony to make a determination of the whether or not the condition is unforeseen.

Unless changed by statute or a clause in the contract, the basic law is that the contractor absorbs the costs of unforeseen site conditions. This rule is a corollary to the **Basic Premise of Contract Law,** which states that a party must perform its contract or respond in damages. Many courts have stated: "Where one agrees to do, for a fixed sum, a thing possible to be performed, he will not be excused or become entitled to additional compensation, because unforeseen difficulties are encountered..."[41] A party has an obligation to perform its contract whether or not unexpected problems or events occur or the profit is less than anticipated. Some exceptions exist. See discussion of exceptions in the section below entitled "Avenues of Possible, but Not Probable, Recovery for Contractors if No Unforeseen Site Condition Clause Exists." Some jurisdictions,[42] by statute, have placed the risk of unforeseen site conditions on the owner, but the majority of jurisdictions do not.

Transferring Risk of an Unforeseen Site Condition onto the Owner

Two basic methods exist for transferring the risk of an unforeseen site condition onto the owner: through a contract clause doing so and under a claim of superior knowledge.

Most standard contracts contain an unforeseen site condition clause. An **unforeseen site condition clause** transfers the risk of an unforeseen site condition onto the owner. These clauses are common in contracts in the United States and are enforceable.

The theories of **superior knowledge, concealment, misrepresentation,** or **fraud** transfer liability for an unforeseen site condition to the owner when the owner has withheld material facts about the condition of the site and the contractor has been damaged. These theories have often produced a result favorable to the contractor. If the owner knows of the existence of an unforeseen site condition and fails to inform the contractor of it, the owner is liable for damages. This result occurs usually occurs even if the contract contains some language disclaiming the accuracy or completeness of information furnished to bidders and/or requiring bidders to fully inspect the conditions at the proposed job site.

The owner's knowledge is an issue of fact and may be difficult to prove. The contractor may be able to use the discovery process to determine whether or not the owner had knowledge of the unforeseen site condition.

EXAMPLE

The owner's dam construction plans stated that, according to certain borings, the material to be excavated was composed of gravel, sand, and clay. However, the owner's engineer knew that stumps, buried logs, and sandstone conglomerate existed at the site but were not visible on a prebid site inspection nor did they appear in the borings. The owner was held liable for the costs of excavating this debris.[43]

EXAMPLE

On a certain sewer project, the government hired a private firm to make test borings of the site but did not reveal the results of the borings to the contractor. The borings showed the existence of underground water that was unknown to the contractor. The contractor was entitled to damages for an unforeseen site condition.

MANAGEMENT TIP

The contractor should negotiate the inclusion of an unforeseen site condition clause into the contract. This type of clause is not uncommon, and owners routinely include them in contracts.

Avenues of Possible, but Not Probable, Recovery for Contractors if No Unforeseen Site Condition Clause Exists

A contractor faced with an unforeseen site conditions often uses many legal theories in an attempt to obtain recovery from the owner, usually to no avail. The theories discussed in this section are some of the most commonly used. These theories include the following:

- Doctrine of objective impossibility
- Doctrine of practical impossibility or commercial senselessness
- Mutual mistake or no meeting of the minds
- Differing site condition argument

The **doctrine of objective impossibility** states that if a provision of a contract is impossible to perform, failure to perform the provision is not a breach of the contract. No case was found that used this rule in a construction contract setting. This theory did relieve the contractor of liability for breach of contract in *National Presto Industries, Inc. v. U.S.*[44] when the weaponry the plaintiff had contracted to produce did not perform adequately, and it was determined that it was impossible to make the weaponry function as desired by the government.

An alternative legal theory, the **doctrine of practical impossibility** or **commercial senselessness,** states that at some point, when the cost of performance becomes ridiculous, performance is excused. This rule is a corollary of the maxim of law holding that the law abhors waste. Maxims of law are discussed in Chapter 1. This doctrine is much invoked and seldom produces a determination in favor of the contractor. No example of this doctrine actually working in the context of an unforeseen site condition was found.

The **mutual mistake** or **no meeting of the minds** theory attacks the actual formation of the contract by proving that one of the required elements of contract formation, "meeting of the minds," has not been met. The theory states that if there has been no agreement or meeting of the minds by both parties as to the requirements of the contract, no contract has been formed.

EXAMPLE

The contractor and the owner entered into a contract for the paving of a certain area that both agreed was 10,000 square feet. Shortly after the contractor began performance, the contractor determined that the area to be paved was actually 16,700 square feet. Both parties are mutually mistaken about the square footage, and so no contract has been formed. The parties must renegotiate the contract, or the owner can hire another firm to do the work.

The final type of argument used by a contractor is that the unforeseen site condition is actually a differing site condition. The contractor attempts to use the rules of contract interpretation to prove that the contract has given the contractor incorrect information about the site. The rules of contract interpretation are discussed in Chapter 9.

EXAMPLE

Connor, the general contractor, entered into a contract with the government for renovation of a hospital on an Army base. Part of the renovation included replacement of ceilings, ductwork, grilles, and diffusers. The contract contained a prebid site inspection clause (FAR 52.236-27) that required the contractor make a reasonable site inspection. The contractor claimed that it had made a reasonable site inspection given the difficult conditions at the site which included the following: the hospital was in use at the time of the inspection, many of the ceilings were locked in place and not subject to view unless demolished, some accessible ceiling areas were dark, and some accessible ceiling areas were obstructed by wires and other objects. When the work on the ceilings was begun, it was discovered that the grilles and diffusers were hard fabricated to the ductwork and not attached by flexible ductwork. This caused an increase in cost of approximately $700,000 for the contractor.

The court did not accept the contractor's attempt to classify this as a differing site condition. The court said that if the contractor believed it could not perform a reasonable site inspection, it should have informed the owner before submitting its bid. The contractor was not entitled to an additional sum.[45]

■ Endnotes

1. *Kenneth E. Curran, Inc. v. State*, 215 A.2d 702 (N.H. 1965).

2. Example based on *Balaban-Gordon Co., Inc. v. Brighton Sewer Dist. No. 2*, 67 Misc. 2d 76, 323 N.Y.S.2d 724 (NY Sup. 1971).

3. Example based on *James Cape & Sons Co. v. Mulcahy*, 268 Wis. 2d 203, 672 N.W.2d 292 (Wisc.App.2003).

4. *Moffett, Hodgkins and Clark Co. v. Rochester*, 178 U.S. 373 (1899).

5. *Hudson Structural Steel Co. v. Smith & Rumery Co.*, 85 A.384 (Me. 1912).

6. *Ex parte Perusini Construction Co.*, 7 So.2d 576 (Ala. 1942). See also *M. F. Kemper Construction Co. v. City of Los Angeles*, 235 P.2d 7 (Ca. 1951).

7. *Kutsche v. Ford*, 222 Mich. 442, 192 N.W. 714 (Mi. 1923).

8. *School District of Scottsbluff v. Olson Constr. Co.*, 45 N.W.2d 164 (Neb. 1950).

9. *Intertech Assocs., Inc. v. City of Patterson*, 255 N.J.Super. 52, 604 A.2d 628 (N.J.App.Div. 1992).

10. *Buffalo Mun. Hous. Auth. v. Gross Plumbing & Heating Co.*, 172 A.D.2d 1030, 569 N.Y.S. 2nd 289 (N.Y.Sup. 1991).

11. Ohio Rev. Code Ann. §9.31.

12. Pa. Stat. Ann. Tit. 73 §1601 *et seq.*

13. *James T. Taylor & Son, Inc. v. Arlington Indep. School Dist.*, 160 Tex. 617, 335 S.W.2d 371 (Tex. 1960).

14. *Santucci Constr. Co. v. County of Cook*, 21 Ill.App.3d 527, 315 N.E.2d 565 (Ill. App. 1974).

15. *Alaska International Constr. Inc. v. Earthmovers of Fairbanks, Inc.*, 697 P.2d 626 (Alaska 1985).

16. *Anco Constr. Co., Ltd. v. City of Wichita*, 660 P.2d 560 (Kan. 1983).

17. *Reformation by United States Court of Claims of Government Contract*, 19 A.L.R. Fed. 645, sec. 14. *Ruggiero v. U.S.*, 190 Ct. Cl. 327, 420 F.2d 709, 713 (Ct. Cl. 1970), *C & L Constr. Co. v. U.S.*, 6 Cl. Ct. 791, 800 (Ct. Cl. 1984), aff'd mem., 790 F.2d 93 (Fed. Cir. 1986), *BCM Corp. v. U.S.*, 2 Cl. Ct. 602, 606 (Ct. Cl. 1983).

18. *BCM Corp. v. U.S.*, 2 Cl. Ct. 602, 31 Cont. Cas. Fed. (CCH) P71,110 (Ct.Cl. 1983).

19. Example based on *Fadeley v. U.S.*, 15 Cl. Ct. 706, 35 Cont. Cas. Fed. (CCH) P75,583 (Ct.Cl. 1998).

20. *48 C.F.R. § 14.407-4(b)(2).*

21. "Relief of Bidders," Cal. Pub. Cont. Code §5100 *et seq.*

22. *State Bd. of Control v. Clutter Constr. Corp.*, 139 So.2d 153 (Fla. Dist. Ct. App., cert den. 146 S.2d 374 (Fla. 1962).

23. *Lassiter Constr. Co. v. School Bd.*, 395 So.567 (Fla. Dist. Ct. App. 1981).

24. Mass.Gen.Laws.Ann.ch.149 §44B.

25. *Drennan v. Star Paving Co.*, 333 P.2d 757 (Cal. 1958).

26. *James Baird Co. v. Gimbell Bros.*, 64 F.2d 344 (2nd Cir. 1933).

[27] The name "*Drennan* rule" is based on the case of *Drennan v. Star Paving Co.*, 51 Cal. 2d 409, 333 P.2d 757 (Cal. 1958). Drennan, the general contractor, was preparing a bid for a public school job and included Star Paving's bid of approximately $7,000 for paving work. The next day, Star Paving informed Drennan that the bid contained a mistake, and the cost of the paving was $15,000. Drennan hired another subcontractor to perform the work, sued Star Paving for the difference, won because Drennan had justifiably relied upon Star Paving's bid.

[28] Example based on *Citiroof Corp. v. Tech Contracting Co., Inc.*, 159 Md. App. 578, 860 A. 2d 425 (Md. App. 2004).

[29] *Double AA Builders, Ltd. v. Grand State Constr. L.L.C.*, 210 Ariz. 503, 114 P.3d 835 (Ariz. App. 2005).

[30] *Ray Bell Construction Co., Inc. v. School District of Greenville County*, 331 S.C. 19, 501 S.E.2d 725 (S.C. 1998).

[31] Richard A. Posner, *Economic Analysis of Law* (3d ed., Little Brown and Co., 1986).

[32] Richard A. Posner, *Economic Analysis of Law* (3d ed., Little Brown and Co., 1986).

[33] *Havens Steel Co. v. Randolph Engineering Co.*, 613 F. Supp. 514 (W.D. Mo. 1985).

[34] Example based on *The Sherman R. Smoot Co. v. Ohio*, 136 Ohio App. 3d 166, 736 N.E. 2d 69 (Oh.App. 2000). The case was remanded for further determination on the issue of differing site conditions.

[35] Problem based on example in the *Construction Claims Training Guide*, Sept. 1989.

[36] *Ruff v. U.S.*, 96 Ct. Cl. 148 (Ct.Cl. 1942). *Air Cooling & Energy v. Midwestern Const.*, 602 S.W.2d 926 (Mo. App. 1980).

[37] *Appeal of Rottau Elec. Co.*, 76-2 B.C.A. (CCH) P12,001 (A.S.B.C.A. June 29, 1976).

[38] *Asphalt Roads & Materials Co., Inc. v. Virginia*, 257 Va. 452, 512 S.E.2d 804 (Vir. 1999).

[39] *KGM Contractors, Inc. v. Cass County, Minnesota*, 2006 Minn. App. Unpub. LEXIS 626 (Minn. App. 2006, unpublished opinion).

[40] *Renda Marine, Inc., v. U.S.*, 66 Fed. Cl. 639 (Ct. Cl. 2005).

[41] *U.S. v. Spearin*, 248 U.S. 132, 136, 39 S. Ct. 59, 61, 63 L. Ed. 166, 54 Ct. Cl. 187 (1918).

[42] State of Washington.

[43] Example based on *Christie v. U.S.*, 237 US 234, 59 L Ed 933, 35 S Ct 565 (1915).

[44] *National Presto Industries, Inc. v. U.S.*, 167 Ct. Cl. 749, 338 F.2d 99 (Ct.Cl. 1964).

[45] Example based on *Conner Brothers Const. Co. v. U.S.*, 65 Fed. Cl. 657, 2005 U.S. Claims LEXIS 159 (Fed. Ct. Cl. 2005).

CHAPTER 3

Employment and Labor Law

■ Introduction

Historically, little law existed to control the employment relationship, and that relationship was basically under the complete control of the employer. The employee's only choice was to terminate the employment. However, as power to control governmental process has increasingly been transferred to a larger and larger percentage of the voting population, the legal landscape regarding the employment relationship has changed dramatically. Today the term **employment law** generally refers to law relating to the employment relationship, and the term **labor law** refers specifically to law relating specifically to unions and the right to unionize.

This chapter explains the hiring and firing of employees, federal statutory protections of employees, employment torts, personnel practices, and union contracts.

■ Hiring and Firing

Issues relating to hiring and firing revolve around acceptable and unacceptable forms of discrimination in the hiring and firing process. Historically employers were free to make determinations regarding hiring or firing based on any reason or no reason. This principle is known as the **employment-at-will doctrine** and is applicable today but with modifications. The employment-at-will doctrine states that the employer may hire or fire an employee for any reason or no reason. In addition, the doctrine holds that the employee may quit for any reason or no reason. Today, many exceptions exist to this doctrine.

Illegal Discrimination

Federal law prevents any employer providing public accommodation or transportation and employers with 15 or more employees from certain types of discrimination, referred to as illegal discrimination. **Illegal discrimination** is discrimination based on race, creed, sex, religion, national origin, citizenship, disability, pregnancy, union membership, or age. The term **protected class** is sometimes used to refer to the group of people with any of these characteristics. **Legal discrimination** is discrimination based on any other reason such as discrimination based on experience, education, family relationship, congeniality, or any other valid reason for discriminating between employees or potential employees.

State law may expand the number of covered employers to virtually all employers in the state. Some states may have additional categories of illegal discrimination. For example, Michigan prevents discrimination on the basis of height and weight.[1]

Most employers use some type of pretext to hide illegal discrimination. If an employer is alleging that the decision to hire or fire was *not* based on illegal discrimination, and the employee or potential employee is claiming that the decision to hire or fire was based on illegal discrimination, a factual issue arises, and a jury must decide which of the two positions is correct.

EXAMPLE

Cynthia, an African American woman with a college degree, worked at a deli in a grocery store. More than a year later, the owner of the store promoted a Caucasian woman to the position of deli manager. The Caucasian woman had worked in the deli for only three months, had only a sixth-grade education, and could not calculate prices or read recipes.

Although the owner gave various reasons for promoting the white woman instead of Cynthia, the court held that these reasons could be pretexts for hiding illegal discrimination based on race and the matter would have to be decided by a jury. The matter was ordered to trial for a determination of whether or not illegal discrimination occurred.[2]

In addition to the illegal discrimination described above, the law forbids what is called disparate impact discrimination. **Disparate impact discrimination** is some employment-related requirement or discrimination that on its face appears to be neutral but actually affects those in a protected class differently or disproportionately. This type of discrimination is subject to legal scrutiny for its validity. The discrimination will only be upheld if some valid reason for the employment-related requirement exists.

EXAMPLE

An employer has both indoor and outdoor maintenance/cleaning job positions. The indoor jobs are more desirable than the outdoor jobs. The employer has a requirement that only persons with a high school degree can have the indoor jobs. This results in the majority of indoor jobs going to Caucasians and the majority of outdoor jobs going to African-Americans. Because the requirement of a high school education is irrelevant to the ability to do the job, it is a form of illegal disparate impact discrimination. The employer may not require a high school diploma for the inside maintenance/cleaning jobs.

Employers may discriminate on race, religion, national origin, or sex if that characteristic is a bona fide occupational qualification. A **bona fide occupation qualification (BFOQ)** is a legitimate requirement that the person holding the job have some characteristic that would normally be classified as illegal discrimination. A BFOQ is a limited exception to antidiscrimination law and allowable only in special circumstances.

EXAMPLE

A Catholic church can discriminate and hire only Catholic priests. A hospital can have a policy to hire only women mammogram technicians. A women's prison can have a policy that only female guards may supervise female prisoners in the shower. All of these qualify as BFOQs.

EXAMPLE

Fragante applied for a clerk's job at the Department of Motor Vehicles but was not chosen because of a heavy Filipino accent that made him extremely difficult to understand. Because being able to communicate clearly with English-speaking customers was an important part of the job, he was not hired. The court determined that this was an acceptable form of discrimination.[3]

EXAMPLE

Johnson announced a policy that women, except those whose infertility was medically documented, could *not* work in jobs involving actual or potential exposure to lead. All of these jobs were higher paying than other jobs at the company. The company stated that it feared lawsuits by pregnant women who were exposed to lead. However, exposure to lead can also harm males and can cause impotence. The court held that the male-only requirement for the high-paying jobs was *not* a BFOQ. While it was true that medical harm could result from the exposure to lead, the court said the way to handle the problem was to protect all workers in those jobs from exposure to dangerous levels of lead.

The federal government and some states prevent employers from discriminating against whistleblowers. See the discussion of whistleblower protection in the section "Public Policy Exception to Employment at Will."

Public Policy Exception to Employment at Will

The terms **wrongful discharge** or **wrongful termination** are used to cover exceptions to the employment-at-will doctrine adopted under state laws but not in the category of illegal discrimination. These additional exceptions to the employment-at-will doctrine are often called the **public policy exceptions** to the employment-at-will doctrine. Each state has its own public policy exceptions and wrongful discharge law. The Model Employment Termination Act is an attempt to unify this area of the law, but no state has adopted it.

Typical public policy exceptions to the employment-at-will doctrine include the following:

- Refusing to break the law
- **Whistleblowing**, or informing government authorities when the employer has broken the law
- Exercise of a public right, such as voting or filing a claim
- Exercise of a public duty, such as jury duty

EXAMPLE

Hauck, a deckhand for Sabine Pilot, was instructed that one of his duties each day was to pump the bilges of the boat into the bay. He observed a placard posted on the boat that stated it was illegal to pump the bilges into the bay. He called the U.S. Coast Guard, and an officer confirmed that pumping bilges into the bay was illegal. Hauck refused to pump the bilges into the bay and was fired. He filed a lawsuit against Sabine Pilot for wrongful discharge. Sabine Pilot claimed the employment-at-will doctrine allowed it to fire Hauck for any reason or no reason, and therefore it could fire Hauck for refusing to break the law. The Texas Supreme Court declined to dismiss the case and so recognized the legal claim of wrongful discharge. The court said, "The sole issue for our determination is whether an allegation by an employee that he was discharged for refusing to perform an illegal act states a cause of action....Upon careful consideration of the changes in American society and in the employer/employee relationship....We now hold that public policy requires a narrow exception to the employment-at-will doctrine....That narrow exception covers only the discharge of an employee for the sole reason that the employee refused to perform an illegal act. We further hold that in the trial of such a case, it is the plaintiff's burden to prove by a preponderance of the evidence that his discharge was for no reason other than his refusal to perform an illegal act."[4]

The law in the United States has become more conservative, favoring business and the government over individuals, in recent years, and the Texas Supreme Court refused to grant protection to an employee fired for *reporting* a possible crime. This result may not be the law in all states.

EXAMPLE

Claude D'Unger was an officer and director of the Ed Rachal Foundation, a charitable organization that owns a ranch in Webb County, Texas and used for wildlife and farming research studies. The ranch covers more than 100 square miles, including five miles along the Rio Grande. Due to its location, migrants from Mexico frequently cross the ranch on foot. D'Unger became concerned that the ranch's foreman, Ed DuBose, was harassing migrants, and he reported his concerns to Paul Altheide, the foundation's chief executive officer. According to D'Unger, Altheide told him "to drop it," which he took as an instruction not to report DuBose's activities to any law enforcement officials.

On September 17, 1997, DuBose apprehended three teenage Mexican nationals at the ranch, handcuffed them, and claimed to have turned them over to border patrol agents. When D'Unger saw a ranch report of the incident, he contacted border patrol agents, who told him they had no knowledge or record of the incident. Concerned that a crime might have been committed, D'Unger subsequently contacted a congressman, two sheriffs, the Texas Attorney General's office, a senator, the IRS, a district judge, and the Mexican Consulate about the matter. When Altheide learned of D'Unger's activities, he first suspended him and then fired him when he refused to resign.

D'Unger sued for wrongful termination. However, the court dismissed his case for failure to state a cause of action, holding that the public policy exception outlined in the Sabine Pilot case is narrow and only protects employees who are fired for refusing to commit a crime. "Sabine Pilot protects employees who are asked to *commit* a crime, not those who are asked not to *report* one."[5]

MANAGEMENT TIP

While the law of the United States does not require just cause for terminating an employee, and recent court decisions have tended to favor employers over employees, hard feelings and lawsuits can be avoided by documenting the cause for an employee's termination and making that documentation available to the employee.

A summary of the exceptions to the employment-at-will doctrine is given in Table 3-1.

TABLE 3-1

Exceptions to the employment-at-will doctrine	
Illegal discrimination	**Public policy exceptions**
Race	Refusing to break the law
Creed	Whistleblowing, or informing government
Sex	authorities when the employer has broken
Religion	the law
National origin	Exercise of a public right such as voting or
Citizenship	filing a claim
Disability	Exercise of a public duty such as jury duty
Pregnancy	
Union membership or attempt to	**Americans with Disability Act exception**
form union	
Age	Employers must make reasonable
	accommodations for disabled employees

Affirmative Action

Affirmative action is a particularly hot topic at the present time and subject to radical change by proposed legislation at both the federal and state levels. Currently, private businesses may engage in voluntary affirmative action programs but only for remedial purposes such as to overcome underrepresentation of a protected class or past discrimination. A case that has been watched by the construction industry for over a decade is the *Adarand* litigation.[6] Adarand was the low bidder on a federal highway project, but the job was given to a disadvantaged business owner as required by statute. Adarand filed suit, claiming unlawful discrimination. The case has been to the U.S. Supreme Court three times, and still the Supreme Court has upheld affirmative action in limited circumstances.

EXAMPLE

Transportation Company, Inc. voluntarily adopted an affirmative action program to achieve a better racial and gender balance in the workforce. The company advertised a position for a road dispatcher. No women held jobs in that category. Two applicants, one male and one female, were the best candidates for the job. Although the woman scored two points lower on her test than the man, she was given the position because her gender was taken into account. The man sued, alleging employment discrimination based on sex. It was held that gender can be taken into account in moderate, flexible, affirmative action plans. The employer was free to hire the woman.

Interviewing

Employers must take care when interviewing potential employees. Because the law prevents discrimination based on race, creed, sex, religion, national origin, citizenship, disability, pregnancy, union membership, or age, an interviewer must be careful not to ask questions in the hiring process that could be seen as an attempt to engage in illegal discrimination.

EXAMPLE

An interviewer asked an applicant what year the applicant graduated from high school. This is a subtle way to find out the applicant's age and could be seen as an attempt to engage in illegal age discrimination. If an employer requires that a photo be attached to the job application, this could be an attempt to engage in illegal discrimination based on race or national origin.

MANAGEMENT TIP

It is best for an interviewer to have a printed set of questions that are asked of all applicants. The questions should be reviewed to make sure that no illegal forms of discrimination occur in the interviewing process. Photos should not be required or taken of applicants.

The following types of questions are unacceptable:

1. How old are you?
2. Are you married?
3. Do you have children?
4. Where do your ancestors come from?
5. What is your religion? Any questions relating to religion should not be asked.
6. Are you disabled?
7. Are you pregnant or planning on becoming pregnant in the near future?
8. What is your opinion about unions?

Interview questions should relate only to the applicant's ability to do the job and meet the job requirements.

Hiring Immigrants and Day Laborers

It is illegal to hire persons who are not legally in the United States, and pursuant to federal law, all employees must fill out an I-9 Employment Eligibility Verification form. The reality is that the construction and farming industries frequently use illegal labor. Employers hiring illegal labor should be aware that employment, labor, and tax law apply to illegal immigrants. That is, taxes and Social Security payments are to be deducted and paid to the appropriate government agency. Some contractors pay day laborers, particularly illegal day laborers, under the table and do not withhold taxes and Social Security from the sums paid.

EXAMPLE

A contractor hires illegal day labor to complete a construction project. The workers are covered by the Fair Labor Standards Act and all other laws. To do otherwise would encourage the use of illegal labor.

In addition, it is illegal to threaten or retaliate against illegal immigrants who are seeking their rights under applicable employment or labor law such as the minimum wage law.[7] In the case of *U.S. v. Kozminski*, 487 U.S. 931 (1988) the court said "it is possible that threatening...an immigrant with deportation could constitute the threat of legal coercion that induces involuntary servitude" in violation of federal criminal statutes. The Alien Tort Claims Act[8] gives aliens a tort cause of action in U.S. courts for torts such as this committed against them. Few illegal immigrants will attempt to protect their rights, and if they do, their illegal status can be used to negotiate a settlement more favorable to the employer than would otherwise be the case if the employee were legal.

EXAMPLE

A former domestic worker from the Philippines sued her "employer" and alleged involuntary servitude.[9] Former landscaping employees from Mexico sued, alleging involuntary servitude.[10] The plaintiffs' status as legal or illegal aliens was irrelevant to the litigation.

Illegal immigrants may not be able to obtain some types of relief or recovery. In *Hoffman Plastic Compounds, Inc. v. NLRB*[11] the U.S. Supreme Court refused to award back pay to an illegal immigrant who was still living in the United States **Back pay** is a sum to compensate an employee for wrongful termination—in other words, pay for work that has not actually been performed.

it is not uncommon for illegal immigrant labor to be treated illegally by those who hire them. Legislation called the "Day Labor Fairness and Protection Act" has been introduced into the U.S. Congress but has not yet been passed. This act affords the following:

- Ensures safe and healthy work environments for all day laborers
- Protects and expands the wage and hour rights of day laborers
- Outlaws retaliation for those seeking to enforce rights under the act
- Hold day labor employers accountable
- Prohibits the use of day laborers as strike breakers

Ability to Terminate Employment

The employment-at-will doctrine generally allows an employer to terminate an employee for any reason or no reason.

EXAMPLE

Employees can be terminated if business is slow, the employer wishes to hire her nephew, the employee wears too much makeup, the employee wears inappropriate clothing, or the employee has a tattoo or piercing. As long as the firing is not a form of illegal discrimination and does not violate any public policy exception, the firing is legal.

The same laws that apply to hiring apply to firing. That is, employers may not fire someone based on race, creed, sex, religion, national origin, citizenship, disability, pregnancy, union membership, or age, as these are all forms of illegal discrimination. As with hiring discrimination, the employer typically gives some other reason for firing the employee without admitting to illegal discrimination. This results in an issue of fact that must be decided by a jury.

EXAMPLE

Santillo, a bus driver, began to unionize the drivers. Evidence showed that Santillo's supervisor was angry at him for engaging in unionization activity. After a few days, Santillo was fired for leaving his keys on the bus and taking unauthorized breaks. Santillo was not given a warning. Evidence showed that the supervisor had not enforced these rules against other employees who had left their keys on the bus or taken unauthorized breaks. Santillo claimed that he was fired for attempting to form a union and filed a claim. The situation raised an issue of fact: whether or not the employer had used a pretext to fire Santillo. If the employer did fire Santillo for attempting to form a union, it engaged in an unfair labor practice.[12]

Employee handbooks may contain detailed explanations of the procedures that a company must use prior to disciplining or terminating an employee. The company must follow all procedures as outlined in the employee handbook.

EXAMPLE

An employee was hired by a company and fired without warning. The employee handbook stated that the employer would only fire someone after giving them warnings and following certain disciplinary procedures. The court held that the employer was required to follow the procedures set out in the handbook.[13]

Employment handbooks, contracts, or even correspondence between the parties may contain a representation that the employee will be terminated only for just cause. These representations are generally upheld by the law.

EXAMPLE

Laura Berent was hired by a company and fired eight months later. She claimed that when she was hired, she was assured that she would remain employed as long as she did a good job. The court held this created a contract that her employment would be terminated only for just cause.[14]

MANAGEMENT TIP

In order for an employer to retain its ability to fire an employee for any reason or no reason under the employment-at-will doctrine, an employer should have language on its application and other forms informing potential employees that they are "at-will" employees. The employer must be careful that any employee handbooks are consistent with this language.

I understand that if I am hired, I will be an at-will employee, and my employment is subject to termination for any reason or no reason. I understand _____ (name of company) can terminate my employment at the discretion of the employer. The only exception to this is in the event that a written contract is entered into between myself and the company altering this provision.

Signed: _____ (signature of applicant).

■ Noncompete and Nondisclosure Agreements

Many employers are beginning to use noncompete and nondisclosure agreements, which limit an employee's options for new employment upon termination of the prior employment. Nondisclosure agreements attempt to prevent employees from using trade secrets of their former employers when they take a different position. Trade secrets are discussed in Chapter 10 in the section entitled "Intellectual Property."

Noncompete agreements are upheld by the law as long as they are for a reasonable time period and are limited to a reasonable geographical location. An agreement preventing a former employee from ever engaging in competing work upon termination from employment would not be upheld.

EXAMPLE

Ackerman was employed by Kimball International and signed a noncompetition agreement that prevented him from "directly or indirectly engaged in inventing, improving, designing, developing or manufacturing, any products directly competitive with the products of Employer." The court issued a preliminary order (injunction) preventing Ackerman from accepting or commencing employment with a competitor for a period of one year. The noncompetition agreement was held to be valid.[15]

MANAGEMENT TIP

Employers with specialty information, processes, customer lists, or any other type of information that gives them an advantage over a competitor should have noncompete and nondisclosure agreements with employees if possible.

In consideration of the Company's continued employment of Employee, Employee agrees as follows:

1. Employee promises that following the termination of employment with the Company, for any reason whatsoever, Employee shall not disclose to any person the business or trade secrets ("Business Secrets") of the Company.

2. Employee promises that during or after Employee's employment with the Company, the Employee will not take from the business premises of the Company or in any way convert to Employee's own use the Company's books, lists, files, computer software, computer disks, or other storage of data or record, records or documents or compilation of information of any kind (or copies, transcriptions, or other excerpts of the same) related in any way to the business of the Company, even if such materials would not be considered Business Secrets. Employee also promises that, upon termination of employment, Employee shall surrender all such material to the Company, but Employee may retain copies of items that relate solely to Employee's personal affairs and copies of any items that are publicly available or generally available to those in the trade. Employee agrees to return all originals of such documents to the Company.

3. If Employee's employment with the Company is terminated, the Employee may establish or accept employment with a competing business and may compete with the Company or aid the competing business to compete with the Company. However, upon the termination of Employee's employment with the Company, for any reason whatsoever and at any time, regardless of whether the Company has complied with its obligations to Employee, Employee promises that for one (1) year after the date of termination of employment, the Employee will comply with the following limitations:

(a) Employee shall not for himself or for another solicit or aid any other person to solicit any person who is a client of the Company at the time of termination or who has been a client at any time within 180 days prior to termination either to terminate the client's private security business services relationship with the Company for a project located in the Territory, or to purchase private security business services from Employee or from another for any other project in the Territory.

(b) Employee shall not solicit or aid any other entity to solicit any person who is at the time Employee's employment is terminated, or who has been at any time within 180 days prior to or after Employee's termination, an employee of the Company.[16]

■ Other Statutory Protections

With the decline in the importance of unions in the workforce since the 1950s, the law has become more complex. The law is now is the major regulator of the employment relationship rather than any union contract. Both federal and state law heavily regulate the employment relationship. The following major federal statutory protections afforded employees are discussed below:

- Fair Labor Standards Act and the Davis-Bacon Act
- Family Medical Leave Act
- OSHA
- Americans with Disabilities Act

Other major federal statutory protections not discussed in detail here include the following:

- National Labor Relations Act (NLRA). This act protects the right of workers to unionize and requires employers to negotiate with union representative.
- Employee Retirement Income Security Act (ERISA). This act regulates pension and employee welfare benefit plans.
- Worker Adjustment and Retraining Notification (WARN) Act. This act requires employers with 100 or more employees to provide at least 60 days' advance written notice to employees who will suffer an employment loss by virtue of a plant closing or a mass layoff.
- Family and Medical Leave Act (FMLA). This act requires employers with 50 or more employees to permit employees to take up to 12 weeks each year of unpaid leave in order to care for a new child, a family member with a serious health condition, or because of the employee's own serious health condition.

Fair Labor Standards Act and the Davis-Bacon Act

The **Fair Labor Standards Act (FLSA)** of 1938 requires employers to pay employees a minimum wage and to pay certain employees overtime pay for work in excess of 40 hours per week. Blue-collar workers or those who do repetitive manual labor are covered. All employees who make less than $455 per week are covered no matter what their job title or classification.

FLSA does not apply to executives, administrators, professionals, or those with "advanced knowledge" unless they make less than $455 per week. Persons who are commonly called "white-collar workers" are therefore not covered by this law.

Recent changes in the FLSA have expanded the group of persons *not* covered by the FLSA. Persons who make in excess of $65,000 are not covered no matter what their job title or duties. In addition, executive secretaries, computer technicians, designers, and drafters are not covered. Historically, the test was whether or not the employee exercised "independent judgment"—employees who exercised independent judgment or managed others in their job were not covered. The independent judgment test has been eliminated, and now many workers previously protected by the FLSA are no longer covered. However, state laws may cover these employees.

The **Davis-Bacon Act** requires federally funded construction projects over $2,000 to pay the prevailing wages of the locality in which the project is located. **Prevailing wages** are the standard wages in the area and are determined by the federal government. The prevailing wages are determined by the Secretary of Labor and are available online at **http://www.wdol.gov**. Many states have a similar law with regard to state-funded construction projects.

Family Medical Leave Act

The **Family Medical Leave Act (FMLA)** requires employers to give employees up to 12 weeks of unpaid leave per year for various family-related problems such as the birth of a child or the serious illness of a child or other family member. FMLA only applies to employers with 50 or more employees within a 75-mile radius of the principle place of business or worksite.

OSHA

The **Occupational Safety and Health Act (OSHA)** of 1970 was passed to improve workplace safety and to encourage healthful working conditions. In furtherance of its goals, OSHA has passed thousands of regulations related to workplace safety and health. These regulations are not discussed in detail here.

Historically, construction was one of the most deadly types of work, and for this reason, OSHA tends to keep a close watch on the construction industry. Every person involved in the industry must have an understanding of workplace safety requirements.

Contractors and subcontractors must be careful not only of their own OSHA violations but must also exercise care for OSHA violations committed by other contractors and subcontractors on the job. OSHA's current policy[17] is to issue a violation against the following:

- Any employer whose employees are *exposed* to hazards, whether or not caused by their employer
- The employer who actually causes or creates the hazard
- The employer who has the general supervisory authority over the worksite
- The employer who has the responsibility for safety and health conditions at the worksite

EXAMPLE

A recent example of an OSHA case arose out of an incident in March 2000, when a scaffold collapsed during the construction of a theatre in Providence, Rhode Island. OSHA cited the general contractor for a total of $14,400 in fines for four alleged serious violations under the construction standard: loading a scaffold in excess of its rated capacity; failing to have a competent person inspect the scaffold for defects prior to each work shift; failure to provide fall protection; and improper bracing and unstable cribbing. The sheet rock contractor was cited for $62,500 in fines for three alleged repeat violations: loading a scaffold in excess of its rated capacity; failure to have a competent person inspect the scaffolding before each shift, and failure to provide fall protection. It was also cited for one serious violation for failure to situate scaffold legs on base plates. Further, it was cited with another serious violation for failure to train employees. The contractor that erected the scaffold faced $19,500 in fines for four alleged serious violations similar to those brought against the general contractor.

continued

> A further example was the citation in February 2000 in New Rochelle, New York, of three construction contractors—a general contractor, a scaffolding contractor, and a plastering contractor—for a total of two willful, five repeat, ten serious, and one other-than-serious violation of OSHA standards with proposed penalties totaling $133,000.[18]

Failure to follow OSHA regulations can lead to criminal liability for corporations, officers, and agents.

EXAMPLE

OSHA standards state "[s]ides of trenches in unstable or soft material, five feet or more in depth, shall be shored, sheeted, braced, sloped, or otherwise supported by means of sufficient strength to protect the employees working within them." When a trench without any type of protection as required by the regulation collapsed, killing a worker, the corporation employing the worker was convicted of criminal willfulness and fined $3,500. The evidence revealed that the responsible corporate agents had all observed the trench prior to the accident and failed to initiate shoring procedures. Because corporations are responsible for the acts and omissions of their agents, the corporation was liable.[19]

EXAMPLE

At a trial in Cook County, Illinois, the president, plant manager, and foreman were found guilty of murder and sentenced to 25 years in prison and fined $10,000 each for a failure to follow regulations in a silver recovery plant that led to the death of an employee.[20]

Americans with Disabilities Act

The **Americans with Disabilities Act (ADA)** prohibits discrimination against, and requires reasonable accommodation of, the disabled in job application, hiring, advancement, discharge, pay, training, and other terms, conditions, and privileges of employments. A person "with a disability who...with or without reasonable accommodation can perform the essential functions" of his or her job cannot be discriminated against. The definition of **disabled** includes a physical or mental impairment that substantially limits one or more of the major life activities of such an individual. The following are not considered disabilities under the act, and therefore people with these characteristics are not protected: drug abuse or addiction, transvestitism, transexualism, pedophilia, compulsive gambling, kleptomania, and pyromania.

If a person qualifies as disabled, the employer must make reasonable accommodations for the disabled person.

EXAMPLE

Ryan Yingling has an injury that makes it extremely difficult for him to climb stairs. His office is on the second floor of a two-story building that has no elevator. The cost to install an elevator is $2 million. The building also has offices on the first floor. A reasonable accommodation is to give the Yingling an office on the first floor. The employer is not required to put in an elevator, because the expense is an unreasonable burden. Prior to the passage of the ADA, the employer could have just fired Yingling under the employment-at-will doctrine, but ADA requires a reasonable accommodation.

If no reasonable accommodation can be made, the employer may terminate or transfer the disabled person. If the employee's disability prevents job performance, even with accommodation, then the employee can be terminated or transferred.

EXAMPLE

Tyndall suffered from lupus, a joint disorder that causes joint pain and inflammation, fatigue, and urinary and intestinal disorders. Her employer made accommodations including sick leave, allowing her to come in late and leave early on occasion, and taking breaks as needed.

After two years, she began to miss more and more work. Her employer suggested that she resign and come back later if she could. At that point, she had missed approximately 40 days in seven months. Tyndall did not resign, and she was fired. The court held that the employer had made a reasonable accommodation, and she could still not do the work. The firing was legal.[21]

EXAMPLE

April Czarnik, a waitress, suffered from panic attack disorder and was unable to handle the pressure of working on the particularly busy evenings of Friday and Saturday. She requested that she only be scheduled to work Sunday through Thursday. She could be terminated because the accommodation requested was not reasonable given the nature of the work.

Greg Hilla has chronic fatigue syndrome and cannot work. He is classified as totally disabled and can be terminated from his employment because he cannot work even with a reasonable accommodation.

A person who is able to correct their condition themselves, such as by buying eyeglasses, is not considered disabled and is not covered by the ADA.

EXAMPLE

Corey Lipar was an applicant for a commercial airline pilot position, but he was rejected because the company requires a minimum 20/100 visual acuity in both eyes. He had 20/200 vision in both eyes without glasses and 20/20 with glasses. He met all the other requirements for the position but was not hired. Since he could correct his vision with eyeglasses, he was not disabled, and the ADA did not cover him. The employer did not need to make a reasonable accommodation and could refuse to hire him.[22]

The ADA does contain a provision allowing an employer to refuse to hire someone or to fire someone who is a "direct threat" to the safety and welfare of other employees or customers. Congress defined a **direct threat** as a "significant risk to the health or safety of others that cannot be eliminated by reasonable accommodation."

EXAMPLE

Sara Lubinski, a bus driver, suffered from a disorder that brought on blackouts. The company fired her because she was a direct threat to passengers in the operation of the bus. The company did not need to make a reasonable accommodation or attempt to find her an alternate position.

Antonio Quinn suffered from bipolar disorder and depression. His employer could not assume that he was a direct threat from the mere fact that he suffered from these disorders. The employer must have had some evidence that he was a threat to the health of other employees.

Recent cases interpreting the ADA have been favorable to employers and limit the rights of employees.[23]

EXAMPLE

Barnett was a cargo handler who had injured his back on the job. He requested a transfer to a mailroom job that would not exacerbate his back injury. However, an employee with more seniority also applied for the mailroom job. The company had a formal seniority policy that would have favored the second, more senior employee. The legal issue was whether or not the ADA's requirement that the employer make a reasonable accommodation trumped the company seniority policy. The Supreme Court held that it did not.[24] The company could legally give the mailroom job to the more senior employee and terminate Barnett, the injured employee.

■ Workers' Compensation

In addition to the above federal statutory schemes, states have laws regulating the relationship between employer and employee. One of the most important laws at the state level is the **Workers' Compensation law,** which provides medical coverage for employees injured on the job. While not all states require employers to have Workers' Compensation insurance, the vast majority do. Workers' Compensation law prevents employees from suing employers for most injuries on the job and provides coverage for any job-related injury no matter how the injury occurred. In a state with optional Workers' Compensation insurance, an injured employee can recover damages only for work-related injuries if the employee can prove that the employer or another employee was negligent.

Damages recoverable under Workers' Compensation laws are limited, often leading the injured party to look for additional sources of recovery. While Workers' Compensation law prevents the employee from suing the employer, unless the employer acts intentionally or with reckless disregard for the safety of employees, it does not prevent the injured employee from suing others who may have contributed to the injury and recovering damages not covered under the insurance policy. In addition, the Workers' Compensation insurance carrier can seek reimbursement from other parties whose acts have contributed to the injury. A commercial general liability insurance policy, also called a comprehensive liability policy will usually provide coverage to a party sued in such a situation. See Chapter 11 for discussion of insurance issues.

EXAMPLE

A designer is injured on the job site when the designer trips over debris that should have been cleaned up by a subcontractor. The designer can recover from the employer's Workers' Compensation policy and can sue the contractor and subcontractor for damages not paid by the coverage—for example, for pain and suffering. In addition, the Workers' Compensation carrier can sue the subcontractor and the contractor for reimbursement of the benefits paid to the designer.

■ Job-Related Torts against Employers

The doctrine of *respondeat superior* or **vicarious liability** says employers are responsible for the torts of their employees. See Chapter 12 for full discussion of torts. **Vicarious liability** arises whenever one party is legally liable for the actions of another party.

Vicarious liability always leads to issues of joint liability. **Joint liability** is liability by two or more entities for the same injury. The vicarious liability of an employer does not relieve the employee from liability; it merely adds another potentially liable party. All parties, including employees, are liable for the torts they commit even if their employer is also liable. In the real world, an injured party is unlikely to sue an employee for two reasons: the employee has no money or the employee is more likely to testify truthfully as to what happened if the employee is not sued (Figure 3-1).

FIGURE 3-1 Vicarious Liability of Employer for Employee's Acts

EXAMPLE

On a construction site, Gary Brown, who was employed by English Construction, Inc., put a tarp over a hole in the uncompleted second floor. David Parker, also employed by English Construction, Inc., and Chris Schreiber, a county inspector, walked across the tarp while inspecting the area and fell through to the floor below. English Construction, Inc. was vicariously liable for injuries to both Chris Schreiber and David Parker. Gary Brown, the employee, was also liable for his own negligence.

Sexual Harassment

Sexual harassment law is extremely complex. Case law has developed two types of sexual harassment: quid pro quo and hostile work environment sexual harassment.

Quid pro quo sexual harassment is the simplest type of sexual harassment. It is a demand for sexual favors in exchange for a job-related benefit. The employer is liable to the sexually harassed employee under the theory of vicarious liability. The employer is liable whether or not the employer has a sexual harassment policy. This is a tort cause of action. Recoverable damages are discussed in Chapter 12.

EXAMPLE

A manager of an employee demands sexual favors from the employee before the employee will be given a raise. This is quid pro quo sexual harassment.

Hostile work environment sexual harassment law is much more complex than quid pro quo sexual harassment. It can be divided into two categories: the law affecting companies *without* a sexual harassment policy (Figure 3-2) and the law affecting companies *with* a sexual harassment policy (Figure 3-3). Companies *without* a sexual harassment policy are liable for hostile work environment sexual harassment if the employee can prove a hostile work environment, as defined below. The employee need not prove any tangible employment action, as defined below. Companies *with* a sexual harassment policy are liable for hostile work environment sexual harassment if a hostile work environment exists, the company has made a tangible employment action, and either the company failed to take reasonable steps to abate the sexual harassment *or* the employee failed to take advantage of company processes to stop the harassment.

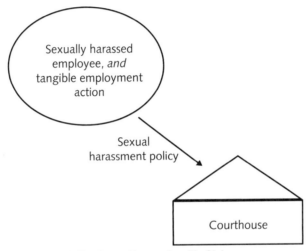

FIGURE 3-2 Sexually Harassed Employee Proceeds with Claim

A **hostile work environment** exists when the work environment is characterized by severe or pervasive sexually offensive conduct. The conduct cannot be merely rude or gross but must rise to a higher level of offensiveness. Because the conduct in such cases can be so varied, each case must be looked at individually to determine if the conduct rises to the level of a hostile work environment. The trier of fact will look at all of the circumstances including the frequency of the discriminatory conduct, its severity, whether or not it is physically threatening or humiliating, as a merely offensive utterance is insufficient, and whether or not the offensive conduct unreasonably interferes with an employee's work performance. A **tangible employment action** or decision includes firing, pay reduction, relocation to a less-desirable job, and similar acts.

Frequently employers fearing a sexual harassment claim will attempt to document transgressions of the sexually harassed employee and then fire the employee based on claimed legitimate employment concerns. The law states that if the tangible employment action is the result of illegal sexual discrimination *and* legitimate employment concerns, the employer is still liable to the employee if the employee can prove sex

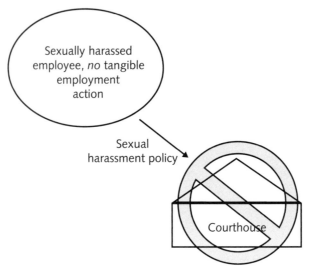

FIGURE 3-3 Sexual Harassment Policy Blocks Claim if No Tangible Employment Action

discrimination was at least a *motivating factor,* even if other legitimate motivating factors were involved. In other words, if a hostile work environment exists and sex discrimination was a motivating factor in the tangible employment action, the employer is liable.

EXAMPLE

Catharina was employed as a warehouse worker and heavy equipment operator for an employer. She was the only woman in this job and in her local Teamsters bargaining unit. She presented the following evidence: (1) she was stalked by a supervisor, (2) she received harsher discipline than men, (3) she was treated less favorably than men in the assignment of overtime, (4) her supervisors stacked her disciplinary record, putting disciplinary items in her file that were not included in the men's files, and (5) her supervisors used and/or tolerated sex-based slurs against her. Her employment file contained many disciplinary notes. Eventually she was terminated for fighting with a male coworker. The male employee was given a five-day suspension, but he did have a clean disciplinary record prior to the incident. The employer claimed that Catharina was fired because of her extensive disciplinary file while the man was not because his disciplinary file was clean. After a trial on the merits, the jury determined that a hostile work environment existed, a tangible employment action had occurred, and the tangible employment action was motivated in part by sex discrimination. The plaintiff was awarded damages.[25]

Employees at a company with a sexual harassment policy must take advantage of complaint procedures in place at the company to stop the sexual harassment. If no specific procedures exist, the employee must take reasonable actions to try and abate the hostile work environment.

EXAMPLE

Suders, a female, was hired by the state police as a communications operator. The state government had a sexual harassment policy and sexual harassment prevention procedures applicable to all state government employees. A copy of this was given to each employee, including Suders. She claimed that some male employees subjected her to a continuous barrage of sexual harassment that ceased only when she resigned from the force. Some of the discussions held in her presence by supervisors, which she claimed contributed to the hostile work environment, included people having sex with animals and how young girls should be given instruction in how to gratify men with oral sex. Acts included sitting across from Suders with spandex shorts and legs opened, making an obscene gesture popularized by television wrestling, a supervisor grabbing his genitals and shouting out a vulgar comment inviting oral sex, and pounding the furniture she was sitting at to intimidate her. Comments made to her included a supervisor rubbing his rear end in front of her and remarking, "I have a nice ass, don't I," "The village idiot could do your job," "A 25-year-old could catch on faster," and calling her "mama." Such acts would occur from 5 to 10 times per shift. At one point, she was accused of taking a missing accident file home. Her supervisors denied some of the acts or said that they were taken out of context.

Suders was required to take a computer skills exam to satisfy a job requirement, and she took it several times. She was told each time that she had failed. One day, she came upon her exams in a cabinet in the women's locker room. Her supervisors had never forwarded the tests for grading, and their reports of her failures were false. She regarded the tests as her property and removed them from the locker room. She was terminated for the theft of her exam forms, but no criminal charges were brought against her.

She filed a hostile work environment sexual harassment suit, and the employer filed a motion to dismiss. The case against the employer was dismissed, because even if all of the allegations made by Suders were true, she had failed to utilize the employer's internal sexual harassment prevention procedures, thereby relieving the employer of liability.[26]

MANAGEMENT TIP

All employers should have a sexual harassment policy and a process that attempts to prevent and correct promptly any sexually harassing behavior. *The Business Owner's Toolkit,* published by CCH, Inc., has sample sexual harassment policies that can be used by businesses. See *http://toolkit.cch.com.* In the horizontal bar across the top, choose Business Tools, scroll down under Employee Management, and click on Sexual Harassment Policy. You can also find forms at *http://entrepeneaur.com.*

Invasion of Privacy and Defamation

With the advent of the technological revolution, an employer's ability to monitor and find information about an employee has been greatly enhanced. For example, a product called SurfWatch® can monitor employee Internet use. It is possible to put small television cameras in any location and record what an employee does at her desk all day. The law has historically found that employees have a reduced level of privacy in the workplace, and employers have been allowed to search lockers, desks, and e-mail. A current issue in the law considers the extent to which an employer can monitor an employee with the newly available software.

The Electronic Communications Privacy Act (ECPA) of 1986[27] prohibits the intentional interception and disclosure of information obtained by means of wiretapping, e-mail, cellular telephone, and all other forms of electronic communication.

EXAMPLE

Konop was a pilot for Hawaiian Airlines and maintained a personal Web site where he posted information critical of his employer. The site was accessible only with a user name and password. In addition, users had to agree not to divulge the information. Davis, a vice president of Hawaiian Air, obtained access through another employee. When Konop found out, he sued Hawaiian Air for invasion of privacy. The appeal court held that Hawaiian Air's access had been an unauthorized interception and remanded the matter for trial on the tort of invasion of privacy.[28]

Employers may monitor business-related electronic communications.

EXAMPLE

An employer told employees that e-mail would be private and not monitored. However, after Smyth sent threatening e-mails, he was terminated. The court upheld the firing. The court weighed the employee's expectation of privacy with the employer's interest in preventing inappropriate e-mail, and the employer's was the stronger criteria.[29]

Employers cannot monitor personal communications, although employees can be required to sign release forms allowing the employer to do so.

Employers are liable for torts committed against their employees. Two torts that may arise are invasion of privacy and defamation. As shown in Figure 3-4, invasion of privacy actually includes four different torts: intrusion into seclusion, false light, public disclosure of private facts, and appropriation of another's name or likeness. These and other torts are discussed in Chapter 12.

MANAGEMENT TIP

Always inform employees in advance of the types of monitoring they will be subject to in the workplace. Do not engage in secret videotaping or audiorecordings of employees at any time. Have a company policy on electronic communications. If appropriate, have employees sign a release allowing the company to monitor personal communications on company hardware.

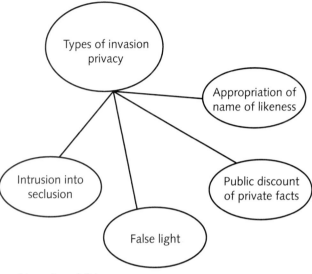

FIGURE 3-4 Types of Invasion of Privacy

Employee Files, Employee Handbooks, and Job Descriptions

For each new employee, a file should be started that contains the following items:

- Job advertisement, if any, through which employee was hired
- Job description
- Job application and any attachments such as resumes
- Contracts of employment, if any
- INS form I-9 and supporting documents
- IRS Form W-4
- Signed receipt for any employee handbook or policy manual
- Performance evaluations
- Awards or commendations, if any
- Complaints, if any
- Warnings or disciplinary documentation, if any
- Documentation of absences, late arrival or similar

Employee files must be kept confidential. State law probably gives the employee the right to review her file.

Employee handbooks can be helpful in outlining the expectations of the employer and the responsibilities of the employee. Be aware that the law treats employee handbooks as binding on the employer even though they may not technically be contracts. The following items are usually included in employee handbooks:

- Normal working hours and possibly pay or pay scales
- Explanation of benefits
- Drug and alcohol policies
- Sexual harassment policy
- Employer expectations and policies regarding job attendance, workplace civility, dress code, smoking, e-mail, and Internet usage

Employment and Labor Law

- Complaints, discipline, and disciplinary procedures
- Safety disclosures regarding workplace hazards
- Ethics statement

The employer should always get a receipt from the employee noting that the employee has been given a copy of the handbook, has read it, and understands it. The receipt should tell the employee who to contact if the employee has questions.

FORM

Receipt and Acknowledgment of Employee Handbook

Welcome to _____ (name of company). We are glad to have you join our team. This Employee Handbook contains company policies that apply to our employees. Please read it carefully. By signing below, you are acknowledging that you have received a copy of the Employee Handbook, have read it, and understand it.

This handbook does not contain all of the information you will need to do your job. You will be given specific information and requirements related to your particular job. In addition, this handbook is updated and changed from time to time, and only the most current employee handbook is applicable to your employment.

This Employee Handbook is not an employment contract. Unless you have a written contract signed by an officer of the company, you are an at-will employee, and this means that either you or the company can terminate your employment at any time for any reason or no reason.

No oral statements or promises regarding the terms and conditions of your employment are valid. If you believe some oral statements or promises regarding your employment have been made, and that they contradict anything in this Employee Handbook, do not sign this form but discuss these questions with the Personnel Office immediately.

Making Recommendations or Giving References

Employers must exercise caution when giving a recommendation or reference for a former employee. If the recommendation or reference contains false information, the employer could be liable for the tort of defamation.

EXAMPLE

Stanbury worked as a project manager for Sigal Construction for slightly over a year. He was told that he was terminated because of a slowdown in the company, but the personnel director testified that she had been told Stanbury's performance had been inadequate. Stanbury contacted a previous employer, Daniel Construction, to find out whether or not any work was available and was offered a position pending a review of his references by Janes. Janes spoke to Littman, an employee of Sigal. Littman told Janes not to hire Stanbury for the following reasons:

1. He seemed detail-oriented to the point of losing sight of the big picture.
2. He had a lot of knowledge and experience on big jobs.
3. With a large staff, he might be a very competent PM.
4. Obviously, he no longer worked for Sigal and that might say enough.

EXAMPLE

At trial, Littman acknowledged that he had made these statements without having supervised, evaluated, read an evaluation of, or even worked with Stanbury. His sole contact with Stanbury was seeing him in the hall on occasion, and they had spoken to each other only once. At trial, Littman said that he made the recommendation based on hearing others talk about Stanbury. Janes testified that Littman had told him he worked with Stanbury on a project. Janes also testified that Daniel Construction did not hire Stanbury because of the negative recommendation of Littman. The jury awarded Stanbury $370,440 for defamation. The award was reduced by the court to $250,000. The reduced award was upheld on appeal.[30]

Because potential employers may rely upon these recommendations, recommendations must be reasonable given the circumstances.

EXAMPLE

Herrara was an employee of Dona Ana County and worked at the jail. He was accused of sexually harassing female inmates, including demanding and receiving sex in exchange for privileges. After reports from female inmates, he was investigated, and although he denied the accusations, his desk contained a pornographic video, condoms, and underwear belonging to a juvenile. The investigation reported his conduct as "questionable" and "suspect." He resigned from his position at the jail rather than attend a hearing and go through the disciplinary process.

continued

Shortly thereafter, he sought alternative employment and requested a letter of recommendation from the Dona Ana County jail. The supervisor of the jail gave him the following letter:

To Whom It May Concern:

This letter will introduce to you Joseph V. Herrera. I have had the distinct pleasure of working with Herrera for the past two years. In my opinion, he is an excellent employee and supervisor for the Dona Ana County jail. In developing social programs for the inmate population, he displayed considerable initiative and imagination. Herrera was instrumental in the Department's maintenance program and was involved in remodeling projects.

I know that this department will suffer for his leaving. Employees of his caliber are difficult to find. I am confident that you would find Herrera to be an excellent employee. Should you need verbal confirmation of his ability, I would deem it a pleasure to respond to any inquiries that you may have.

Sincerely, [Signed] Detention Administrator, Dona Ana County Jail

Based upon this letter of recommendation, Herrera was hired as a mental health technician at a psychiatric hospital, where he allegedly sexually and physically abused a female patient. The female patient sued, among others, Dona Ana County for negligent misrepresentation. The county filed a motion to dismiss the case, claiming that the patient had no legal cause of action against the county as the county had no duty to the patient. The court disagreed and stated that the county had a duty to act reasonably when making employment recommendations; the employment recommendation could possibly be a negligent misrepresentation. The matter was referred to trial for a jury to decide but was settled without further public record.[31]

MANAGEMENT TIP

If you cannot give a good recommendation to a former employee, limit the recommendation letter to factual information such as the dates of employment, job title, and job description. If you are pressed for additional information, say something to the effect that you are sorry but you cannot give any further information.

Arbitration and Mediation of Employment Disputes

It is becoming more common for employers to include arbitration and mediation requirements in employment contracts. All types of claims can be made subject to these forms of alternative dispute resolution, thus relieving the employer from having to engage in litigation with the employee. The types of claims subject to alternative dispute resolution include discrimination, sexual harassment, retaliation, and wrongful discharge. While a complete waiver by the employee of the right to engage in litigation against the employer would probably not be upheld, the Federal Employment Arbitration Act states that employment arbitration agreements are valid. As with all things, the arbitration agreements must be reasonable. Case law has refused to uphold unconscionable or one-sided arbitration agreements.

EXAMPLE

An employer presented employees with an agreement to arbitrate and a waiver of all rights to a jury trial for most claims that could arise between the employer and the employee. The court found the agreement to arbitrate invalid because it was unconscionable, against public policy, and a voidable adhesion contract. The agreement incorporated by reference certain rules which in actuality contained the meat of the agreement. The rules were extremely one sided, favored the employer, were ambiguous, and in some places were contradictory. The rules and therefore the agreement allowed the following:

the employer could change the rules at any time and without notice to the employee, employees were stripped of most statutory rights, damages were extremely limited, the employer chose the arbitrators and the employee could not object to them, discovery was severely limited, judicial review of the arbitration award was even more limited than the law allowed, employees, but not the employer, had to disclose witnesses and evidence prior to the arbitration, the employer, but not the employee, could cancel the arbitration and proceed with a lawsuit, it appeared the arbitrator's decision was binding only on the employee but not the employer, and the rules appeared to make the rules controlling rather than the law—for example, the rules could redefine what sexual harassment was or was not. The court said, "Unconscionability has generally been recognized to include an absence of meaningful choice on the part of one of the parties together with contract terms which are unreasonably favorable to the other party.... Unconscionability is measured at the making of the contract, and the test is 'whether the terms are so extreme as to appear unconscionable according to the mores and business practices of the time and place'....It seems to be well established in this State that contracts having for their object anything that is obnoxious to the principles of the common law,...or repugnant to justice and morality, are void; and the Courts of this State will not lend aid to the enforcement of contracts that are in violation of the law or opposed to sound public policy (all internal citations omitted)."[32]

Alternative dispute resolution is generally beneficial to both the employer and the employee, although the employee does give up substantial rights if she agrees to mandatory arbitration. Arbitrators may or may not understand or follow the law to the same extent that a court would. Arbitrator's decisions are not reviewed for legal validity. On the other hand, the process is generally much faster and less adversarial. Alternative dispute resolution processes are discussed in Chapter 1 in the section entitled "Dispute Resolution."

EXAMPLE

Leslie Balanger is employed by Venegoni and Associates, Engineers, and as part of the employment contract, she signs an agreement that all employment disputes between employee and employer are subject to binding arbitration. This means that Balanger can never file an employment-related lawsuit against the employer. Balanger was terminated and claimed she was fired for illegal discrimination. The mandatory arbitration agreement is valid, and the matter will be decided in a private arbitration, not in a court of law.

Some exceptions to this situation, such as fraud, do exist, but their applicability is rare. If the agreement is totally one sided, in which the employee gives up her rights but the employer does not, the law is likely to disregard the agreement to arbitrate and allow the employee to file a lawsuit.

MANAGEMENT TIP

All employers need to realize that disputes will arise and should have policies for handling disputes so they do not escalate into lawsuits. A simple requirement in an employee contract or handbook requiring mediation before a lawsuit can be filed will usually prevent litigation. In addition, the employer should think seriously of requiring that disputes be subject to binding arbitration. Such a requirement requires legal advice.

Union Contracts

The rights of employees represented by a union include the common law and statutory rights of nonunion employees, as discussed previously, and also the rights as outlined in the union contract. Union contracts generally contain sections outlining wages, benefits, employee evaluation procedures, grievance procedures, what information is kept in the personnel file, and other information. In nonunionized environments, such information may be outlined in an employee handbook.

If a question arises as to the interpretation of the union contract, the rules of contract interpretation are applicable. See Chapter 9 for a discussion of the law of contract interpretation. However, since most union contracts contain mandatory arbitration procedures, and the arbitrators may not understand or follow the law, and the arbitration decision is not usually reviewable for legal validity, the applicability of the law in an arbitration is not as certain as it is in litigation. Arbitration is discussed in Chapter 1 in the section entitled "Arbitration—Contractually Required."

It is illegal for an employer to discriminate against employees who are attempting to form a union or who belong to a union. The reality of the situation is that employers use many tactics including hiring specialized union-busting firms to thwart employees' attempts to unionize. Attempting to unionize in the present legal climate in the United States is difficult, although some legislation to make it easier has been introduced in the Congress.

■ Endnotes

1 See Michigan S.A. 3.548(202).

2 Example based on *McCullough v. Real Foods, Inc.*, 140 F.3d 1123 (8th Cir. 1998).

3 Example based on *Fragante v. City and County of Honolulu*, 888 F.2d 591, 104 A.L.R. Fed. 801 (9th Cir. 1989).

4 Example based on *Sabine Pilot Service, Inc. v. Hauck*, 687 S.W.2d 733 (Tex. 1985).

5 Example based on *Ed Rachal Found. v. D'Unger*, 207 S.W.3d 330 (Tex. 2006).

6 Most recent U.S. Supreme Court case involving *Adarand Constructors, Inc. v. Mineta*, 534 U.S. 103 (2001).

7 See *29 U.S.C. 216*(b) (2000) (stating that an employer who violates antiretaliation provision of FLSA "shall be liable for such legal or equitable relief as may be appropriate") and *42 U.S.C. 2000e-5*(g) (listing remedies available when employer engaged in unlawful employment practice).

8 *28 U.S.C. 1350* (2000), establishing cause of action and district court jurisdiction over "any civil action by an alien for a tort only, committed in violation of the law of nations or a treaty of the United States."

9 Example based on *Manliguez v. Joseph*, 226 F. Supp. 2d 277 (E.D.N.Y. 2002).

10 Example based on *Castillo v. Neave*, No. O3 Civ. 0763 (S.D.N.Y. filed Feb. 3, 2003).

11 *Hoffman Plastic Compounds, Inc. v. NLRB*, 535 U.S. 137 (2002)

12 Example base on *NLRB v. Transportation Management Corp.*, 462 U.S. 393; 103 S. Ct. 2469; 76 L. Ed. 2d 667 (1983).

13 Example based on *Lukoski v. Sandia Indian Management Co.*, 106 N.M. 664, 748 P.2d 507 (N.M. 1988).

14 Example based on *Hetes v. Schefman & Miller Law Office*, 152 Mich. App. 117, 393 N.W. 2d 577 (Mich. Ct. App. 1986).

15 Example based on *Ackerman v. Kimball Int'l, Inc.*, 652 N.E.2d 507, 510-511 (Ind. 1995).

16 Example based on *International Security Management Group, Inc. v. Sawyer*, 2006 U.S. Dist. Lexis 37059 (M.D. Tenn. 2006).

17 OSHA Field Operations Manual, Ch. V, F.1.a., and OSHA Instruction, CPL 2-0.124, Multi-Employer Citation Policy, Oct. 12, 1999.

18 Example from Johnstone, Mayhew, & Quinlan, *Outsourcing Risk? The Regulation of Occupational Health and Safety Where Subcontractors Are Employed*, 22 Comp. Lab. L. & Pol'y J. 351 (2001).

19 Example based on *U.S. v. Dye Construction Co.*, 510 F.2d 78 (10th Cir. 1975).

20 Example based on *People v. O'Neil*, 194 Ill. App. 3d 79, 97 (Ill. App. Ct. 1990), case remanded on appeal for inconsistent verdicts.

21 Example based on *Tyndall v. National Education Centers*, 31 F.3d 209 (4th Cir. 1994).

22 Example based on *Sutton v. United Air Lines*, 527 U.S. 471, 475 (1999).

23 *Barnes v. Gorman*, 122 S. Ct. 2097 (2002), *Chevron U.S.A., Inc. v. Echazabel*, 122 S. Ct. 2045 (2002), *Toyota Motor Mfg., Ky., Inc. v. Williams*, 122 S. Ct. 681 (2002), *U.S. Airways, Inc. v. Barnett*, 122 S. Ct. 1516 (2002).

[24] Example based on *U.S. Airways, Inc. v. Barnett*, 122 S. Ct. 1516 (2002).

[25] Example based on *Desert Palace, Inc. v. Costa*, 539 U.S. 90, 123 S. Ct. 2148, 156 L. Ed. 2d 84 (2003).

[26] Example based on *Pennsylvania State Police v. Suders*, 42 U.S. 129, 124 S. Ct. 2342, 159 L. Ed. 2d 204 (2004).

[27] 18 U.S.C. §2510-2521.

[28] Example based on *Konop v. Hawaiian Airlines, Inc.*, 236 F.3d 1035 (9th Cir. 2001).

[29] Example based on *Smyth v. The Pillsbury Co.*, 914 F. Supp. 97 (E.D. Pa. 1996).

[30] *Sigal Const. Corp. v. Stanbury*, 586 A.2d 1204 (D.C. App. 1991).

[31] Example based on *Davis v. Dona Ana County*, 127 N.M. 785, 987 P.2d 1172 (N.M. 1999).

[32] Example based on *Hooters v. Phillips*, 39 F. Supp. 2d 582 (D.S.C. 1998).

CHAPTER 4

Forms of Doing Business, Project Delivery Methods, and Contractual Relationships Common in Construction Projects

■ Forms of Doing Business

Prior to the 1990s, the traditional forms of doing business were the sole proprietorship, the partnership, and the corporation. During the 1990s, and before in some states, alternative forms of doing business were recognized. At the end of the decade, business forms included the following:

- Unlimited liability business entities
  - Sole proprietorship
  - Partnership and joint venture
- Limited liability business entities
  - Corporation
  - Professional corporation
  - Limited liability company and limited partnership

This chapter discusses the above business entities. The term **business entity** includes all of the possible forms of doing business. The term **unlimited liability business entity** means a business entity whose owners have unlimited personal liability for all of the business entity's debts. The term **limited liability business entity** means a business entity whose owners have no personal liability for the business entity's debts. The term **entity** is broader and can refer to a person, partnership, joint venture, limited partnership, corporation, nonprofit corporation or group, church, limited liability company, professional corporation, government, or part of a government.

In addition to the above forms of business entities, this chapter discusses the concept of "doing business as" or the "fictitious business name statement."

Sole Proprietorship

A **sole proprietorship** is a business entity operated by one person. If the person is married, the spouse may have community property or marital property[1] interests in the business should the parties divorce or separate. The legal concepts of community property or marital property law are not discussed herein.

No special legal formalities are required to form a sole proprietorship. A single person operating any type of business venture, profitable or not, is considered to be operating a sole proprietorship.

Sole proprietorships do not file income tax returns; all income and deductions are taken on the individual income tax form of the individual operating the business. A federal income tax Schedule C or Schedule C-EZ is used for this purpose.

The sole proprietor, including all his nonexempt property, is liable for all debts of the business. All states have lists of property that cannot be seized to cover the debts of a person, and the property on that list is called **exempt property**. Exempt property may include some amount of equity in a home, car, tools, retirement plans, and other assets. **Nonexempt property** is all of the property owned by the individual that does *not* appear on the exempt property list.

Debts arise from three sources: contract, tort, and tax liabilities. Owners of sole proprietorships are vicariously liable for the torts of their employees committed within the scope of employment. A **tort** is a noncontract injury such as negligence or defamation. Torts are discussed in Chapter 12, Torts and Warranties. Vicarious liability is discussed in Chapter 3, Employment and Labor Law. A sole proprietor considering hiring employees should consider an alternative form of business entity such as the limited liability company.

EXAMPLE

Herb Wray is a high school teacher, but he washes and repairs decks as a side business, particularly in the summer when he is not teaching. He has started a sole proprietorship. He decides to hire Matt Tice to help him. One day, Matt is alone on a job and fails to nail down some boards on a deck he is repairing. Amy Svorec, a neighbor visiting the owner of the deck, walks on the deck, falls through, and is injured. If Matt is negligent, Herb Wray is personally liable for the injuries. Matt and the business entity are also liable.

EXAMPLE

Herb buys equipment costing $5,000 but only makes $2,000 from his business. He must still pay the entire $5,000 out of the other income or assets he has available.

Should the business become bankrupt, the sole proprietor must file bankruptcy, and all of the sole proprietor's nonexempt assets are subject to be sold to satisfy the businesses' debts.

EXAMPLE

Amy Svorec, the neighbor who was injured in the previous example, was a professional dancer, and her career has been ruined because of the injury she sustained when she fell through the deck that was negligently repaired by Mike. Her damages are estimated at over $1,000,000. Herb decides to file bankruptcy because he has no insurance and does not have assets to cover the damage to Amy. He must file a personal bankruptcy, and any nonexempt assets he owns will be liquidated and used to pay his debts, including the debt to Amy.

MANAGEMENT TIP

Sole proprietorships should only be operated in limited circumstances. The limited liability company, discussed below, is usually preferable to the sole proprietorship.

A sole proprietorship terminates upon the death of the sole proprietor, and all assets of the business are distributed according to the will of the deceased. If no will has been made, the property will be distributed by intestate succession. **Intestate succession** is the state law outlining who receives a person's property if they die without a valid will. Intestate succession law typically distributes the assets to the spouse and children of the deceased. If there is no spouse or children, state laws distribute the property to parents, siblings, cousins, and so on.

EXAMPLE

After many years, Herb has built up his deck repair business into a small construction company with a good reputation and assets including trucks and tools. Herb's son has been working for him for several years in the business. Herb has a will that gives his estate to his wife upon his death. Herb dies, and his son would like to continue with the business. However, Herb's wife wants to liquidate all of the assets of the business and use the money to buy a condo in Florida. In this case, the son can buy the assets from his mother but has no legal rights to the business.

General Partnership and Joint Venture

A **general partnership** or **joint venture** is a business entity operated by two or more entities for profit. Any two or more business entities can form a partnership, and no special paperwork, legal formalities, or written agreements are needed. Two people or two corporations can form a partnership. A person and a government entity can form a partnership. Working together forms the partnership or joint venture.

The term **partnership** is often used to mean a general partnership as compared to a limited partnership, discussed below, and is used herein to refer to a general partnership. The difference between a partnership and a joint venture is usually the length of the operation. When the parties contemplate operating their business for an indefinite period of time, their relationship is a partnership. A **joint venture** is a temporary business relationship between two entities. The legal liability of both is the same.

The legal liability of partners and joint venturers is so great that it has been questioned whether or not it is legal malpractice for a lawyer to advise a client to form either. Almost any other business form except the sole proprietorship, which is unavailable to two or more entities, is a better choice.

MANAGEMENT TIP

Partnerships should only be operated in limited circumstances, and the limited liability company, discussed below, is usually preferable to the partnership. Joint ventures should only be entered into by limited liability entities such as corporations, limited liability companies, and limited partnerships. The partnership or joint venture agreement should always be in writing.

EXAMPLE

Joel Tubergen has been engaged in the practice of engineering for many years and has formed a professional corporation called Joel Tubergen, Inc. Kelly Sharp is a newly licensed engineer. Joel Tubergen, Inc. and Kelly Sharp form a partnership to provide engineering services. No special paperwork is needed, and no forms need to be filed. The parties should have a written partnership agreement.

Partnerships file an informational tax return (IRS Form #1065) with the IRS and send the partners a Schedule K. Joint ventures may also do so but, depending on the complexity and length of time of the joint venture, the parties may simply report the income and expenses on the business entity's return without preparing the informational return. The individual partners and joint venturers then report the share of the income and liabilities on their personal tax returns. Similar forms may also be required by the state.

EXAMPLE

Advanced Architectural Design, a partnership owned and operated by Maria and Tanya, must file Form 1065 with the IRS as well as Schedule Ks. All of the income and expenses of Advanced Architectural Design are reported on Form 1065 and in the Schedule K. Maria and Tanya then include the income and expenses from the Schedule Ks in their own personal income tax returns.

Partners are jointly and severably liable for all debts of the business. Debts of the business arise from contract, tort, and tax liabilities. **Joint liability** means each partner/joint venturer is individually liable for a *percentage* of the debt equal to their ownership share of the business. **Severable liability** means each partner/joint venturer is individually liable for the *entire* debt should any of the other partners/joint venturers not have sufficient assets to cover their portion of the debt. **Joint and severable liability** therefore means that each partner is individually liable for his/her/its individual proportion of the debt and is also individually liable for all of the debt should the other partners/joint venturers be unable to pay.

EXAMPLE

Advanced Architectural Design, the partnership equally owned and operated by Maria and Tanya, provides architectural services. Maria designs a new home project, but the spaces between the rails on the balcony are too large and do not meet code. After the home is occupied, a child falls through the rails and is injured. This is a tort liability of the business entity. Maria and Tanya are individually liable for half of the damages, but if only Tanya has any assets, Tanya's assets are liable for the entire debt.

Historically, if any partner/joint venturer declared bankruptcy, the other partner/joint venturer had to declare bankruptcy and the partnership/joint venture was terminated. The Revised Uniform Partnership Act, enacted by every state in the United States, no longer requires this. Now, if one partner/joint venturer files for bankruptcy, the partnership need not be dissolved.

Under the Revised Uniform Partnership Act, the partnership continues upon the death or dissociation of a partner. If one partner wishes to leave the partnership or dies, the partnership is required to buy out the partner, but the partnership itself is not terminated.

MANAGEMENT TIP

Partners should have partnership agreements and should seek estate-planning advice. Since the law may require the partnership to buy out the share of the deceased partner, the partnership should consider buying life insurance on each of the partners to provide cash for this purpose.

Corporation

A corporation is a legal entity created by a state government and is considered a separate and legally recognized "person" for most laws. This form of business entity has existed for more than 2,000 years, having been invented by the ancient Romans. It has proved to be the most versatile business form for both large and small businesses. Because ownership of the business entity is evidenced by shares of stock, transfer and sale of the business entity is easy. The corporation pays its own taxes and is liable for its own debts. States generally recognize both for-profit corporations and nonprofit corporations. Corporations are owed by persons called shareholders who elect a board of directors to run the corporation.

A **publicly traded corporation** is one whose stock is bought and sold on exchanges such as the New York Stock Exchange. A ready market exists for the shares. The federal government extensively regulates publicly traded corporations.

A **closely held corporation** is one whose stock is owned by a small group of people. The stock of a closely held corporation is not available for sale to the general public. It is often held by family members and/or people who wish to operate a business together but do not wish to form a partnership because of the liability or tax consequences. Federal regulations place limits on the numbers of owners of close corporations. A person can be both an owner (shareholder) and an employee of a corporation.

Most closely held corporations have shareholder agreements that prevent the sale or transfer of the stock to outsiders and outlines what happens to the stock on the death of the shareholder. It is common for the shareholder agreement to require that if a shareholder wants to sell his shares, the shareholder must sell to existing shareholders or the corporation only.

MANAGEMENT TIP

Care must be exercised when selling stock in a close corporation. Owners of close corporations should seek legal, tax, and estate-planning advice.

All corporations are started by one or more **incorporators,** or persons who file the required forms with the state. The forms include copies of the bylaws that outline how the corporation will function. Shortly after the state certifies the corporation, the incorporators have the initial meeting at which the shares of stock of the corporation are issued and the officers are appointed. From that point on, the corporation is run by the officers. A board of directors may be formed, and the corporation is operated according to the instructions in the bylaws.

EXAMPLE

Georgia, Harriet, and Phillipe decide to form a corporation called Ace Construction, Inc. They file the necessary paperwork as the incorporators, have a shareholders meeting, and elect Georgia as the president, Harriet as the vice president, and Phillipe as the secretary/treasurer. Each contributes $25,000, and a bank account in the name of the corporation is opened. The corporation issues 100 shares of stock to each of them. They enter into an agreement stating that in the event any of them want to sell their shares of stock, the stock must be sold to the corporation at one-third of the book value of the assets. In addition, the shareholder agreement states that upon the death of any of the shareholders, the corporation will buy the shares of stock at one-third of the fair market value of the assets. The corporation buys insurance to provide funds for the deceased shareholders' shares. This is a closely held corporation because the shares are held by a small number of people and the shares cannot be sold to others.

Some states require a minimum of three people to form a corporation, but some allow single-owner corporations. A state allowing single-owner corporations may not have recognized the limited liability company business form because the single-owner corporation fulfills that purpose.

The main advantage of the corporate form of business is the corporation becomes a separate legal person responsible for its own debts, contracts, torts, and taxes. The shareholders, directors, and officers are not liable for the debts of the corporation unless they personally guarantee a debt or have personally engaged in some tortuous conduct. Some jurisdictions do place legal liability for unpaid taxes on officers.

EXAMPLE

Ace Construction, Inc., orders $5,000 of flooring from supplier. Only the corporation is responsible for the debt, not the shareholders or officers.

EXAMPLE

Georgia, a shareholder of the corporation but also an employee of the corporation, injures a pedestrian while going to pick up flooring for a job. Georgia was driving in a negligent manner. Both the corporation and Georgia individually are liable to the injured pedestrian. None of the other shareholders, officers, or employees is liable to the injured pedestrian.

Even though a corporation is personally responsible for contracts, few lending institutions will lend small corporations money or extend them credit without a personal guarantee by some individual involved in the corporation. If a shareholder gives a personal guarantee to a lending institution for a loan to a corporation, that shareholder is personally responsible for the debt.

Corporations require bookkeeping and tax filings generally beyond the capability of the average person. Professionals such as accountants and lawyers are usually retained by the corporation to make sure that the corporation is in compliance with state and federal laws.

The corporation is considered a person under the law, must get state and federal tax identification numbers, and must pay taxes on profits. After the corporation has paid the taxes on its profits, it can do the following:

- Keep the profits in the business: use them to expand the business, pursue other business opportunities, or save them
- Pay out the profits as salary and bonuses to employees
- Pay out the profits, or a portion of them, to the shareholders as dividends

If the corporation pays dividends to the shareholders, the shareholders must pay income tax on these dividends. For this reason, closely held corporations rarely pay dividends but only pay salaries and bonuses to the shareholders who are also the employees of the corporation. Salaries are deductions on corporate income and reduce the amount of tax owed. A dividend is not a deduction and does not reduce the tax owed.

A **subchapter S corporation** is not a special type of corporation but a federal tax status. It gives the corporation the tax advantages of a partnership but the protection of a corporation. It has declined in use, since most states have allowed the formation of limited partnerships, limited partnerships, and limited liability companies that offer the same benefits.

MANAGEMENT TIP

Companies making a significant profit may wish to incorporate, because corporations are usually taxed at a lower rate than individuals. Limited partnerships, limited partnerships, and limited liability companies should review their income liability with a tax professional to determine if the corporate form of business is more advantageous.

Corporations do not die and can remain in existence indefinitely as long as the fees and taxes are paid. If fees or taxes are not paid, the state can dissolve the corporation. The bylaws of the corporation usually provide mechanisms for dissolution, should the shareholders or board of directors decide to embark on such a course of action. Shareholders cannot usually force the dissolution of the corporation; they can only sell their shares.

The shares of the corporation are an asset distributed per the will of the deceased shareholder. If the shareholder leaves no will, the shares are distributed to the surviving relatives according to state law of intestate succession. The corporation continues unimpeded by the death of any shareholder.

It is very common for shareholders of small corporations to have shareholder agreements that outline what is to happen with the shares of stock upon the death of the shareholder. The corporation or the other shareholders often purchase insurance to buy the shares upon the death of the deceased shareholder.

Professional Corporation

A professional corporation is one for which the shareholders must have some type of license in order to operate. For example, two or more people who want to start an architecture firm, but do not want to form a general partnership, may establish a professional corporation in most states. This allows the owners to receive the protection of the corporate shield.

MANAGEMENT TIP

With the recent influx of new business forms such as limited liability companies and limited partnerships, individuals should investigate whether or not a limited liability company or limited partnership is preferable under state law.

As in other areas of the law, individuals of professional corporations are liable for their own negligence. Under the Model Professional Corporation Supplement, an individual performing professional services as an employee of a professional corporation is liable only for his own negligence or wrongful acts. He is not liable for the acts of other employees of the professional corporation. Under the doctrine of *respondent superior,* the professional corporation is liable for the negligence or wrongful acts of all of its employees within the scope of employment.

EXAMPLE

Southland hired Richeson Corporation to design a retaining wall. The wall later cracked, bulged, and was near collapse due to an employee's inappropriate design. The design did not comply with professional engineering standards. Both the employee and the corporation were held liable.[2]

Limited Liability Company and Limited Partnership

Many states recognize specialized forms of sole proprietorships and partnerships that give the sole proprietor or partners the tax advantages of a sole proprietorship or general partnership and the limited liability of a corporation. The most recent form—which is becoming the most common because of the ease of its formation—is the limited liability company or LLC. Some limited partnerships may still exist, but the LLC is quickly replacing the limited partnership.

These business forms require filing paperwork with the state in order to be validly formed. Failure to follow the proper procedures may result in the formation of a sole proprietorship or partnership, thus subjecting the sole proprietor/partners to personal liability, as discussed above.

Start-up businesses frequently experience losses, and these business forms allow those losses to flow into an entity's tax return and reduce the taxes on other income received by the entity. These business forms must file the same IRS Form 1065 as a general partnership and issue Schedule Ks to the partners or sole proprietor. The income and deductions of the entity are then taken on the owner's income tax returns.

EXAMPLE

Hakan and Andrea Erden are the sole shareholders in a corporation called Erden Enterprises, Inc. This corporation owns several rental properties. Erden Enterprises, Inc. and Sung Lee want to buy an apartment building. Erden Enterprises, Inc. and Sung Lee form a limited liability company called Live Oaks, LLC. Each entity, Erden Enterprises, Inc. and Sung Lee, owns 50% of Live Oaks, LLC, and all of the required paperwork to form Live Oaks, LLC is filed.

Live Oaks, LLC incurs $10,000 in losses due to the depreciation on the property. At the end of the year, Live Oaks, LLC files an informational return, IRS Form 1065, with the federal government and any required state tax forms, reporting $10,000 in losses. It sends a Schedule K to Erden Enterprises, Inc. and Sung Lee, who each deduct $5,000 on their income tax forms. During the year, the property increases in value by $15,000, but this has no effect on the partnership income or loss until the property is sold.

The owners of these business entities are not personally liable for the business's debts. Some states may require a limited partnership to have a general partner who is personally responsible for the debts of the limited partnership. This general partner can be a corporation, thereby relieving any individuals from personal liability for the debts of the business entity.

EXAMPLE

Bill and Jill are the limited partners of Red Oak, LP, which owns an apartment building. The general partner of Red Oak, LP is Red Oak, Inc. Bill and Jill each own 50% of the stock of Red Oak, Inc. Red Oak, LP hires a manager to handle the property. A tenant falls on a broken stair on the property and is injured. Red Oak, LP and Red Oak, Inc. are liable to the tenant for injuries. Neither Bill nor Jill is liable to the tenant. It is common for entities to purchase insurance to cover this situation.

What happens if the owners do not file the requisite paperwork to be recognized as a limited liability company or limited partnership? Depending on the exact nature of the failure, the law may say that the owners are operating a sole proprietorship or general partnership and subject the owners to joint and severable liability for all business debts.

EXAMPLE

Ian Gorski and Spencer Tolas want to form a limited liability company to operate a construction firm, but they never get around to filing the required paperwork. They do call their company Gorski and Tolas, LLC, and all of the paperwork contains this name. A customer in a strip mall they construct is injured when the roof caves in. If it is determined that the roof was negligently constructed, Gorski and Tolas are individually liable, not the business entity called Gorski and Tolas, LLC, as Gorski and Tolas are in a general partnership.

State law varies in this area, and care must be taken to understand how long the state allows these entities to exist. The state law may allow the entity to exist only for a limited period of time, such as 20 years. State law may require the business entity to be dissolved on the death of one of the owners.

EXAMPLE

Ahmed dies and leaves his limited partnership interest in Sequoia, LP to The Society for the Prevention of Cruelty to Animals. State law does not require the dissolution of the LP upon the death of a partner. If the LP has no provisions requiring the remaining limited partners to buy Ahmed's share, it goes to the SPCA. The SPCA may try to sell the limited partnership interest, but there may be few buyers.

MANAGEMENT TIP

Owners of LLCs and LPs should seek legal, tax, and estate-planning advice. They should have written agreements outlining what happens in the event that one party wishes to leave the business entity or dies. The business entity may wish to buy life insurance to fund the purchase of a deceased partner's interest. Since a limited market exists for these business interests, this is usually beneficial to the heirs. Having an agreement will also ensure that unknown parties, such as children or the spouses of deceased partners, are not allowed to run the business.

EXAMPLE

Arturo and Arnold are partners in a limited partnership called AA Engraving, LP. The general partner is AA, Inc., and Arturo and Arnold each own 50% of the stock of AA, Inc.

Their partnership agreement requires that upon the death of Arturo or Arnold, AA Engraving, LP must buy, and the estate of the first to die must sell, the deceased partner's interest. A way to calculate the purchase price is included in the agreement. AA Engraving, LP has insurance on both Arturo and Arnold to fund the purchase. When Arnold dies, his wife wants to be part of the business, but Arturo does not want to work with her. The wife has no legal right to work in the business. AA Engraving, LP will pay the money owed to Arnold's estate, and the money will be distributed to Arnold's heirs according to his will or by intestate succession if he has no will.

Doing Business As or Fictitious Business Name Statement

Any business entity may do business under any name it chooses. However, if the business entity does business under any name other than its legal name, it must file a certificate called a doing business as (DBA) certificate or a fictitious business name statement. A **doing business as certificate** or a **fictitious business name statement** is a government document, usually filed in the office of the county recorder where the business entity is doing business, and it records the name under which the entity is doing business *and* the legal name of the business entity. The statement will also require the listing of the entity or agent that is empowered to receive **process**. The word *process* in this context means the person who can receive forms of legal notice, such as a lawsuit, on behalf of the entity. Once process is given to the agent, the business entity is legally notified of the matter, and all legal processes can proceed with or without the business entity's involvement.

EXAMPLE

Maria Steuf and Tanya Juen decide to form a business to sell architectural design services. They want to name the business Advanced Architectural Design. In order to legally operate under that name, they must file a fictitious business name statement in the county where their office is located. This statement will put people on notice that the business called Advanced Architectural Design is being operated by Maria Steuf and Tanya Juen. They also decide to designate Maria Steuf as the agent for service of process and give the address where Maria can be served.

Several months later, Maria moves from the address listed in the statement. Advanced Architectural Design is sued for failure to pay rent, and the plaintiff serves the process on Maria at the address in the form. Maria never receives the process. The plaintiff proceeds with the lawsuit and can obtain a default judgment against Advanced Architectural Design for the unpaid rent.

MANAGEMENT TIP

The law requires that fictitious business name statements be kept up to date. In addition, jurisdictions may require fictitious business name statements to be refiled every five years or other time period. Always calendar the date for renewal of the fictitious business name statement, if one applies.

Piercing the Protective Veil

Owners of limited liability business entities such as corporations, LPs, and LLCs must be careful to maintain the separateness of the business entity and to capitalize the business entity adequately at start-up. The owners must not use business accounts for personal reasons. While it is true that owners of a business entity can also be employees of their own business, the business entity must treat them as employees when they are acting as employees. For example, taxes must be withheld from their income.

Failure to keep the business entity separate may subject the owners to personal liability for the debts of the business entity under the doctrine of "piercing the protective veil," also known as the "alter ego theory." The doctrine of **piercing the protective veil** or **alter ego theory** allows the law to disregard the protection (the "veil") of limited liability and in essence transforms the business entity into either a sole proprietorship or a general partnership. Because owners of sole proprietorships and general partnerships are responsible for the debts of the business entity, they are no longer protected. In some cases, this can be financially devastating.

EXAMPLE

Gene and Betsy are the sole shareholders of Platt River Construction, Inc., which sells construction services to homeowners and small commercial businesses. Both Gene and Betsy work in the business. Platt River Construction, Inc. has a separate company bank account, but Gene and Betsy use it to make their house payments and pay their personal credit card bills. When the company checking account has money in it, they write themselves checks but do not take out income or other taxes.

An employee of Platt River Construction, Inc. is negligent and causes damage to the house on which Platt River Construction, Inc. is doing a remodel. Platt River Construction, Inc., Gene, Betsy, and the negligent employee are all liable for the damage. The employee is liable under the theory of negligence, and Gene and Betsy are liable under the alter ego theory. Their corporation is considered a general partnership, and they are vicariously liable for the acts of their employee.

■ Contracts with Government Entities and the Doctrine of Sovereign Immunity

Government entities, such as the United States government, state governments, cities, counties, school districts, and highway departments, frequently purchase construction projects and construction services. In addition, with the growth of tribal casinos and other businesses, tribal law may be applicable to a contract with a tribe or on tribal land.

Special laws not applicable to a private contract may be applicable to a contract with a government agency.

EXAMPLE

A local government entity may require all of the contracts with that entity over $1,000 to be bid competitively.

A local government entity may require licensed engineers on certain government projects.

A local government entity may prohibit design-build projects for government projects or allow them only for school projects.

All of the above are examples of laws that operate in addition to and separate from local building codes and regulations.

Most government entities have complex **procurement laws** and **false claims laws**[3] that are designed to help the government entity obtain competitive rates for contracts, prevent corruption and favoritism, and prevent fraud in government contracting. These laws make it illegal to bribe government entities to obtain contracts. It may be illegal to give gifts and pay for meals and other services for procurement officers. Violation of these laws can result in criminal penalties.[4]

Many government entities have **affirmative action requirements** that must be followed by entities entering into government contracts. These requirements are usually designed with the goal of making the contractor's and major subcontractors' workforce mirror the population in which the contractor works and that government entity operates. Construction contractors on federal projects have slightly different affirmative action requirements than nonconstruction contractors. In connection with construction contracts, only certain areas of the country have affirmative action requirements, and the government entity will have this information for the contractor. The contracting officers will require certification of compliance with affirmative action requirements before the contract is entered into.[5]

Information given to government agencies may also be subject to disclosure under the Freedom of Information Act (FOIA). Designed to prevent secrecy, this act allows the public access to the government and its operations.

MANAGEMENT TIP

Business entities dealing with a government entity should always require the return of any documents, designs, or other paperwork that the business entity does not want to be shared with competitors. A business entity that must share sensitive information with the government must take precautions that its information is not subject to being divulged to competitors under a FOIA request.

The law controlling government contracts and private contracts also differs in one important respect: the applicability of the doctrine of sovereign immunity to government contracts. The doctrine of sovereign immunity is an old common-law doctrine that held the "king could not be sued." The doctrine of **sovereign immunity** has been transformed by the common law into the maxim that the "government cannot be sued."

Most states and the federal government, though not tribal governments, have passed laws that to some extent have abolished sovereign immunity for both contract[6] and tort[7] actions. Some southern states still apply this doctrine and prevent private parties from suing the state without permission of the legislature, at least in some circumstances.[8] In very rare circumstances, equitable doctrines may trump the doctrine of sovereign immunity and allow a party to sue the government.[9]

Many government entities, particularly cities and counties, allow themselves to be sued but have placed limitations on that ability. For example, a county may require 30 days' notice before a suit is filed.

Government entities may have enacted a shorter statute of limitations for filing lawsuits against government entities. The **statute of limitation** is the time period after some event, such as a breach of contract, during which one can file a lawsuit.

MANAGEMENT TIP

If one has a claim against a government entity, it is necessary to obtain legal advice or risk loss of the claim. Many contracts with government entities have elaborate claims procedures outlined in them, and these will control contract claims. If a claim is for a tort (a noncontract-related injury), the claim may not be covered by the contract provisions. However, it may be covered by other separate laws and regulations that greatly limit a business entity's ability to bring a lawsuit against the government.

The protection of sovereign immunity extends to suits against contractors who have built projects for the government. Contractors are considered part of the government. They are protected by the "government contractor defense" and are not liable to subsequent users of the project.[10]

■ Project Delivery Methods

The three most common methods of project delivery are briefly outlined here. The types of contracts are more fully described in the following section, Contractual Relationships Common in Construction Projects. The most common forms of project delivery are as follows:

- Traditional form
- Construction Management or CM
- Design-build

Traditional Form

In the traditional form of construction project delivery (Figure 4-1), an owner enters into an agency or possibly employment relationship with an architect, an engineer, or both to design the project and to oversee the construction of the project. After the project is designed, the owner enters into a principal/independent contractor relationship with the contractor to build the project.

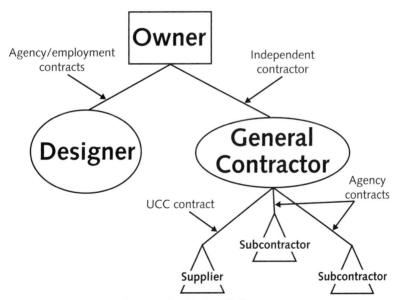

FIGURE 4-1 Traditional Form of Construction Project Delivery

In the traditional form, it is common for the architect/designer to blame the contractor for problems and the contractor to blame the architect/designer. No contractual relationship exists between the general contractor and the designer, and neither can sue the other for breach of contract. They can only sue the owner. Most states limit the ability of the general contractor and the designer to sue each in tort. See Chapter 12, Torts and Warranties.

Construction Management or CM

Figure 4-2 illustrates one form of the Construction Management or CM project delivery method. Many variations are possible. This relatively new type of relationship has developed due to several factors in the industry. As projects become more and more complex, the designer is no longer able or willing to supervise the construction. Another important factor, which can reduce future problems, is that the manager can be involved in the design phase and review the plans and specifications for buildability.

The CM may be either an employee of the owner or an agent. The CM enters into independent contracts with the subcontractors on behalf of the owner/principal, and there is no general contractor. If hired during the design phase, the CM can contribute to the planning and scheduling of the project.

This process is advantageous for the owner because the owner retains more control over the construction. In addition, the CM has a fiduciary duty toward the owner. A party with a fiduciary duty must always operate in the best interest of the other, not in its own best interest. In contrast, the principal independent/general contractor contract creates no fiduciary duty between the parties. The general contractor acts in its own best interest within the confines of the contract.

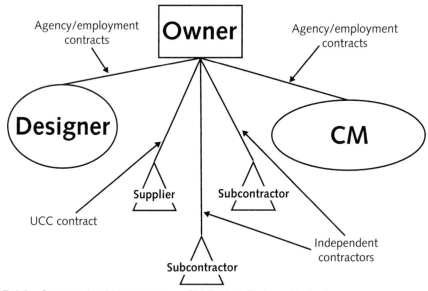

FIGURE 4-2 Construction Management or CM Project Delivery Method

Design-build

Figure 4-3 illustrates the design-build project delivery method where, again, many variations are possible. This is also a relatively new type of relationship and has developed due to several factors in the industry. It has proved to be a very economical form of project delivery from the owner's standpoint. The owner enters into one contract with one company to both design and build the project, thereby avoiding or at least minimizing conflicts between the designer and the general contractor. The design-build firm enters into contracts with designers, contractors, subcontractors, and suppliers as needed to complete the project.

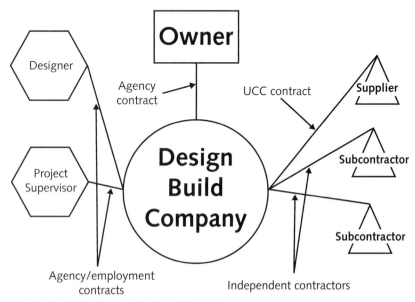

FIGURE 4-3 Design-Build Project Delivery Method

■ Contractual Relationships Common in Construction Projects

The typical contractual relationships encountered on construction projects are as follows:

- Employer-employee
- Principal-independent (aka general contractor)
- Principal-agent
- Buyer-seller-Uniform Commercial Code relationships

Employer-Employee

The employer-employee relationship is a contractual relationship between two parties: the employer or principal hires the employee to perform tasks the employer wants completed, often specific tasks such as painting or hanging drywall. Two major characteristics of the employer/employee relationship are that the employer supervises the work of the employee and the employee is not empowered to enter into contracts for the employer. An entity empowered to enter into contracts for another (the principal) is called an **agent**, and this relationship is discussed below. A person can be both an employee and an agent for the same employer/principal.

> # EXAMPLE
>
> An office assistant is authorized to purchase up to $500 of office supplies per month for the company. When acting in this role, the office assistant is the agent of the employer, and the employer is obligated to pay for the office supplies.

As one of the most important issues to the voting population, the employer/employee relationship is the subject of much government regulation. See Chapter 3, Employment and Labor Law.

Principal-Independent (aka General Contractor)

The relationship between the owner and the general/prime contractor on a construction project is usually that of principal and independent contractor. In this relationship, the principal does *not* oversee the day-to-day operations of the independent contractor or the contractor's employees but contracts for a service or completed project. It is up to the independent contractor to determine how to provide the service or complete the project. The principal-independent contractor is an "arm's length" relationship where both parties are expected to operate in their own best interests within the boundaries of law and ethics. The independent contractor has no fiduciary duty to the principal but must operate in good faith. In the construction industry, the independent contractor frequently has the title of general contractor.

The principal-independent contractor relationship is advantageous to the owner because it shields the owner from liability for the debts and torts of the independent contractor. In this way, it differs significantly from the employer-employee relationship and the principal-agent relationship, in which the employer is liable for the acts of its agent or employee.

> # EXAMPLE
>
> A general contractor is negligent by failing to clean up a certain area of a project site, and an employee of a contractor is injured. The owner is not liable to the general contractor's employee for damages.

One major exception of nonliability of the principal for acts of its independent contractor exists and may be applicable in some construction projects. This exception addresses abnormally or inherently dangerous activities. The principal is strictly liable, that is liable even if not at fault, for injuries caused by abnormally or inherently dangerous activities contracted to an independent contractor.[11] The independent contractor is still liable and therefore an injured party can collect from either the principal, the independent contractor, or both. Transporting or disposing of dangerous or hazardous materials and blasting are examples of abnormally or inherently dangerous activities.

EXAMPLE

A general contractor is negligent when transporting hazardous waste, which is being treated and disposed of according to law, on the owner's property. Some of the waste is spilled, and the property of a neighboring land is contaminated. Both the general contractor and the owner are liable for the damages to the neighboring landowner.

MANAGEMENT TIP

When engaging in any type of hazardous activity, always perform a risk analysis and determine the appropriate actions such as buying insurance or self-insuring.

Most owners are careful to protect the owner-general contractor relationship by not interfering with the general contractor's control over the means and methods of construction. Failure to do so could destroy the principal-independent contractor relationship, making the general contractor the employee of the owner. This action would subject the owner to liability for the debts of the general contractor related to the project.

The principal-independent contractor status has tax advantages for the principal. Though the principal is not required to withhold income taxes from independent contractors, the principal is required to report income over a set amount—currently $600 per year—to the IRS on a Form 1099. In addition, the principal does not pay any Social Security taxes, unemployment insurance taxes, or Workers' Compensation premiums on payments to an independent contractor. Independent contractors are not protected by labor and employment laws.

Because of these advantages, some business entities may erroneously classify employees as independent contractors. The IRS can impose back taxes, penalties, and interest on employers engaging in this illegal conduct.

One of the most commonly used tests to determine if an individual is an employee or not is the "right to control" test. That is, if Entity A has the right to control the work of Person B, Person B is the employee of Entity A. The law in this area is complicated, and the IRS and the courts have developed a Form SS-8 for determination purposes. This form lists the factors to be reviewed when deciding whether a person is an employee or an independent contractor.

MANAGEMENT TIP

Do not try to save money by classifying positions such as secretaries and office employees as independent contractors. Persons who are supervised or who are supplying labor exclusively to a business entity and over long periods of time are probably employees.

The 20 questions asked on the IRS Form SS-8 are as follows:

1. Are instructions provided by the employer?
2. Is training provided by the employer?
3. Are the worker's services integrated into the general operations of the employer?
4. Are the services rendered personally?
5. Does the worker hire, supervise, and pay others who work with him in the performance of services on behalf of the employer?
6. Is there a continuing relationship between the worker and the employer?
7. Does the worker have set hours of work?
8. Does the worker work full time for the employer?
9. Does the worker work at the employer's premises or his own?
10. Does the employer specify the sequence of services to be performed?
11. Is the worker required to regularly furnish oral or written reports to the employer?
12. Is the worker paid on an hourly basis, weekly basis, or by the project?
13. Is the worker reimbursed for business expenses, or must he bear these expenses himself?
14. Does the employer provide materials needed to do the work, or does the worker provide these materials?
15. Does the worker have an investment in the enterprise?
16. Is it possible for the worker to either profit or lose money from his endeavors?
17. Does the worker work for a multitude of clients or just one?
18. Does the worker hold himself out to the public as being self-employed?
19. Does the worker have the right to terminate his services at any time without risking breach of contract?
20. Does the employer have the right to terminate the worker's services at any time?

Source: The Internal Revenue Service Web site: http://www.irs.gov.

Principal-Agent: Owner (Principal)-Designer (Agent) Relationship

Agency law is the term given to the law covering the relationship between a principal and an agent. A **principal** is an entity that gives power to another, called the **agent,** to act for the principal. In the construction industry, the owner frequently has a principal-agent relationship with the designer. That is, the owner gives the agent (the designer) the power to manage the building of the project and make decisions that the owner would otherwise make.

The decisions of an agent are binding on the principal, and the negligence of the agent is attributed to the principal.

EXAMPLE

Duhamel Broadcasting contracted with SST to perform structural work on its broadcasting tower to reinforce the tower so that it could support a digital antenna. SST subcontracted with Mid-Central to install certain horizontal members and stronger diagonal members on the tower. Mid-Central removed some of the bracing without installing temporary bracing. Experts testified that it was necessary to install temporary bracing when removing long-term bracing. The removal of the bracing without the installation of temporary bracing violated the standard of care. In other words, Mid-Central was negligent. The tower collapsed. Duhamel sued SST for damages to the tower.

SST claimed Mid-Central was its independent contractor and therefore SST was not liable for the negligence of Mid-Central. The court held as a matter of law that Mid-Central was the agent of SST and therefore the negligence of Mid-Central was imputed to SST. Mid-Central was the agent of SST because of the control SST exerted over Mid-Central. The contract between Duhamel Broadcasting and SST expressly stated that SST would supervise all work on the project. The contract between SST and Mid-Central gave SST the power to control Mid-Central and supervise Mid-Central's work.[12]

The principal is bound only to acts within the scope of the power given by the principal to the agent. If the agent acts outside of the power given to him, the principal is not liable.

EXAMPLE

An engineer, Jasmine Daoust, is given the "power to interpret the plans and specifications" in her contract with the owner. She tells the contractor that it is not necessary to prepare a written change order but to go ahead and upgrade the fill that will be used on a site. This upgrade increases the project cost by $5,000. The owner is not obligated to pay the contractor the increased fill because Daoust operated outside of the power given to her by the owner. She was not given the power to change or modify the contract.

MANAGEMENT TIP

Contractors must exercise care when the designer authorizes changes to a contract. The contractor must check to make sure that the designer has this power. If in doubt, the contractor should get owner approval of any major change.

The power given to the agent can be express, implied, or apparent.

Express power is the power specifically given to the agent in the agency contract.

> ### EXAMPLE
>
> Express power: "Agent is to hire a contractor to perform construction of the project." Agent has power to hire a contractor for the owner.

Implied power is the power to do the things necessary to carry out the express powers but is not specifically outlined in the contract.

> ### EXAMPLE
>
> Express power as stated in the contract: "Agent is to hire a contractor to perform construction of the project." Agent has the implied power to interview and evaluate several contractors.

Apparent power is the most problematic. Apparent power is the power given to the agent by the principal's acts and is separate and apart from express or implied authority. "Apparent authority is 'entirely distinct from authority, either express or implied,' and arises from the 'written or spoken words or any other conduct of the principal which, reasonably interpreted, causes [a] third person to believe that the principal consents to have [an] act done on his behalf by the person purporting to act for him.'"[13]

> ### EXAMPLE
>
> The contract between the owner and the designer requires that the owner approve, in writing, any changes to the construction contract. However, on a certain project the designer routinely makes changes to the contract without the owner's approval and the owner pays the contractor for the changes. The owner has given the designer the apparent authority to make changes in the contract without owner approval.

EXAMPLE

Brook owned and offered for sale a piece of heavy construction equipment. Lavant was in the market for heavy construction equipment similar to that offered by Brook. Thelen had previously acted as Lavant's agent in the purchase of heavy construction equipment and discovered that Brook had a piece of equipment that Lavant might be interested in purchasing. Thelen contacted Lavant about acting as Lavant's agent in the purchase of the Brook equipment. It was agreed that Thelen would be paid $5,000 by Lavant if he concluded a purchase.

Lavant claimed it authorized Thelen to offer a maximum of $395,000 million for Brook's heavy construction equipment. There is no dispute that Lavant authorized *some* offer to be sent; the dispute is as to the amount of the offer.

Thelen sent Brook a $400,000 offer on June 1, signed by him as Lavant's agent. This offer was accepted by Brook. The offer included the payment to Thelen of $5,000 by Brook. Brook insisted that the offer contain the provision authorizing him to pay Thelen the $5,000 fee. However, when Thelen faxed the offer to Lavant, he "whited out" the provision stating that Brook was going to pay him $5,000. Thelen expected to obtain a fee of $5,000 from *both* Brook and Lavant but without Lavant's knowledge.

On June 4, Lavant and Thelen met with Brook to inspect the heavy construction equipment. Everyone agreed the equipment was in excellent condition. Lavant told Brook he should speak with Thelen to "move the heavy construction equipment." This was the only direct contact between the Lavant and Brook.

On June 5, Thelen faxed to Lavant a copy of the fully executed offer for $400,000, signed by both Thelen and Brook. The purpose of the fax was actually to document the deposit of $20,000 into an escrow account. Lavant claimed this was the first time he had seen the $400,000 figure. However, Lavant did not protest or attempt to retract that $400,000 offer on the grounds that it was not authorized. Lavant transferred $200,000 into an escrow account to begin the process of purchasing Brook's equipment. The closing was set for June 21. On June 18, Thelen informed Lavant via fax that $3.8 million should be transferred as the "balance payable for aircraft" on June 21. Lavant contended that he did not know at that time that Thelen had signed a purchase agreement for Brook's heavy construction equipment. Lavant did not object to the price, and Lavant testified that he knew a signed contract would have to exist before Lavant would make any final payment.

At this point, another company, Theon, entered the picture and began negotiating with Lavant for the purchase of some of its heavy construction equipment near the Brook equipment. Lavant asked Thelen to delay the Brook purchase because it needed time to obtain additional financing. Lavant did not claim it had not authorized Thelen to purchase the Brook equipment for $400,000. Eventually Lavant purchased the Theon equipment and refused to purchase the Brook equipment.

Brook sued Lavant for breach of contract. Lavant breached the contract entered into by his agent, Thelen, with Brook and owed Brook damages. Even though Lavant claimed he did not authorize the $400,000 fee, his actions had given his agent, Thelen, apparent authority to enter into the contract. The fact that Thelen was not acting in good faith did not allow Lavant to void the contract with Brook although Lavant may have had cause of action against his agent for breach of the duty of good faith.[14]

The principal-agent relationship is one of extreme importance in the law. The agent owes the principal a fiduciary duty that is higher even than the duty of good faith. The **fiduciary duty** means that the agent must always act in the principal's best interest and never in the agent's interest. The duty of **good faith** exists in almost all contracts and requires the parties not to engage in acts that prevent the other party from receiving the benefit of the contract.

EXAMPLE

Dixon was the general partner and manager of a limited partnership named Trinity, LP. Trinity, LP bought land in Maryland, developed it, and sold it. Dixon bought a parcel of land in Maryland near others bought by Trinity, LP. He made a profit of $60,000 when he sold it to Trinity, LP. The limited partners discovered this and sued him for breach of fiduciary duty. It was determined that Dixon had breached his fiduciary duty to Trinity, LP and owed the limited partnership $60,000.[15]

Principal-Agent: Owner (Principal)-Construction Manager (Agent) Relationship

In the construction industry, the term Construction Manager or CM is used to describe the relationship between the owner and a person hired by the owner to supervise the construction and/or to keep the owner informed of the progress of the project. Owners who are government entities or business entities may hire a CM to be the voice of the owner. Many owners are using a CM early in the construction process as part of a design team or to engage in preconstruction planning and scheduling.

An owner may choose to bypass the owner-general contractor relationship and use a CM instead. The CM may be the agent or employee of the owner. If the CM is an employee, another person such as a president or officer will enter into contracts for the owner, and the CM will coordinate the construction process with the general contractor or subcontractors. If the CM is an agent, the CM is empowered by the owner to enter into contracts to complete the construction. The principal-agent relationship is discussed above.

Buyer/Seller or Uniform Commercial Code Relationships

If an exchange between a buyer and seller involves the sale of goods such as lumber or appliances, whatever form of Article 2 of the Uniform Commercial Code the state has adopted governs the transaction. The **Uniform Commercial Code** (UCC) or the law of sales is a set of statutes developed by experts, and it has contributed greatly to the standardization of commercial transaction law in the United States. The UCC does not actually govern in any state as it is only a suggested set of laws, and states are free to adopt it or modify it. For ease of discussion, the term UCC is used herein to refer to whatever version of the UCC is applicable in the jurisdiction.

Service contracts, including contracts for construction services, are *not* governed by the UCC, although the UCC has greatly influenced the common law since its adoption.

Although construction contracts are not governed by the UCC, contracts for goods incorporated into a construction project are. A **good** is a piece of tangible property capable of being moved. Examples include appliances, cables, and drywall.

Problems arise in the construction industry because often goods and services are combined, and questions may arise concerning which law applies, Article 2 of the UCC *or* the common law. The basic rule is that whichever aspect, services or goods, is predominant, this aspect governs. The sale and installation of a roof or windows is probably a sale of goods, therefore covered by the UCC, while the plumbing subcontract is a contract for services, not governed by the UCC. Many cases have addressed specific types of contracts. This is an issue of choice of law or what law governs the transaction.

Major differences exist between the UCC—discussed in more detail in Chapter 13, Sales of Materials and Supplies—and the common law, discussed in Chapter 7, Contract Formation, Breach, and Damages and Chapter 8, Changes to the Contract. Contract formation, acceptance of the contract, and modifications to the contract. are areas with different results depending on whether or not the UCC or the common law controls. The battle of the forms is a concept applicable to the UCC and not the common law and is discussed in Chapter 13, Sales of Materials and Supplies. The UCC implies two warranties into a contract: the warranty of merchantability and the warranty of fitness for a particular purpose. Warranties are discussed in Chapter 12, Torts and Warranties.

For example, under common law, a contract must contain the essential terms to be enforceable. Under the UCC, a contract is enforceable even if some of the essential terms are *not* included; Essential terms not included in the contract are automatically added to the contract by the UCC.

■ Endnotes

1 California, New Mexico, Arizona, and Texas are community property states. Maine is a marital property state, which is the same or very similar to a community property state.

2 Example based on *Southland Construction, Inc. v. The Richeson Corp.*, 642 So.2d 5 (994).

3 False Claims Act, *31 U.S.C. § § 3729*–3733.

4 Dietrich, Elizabeth, The Potential For Criminal Liability In Government Contracting: A Closer Look At The Procurement Integrity Act, 34 Pub. Cont. L.J. 52 (Spring, 2005).

5 FAR 22.804-2.

6 The Contract Disputes Act, 4 U.S.C. § § 601–613, waives sovereign immunity and allows contractors to sue the federal government for contract performance claims but the contractor must either appeal a Contracting Officer's final decision within 90 days or file a suit in the U.S. Court of Federal Claims within 12 months.

7 The Federal Tort Claims Act *28 U.S.C. § 1346.*

8 *Federal Sign v. Texas Southern University,* 951 S.W.2d 401 (Tex. 1997).

9 Pugsley, Christopher S., The Game Of "Who Can You Trust?"—Equitable Estoppel against the Federal Government, 31 Pub. Cont. L.J. 101 (Fall 2001).

10 See Ronald A. Cass & Clayton P. Gillette, The Government Contractor Defense: Contractual Allocation of Public Risk, 77 VA. L. REV. 257 (991).

11 *Erickson v. Monarch Indus., Inc.*, 216 Neb. 875, 347 N.W.2d 99, 104 (Neb. 1984).

12 Example based on *Duhamel Broadcasting Enterprises, v. Structural Systems Technology, Inc.*, 2005 U.S. Dist. Lexis 42561 (D.Neb. 2005). Note if reading case: the case refers to a party called "Mid-States" at one point in the opinion. This appears to be a typographical error for "Mid-Central" as no entity called "Mid-States" otherwise appears in the contractual relationships described in the case.

13 *Minskoff v. American Exp. Travel Related Servs. Co.*, 98 F.3d 703, 708 (2nd Cir. 1996) (quoting *Restatement (Second) of Agency § 7* cmt. a (1958)).

14 Example loosely based on *Brook, Inc., v. Thelen Aviation International Inc.*, 241 F. Supp. 2d 246 (SD NY 2002).

15 Example based on *Dixon v. Trinity*, 431 A.1364 (Md.App. 1981).

CHAPTER 5

Bonds, Liens, and Waivers

■ Bonds

A **bond** or **guarantee** is similar to an insurance policy and is purchased by the owner to insure that the contractor performs some task such as completing a project. Contractors can also purchase bonds on a subcontractor's work, and employers can purchase bonds on employees. For ease of discussion in this chapter, it is assumed that the owner is the purchaser of the bond and the general contractor is the entity that is bonded, although the law applies equally to other types of bonds.

A bond is very similar to an insurance policy but is treated differently by the industry. The distinctions are not always readily apparent to purchasers of the products and are not usually relevant to the purchasers. Bonds generally ensure that a business entity or person fulfills some task. Several different types of bonds exist and are discussed below. The bond or guarantee actually provides some relief, such as a fund of money, to aid the owner should the contractor fail to complete the task. Bonds are issued by companies called sureties, and the **surety** is the entity agreeing to provide the relief to the owner should the contractor fail to perform as required by the bond. The bond usually lists several actions the surety is free to choose from in the event of contractor default such as hiring a replacement contractor or paying the owner a lump sum and allowing the owner to hire a replacement contractor. The choice of remedy is under the surety's control, not the owner's. Sureties are allowed to seek reimbursement from the defaulting contractor for any payments it makes on a bond claim. This type of lawsuit is called a **subrogation** lawsuit.

An owner purchases a performance bond from Dwyer Surety on a project for which Frushour Construction, Inc. is the general contractor. Frushour fails to complete the project. Dwyer Surety pays the owner $250,000 to hire a replacement contractor. Dwyer Surety sues Frushour Construction, Inc. for reimbursement of the $250,000. It may be difficult if not impossible to recover from Frushour Construction, Inc., because it has no assets. Since the owners of Frushour Construction, Inc. are not liable for the debts of the corporation but have probably depleted the company of its assets, Dwyer Surety is unlikely to recover anything it pays out on the bond. See Chapter 4 in the section entitled "Corporation."

Insurance provides a fund in the event of a natural disaster or fire that damages some type of property. See Chapter 11 for more information on this topic. The major difference of importance to purchasers between a bond and an insurance policy is that while insurance companies expect to pay claims, sureties do not. "Unlike the insurance industry, sureties are not set up with the expectation of incurring losses. The premium rates charged are regulated by the government and are a function of the bond amount. As a result, any losses incurred cannot be passed to customers through higher rates."[1] For these reasons, while it is not impossible to collect on any type of bond, it is extremely difficult.

The federal government requires performance and payment bonds on all government construction, alteration, or repair projects over $100,000 under the Miller Act.[2] Many states have adopted similar laws to cover state construction projects. State laws are often referred to as "Little Miller Acts."

Bid Bond

In the traditional project delivery method, a project is put out to bid, the owner chooses one of the bids, and awards the contract to a particular contractor. Although rare, a contractor may refuse to perform the contract after award, forcing the owner to hire a different contractor, usually at a higher cost. The **bid bond** assures the owner that the contractor awarded the contract will sign the contract or at least begin to perform on the contract. The bid bond does *not* assure the owner that the contractor will complete the job; the performance bond, discussed below, exists for this purpose. In some circumstances, if the contractor fails to sign the contract or refuses to begin performance after award, the owner is entitled to the value of the bond or other relief as outlined in the bond.

EXAMPLE

Harper Drilling submitted a bid on a state project requiring contractors to demonstrate affirmative steps to utilize minority businesses. Harper knew of this requirement before submitting the bid. After being awarded the contract, Harper was either unwilling or unable to meet the requirement. The city sued on the bond and was awarded a judgment in its favor.[3]

The most likely reason for a contractor to refuse to sign the contract or begin performance is that it has made a major mistake in the bid and realizes it will lose a lot of money, should it be required to perform. In this circumstance, the owner is precluded from collecting on the bond if work has not yet begun. For a more detailed discussion of this topic, see Chapter 2 in the section entitled "Mistakes in Bidding."

EXAMPLE

Thomas Construction submitted a bid and a bid bond to the owner for the construction of a hospital wing. At the bid opening, Thomas noticed a major mistake in its bid, making it approximately 30% below the next-lower bidder. Thomas may rescind its bid and refuse to sign the contract, and the surety is not obligated to pay on the bond. The owner's only relief is to award the contract to the next-higher bidder, and the owner cannot take advantage of Thomas Construction's mistake.

Once the contractor signs the contract or begins performance, the coverage of the bid bond and the liability of the surety on the bid bond ends. Should the contractor fail to complete the project at a later date, the bid bond is not a source of relief for the owner. In practice, the owner often buys a bid bond, a performance bond, and a payment bond at the same time and from the same surety but if no performance or payment bond is purchased no relief is available to the owner under a bid bond for contractor failure to perform the contract or pay the subcontractors.

EXAMPLE

JJ Construction prepared a bid for a school project and submitted a bid bond issued by Sure Surety as required by the bid documents. No performance bond was purchased. JJ Construction was awarded the contract and signed it. Later, JJ Construction did not complete the project. Sure Surety is not obligated to the owner in any way. The owner still has a cause of action for breach of contract against JJ Construction but must proceed through the court system or other alternative dispute resolution process to obtain any relief.

Owners cannot collect on bid bonds if the owner changes the terms of the project.

EXAMPLE

The government solicited bids for the construction of 10 miles of paved highway. After the bids were opened, the government accepted the bid of Neptune Construction but qualified the acceptance and said only 8.5 miles were to be paved because the funding had been decreased. Neptune Construction refused to sign the contract for 8.5 miles of paved highway, and the government sued to recover on the bond. Neither the surety on the bond nor the contractor were liable to the government.

The surety may be liable to the contractor under theories of negligence and/or breach of contract should the bond it has agreed to provide the contractor prove to be defective. This would be similar to a professional malpractice claim. See Chapter 12.

112

EXAMPLE

Defelice Construction had a 25-year relationship with Farmers to provide it with bid and performance bonds on various road construction projects. Defelice was the low bidder on a certain road construction project, but its bid was rejected by the government, because the government said the bid bond provided by Farmers did not comply with state law. Defelice sued Farmers for failure to provide it with a viable bid bond. Farmers filed a motion to dismiss the case. The court refused to dismiss the case, stating that two triable issues of fact existed. The first was whether or not Farmers breached an oral contract with Defelice to provide it with a bond meeting the state's requirements. The second was whether or not Farmers was negligent in the preparation of the bond.[4]

Performance or Completion Bonds

The **performance or completion bond** assures the owner that the contractor will complete the job. A contractor may also require a subcontractor to supply a performance bond.

MANAGEMENT TIP

On private projects, bonds are an option, and owners should assess the risks of failing to purchase such a bond before entering into a construction contract. Naturally, purchasing the bond adds to the cost of the project.

If a claim is made on the bond, the surety has several options, depending on the wording of the contract:

- Hire a new contractor
- Pay the owner the value of the bond
- Negotiate with the existing contractor to complete the job

As mentioned above under bid bonds, the surety always has the right under the bond to seek subrogation that is reimbursement from the contractor for any losses paid under the bond.

EXAMPLE

An owner enters into a contract with Agila Contractors, Inc. for construction of a sports facility and purchases a performance bond from Surety Company. Because of financial difficulties, Agila Contractors, Inc. does not complete the project. The owner files a claim with Surety Company. After evaluating the project, Surety Company decides to hire Best

continued

Contractor to complete the project. The owner is only required to pay Best Contractor the remainder due on its contract with Agila and not any additional amounts attributed to Best Contractor's need to come up to speed on the project or correct errors. Surety Company is required to pay those amounts. Surety Company may sue Agila for any sums it pays on behalf of the owner. Agila will also find it extremely difficult if not impossible to get a bond in the future and is probably no longer a viable business entity.

EXAMPLE

The owner sued the contractor for defects and damages for delay in the construction and made a claim on the performance bond. The contractor claimed the defects and the delays were caused by the owner's excessive change orders. The contract between the owner and the contractor contained an arbitration clause. The contractor filed a motion to compel arbitration, which the court granted. At the arbitration, the arbitrator decided in the contractor's favor: the defects and delay damages were the owner's fault and not the contractor's. The owner was not able to recover on the bond since the delays were the owner's fault.

MANAGEMENT TIP

Declaring default on a performance bond is a major step and not one an owner should enter into lightly. While the owner may no longer have problems with the contractor, the owner will have many problems with the surety, as the surety will not want to pay on the bond. The owner should always inform the surety of its intent to terminate a construction contract before actually doing so. Most sureties have language in the bond requiring this, and the owner's failure to do so will mean that the surety has no obligation to the owner.

FORM

The Surety's obligation under this Bond shall arise after:
The Owner has notified the Contractor and the Surety...that the Owner is considering declaring a Contractor Default and has requested and attempted to arrange a conference with the Contractor and the Surety to be held not later than 15 days after receipt of such notice to discuss methods of performing the Construction Contract.

Payment Bonds

The **payment bond** assures the owner that the contractor will pay the subcontractors and suppliers. Owners need this protection because lien laws, discussed below, allow subcontractors and suppliers to file and foreclose liens on the owner's real estate if they are not paid by the general contractor. In order to remove the lien the owner must pay the subcontractors and suppliers even if the owner has paid the general contractor. In other words, the owner pays twice for the work.

If the owner does not have a bond, the owner can always sue the contractor for reimbursement of any sums it pays to subcontractors or suppliers, but this can be very costly and time consuming. In all likelihood, if the contractor has not paid the subcontractors or suppliers, it probably has no money. Even if the owner wins a judgment, it may be impossible to ever collect.

EXAMPLE

An owner enters into contract with CC Construction to add an additional bathroom and bedroom to the owner's house for a total cost of $25,000. CC Construction subcontracts the plumbing to PP Plumbing and agrees to pay PP Plumbing $5,000. PP Plumbing installs the plumbing, but CC Construction never pays it. The owner pays CC Construction $25,000, and shortly thereafter, CC Construction goes out of business. PP Plumbing files a lien of $5,000 on the owner's property. The owner must pay PP Plumbing the $5,000 in order to have the lien removed. The owner may sue CC Construction for reimbursement of the money paid to PP Plumbing; however, since CC Construction is out of business, it will have no assets to pay the owner.

Other Types of Bonds

Other types of specialized bonds have been authorized by statute or made available to buyers due to market demand. For example, it is possible to buy a licensing bond in some states. A **licensing bond** assures the owner that the contractor will comply with all licensing requirements of the applicable governments. Some states or local authorities may require an improvement bond. An **improvement bond** assures the buyer that the contractor has complied with wetlands improvement or other required improvements.

■ Liens

Lien laws allow general contractors, subcontractors, material suppliers, and, in some states, laborers to file a lien on the owner's property for amounts owned to them for work done on the owner's property. A **lien** is a legally recognized obligation of the owner of a particular piece of property to pay an amount to another, the **lien holder.** If the owner refuses to pay the lien, the lien holder may sell or dispose of the property in satisfaction of the amount of the lien. Lien laws transfer the risk of a contractor failing to pay a subcontractor or supplier onto the owner who has benefited from the work or supplies. In other words, these laws make construction more risky for owners and less risky for subcontractors and suppliers. The owner is free to sue the general contractor for any sums it pays on a lien, but the owner is unlikely to collect.

Each state has different lien laws, forms, and requirements, and these requirements must usually be strictly followed in order for the lien to be valid. The process of enforcing lien rights is often called perfecting a lien. **Perfecting a lien** means to carefully follow the requirements of state law for enforcing the rights under the lien. Failure to follow the law, or perfect the lien, will usually make a lien invalid. Most lien laws require the following:

1. Information must be posted on the construction site by the owner. States generally require the owner to post certain information on the property site giving potential lien claimants the information they need to make it easier to file a lien. Potential lien claimants should get this information before supplying work or materials to the site. See the example in Appendix A.
2. Notice must be given to the owner. The subcontractor or supplier must give notice to the owner, prior to or shortly after they begin work on the project or supply material to the project, that they intend to enforce their rights under the lien law of the state where the project is located should the general contractor fail to pay them. See the example in Appendix B.
3. Filing the lien or claim of lien with the county recorder. If the subcontractor or material supplier is not paid by the contractor, they file a lien on the property with the county recorder for the amount due. A copy is served on the owner. A lien filed in the county recorder's office is a cloud on the owner's title similar to a mortgage. See the example in Appendix C.
4. Filing a lawsuit to foreclose on the lien. If the owner does not pay the lien, the subcontractor or material supplier can then file a lawsuit to foreclose the lien. Most states offer expedited procedures for this type of lawsuit. At the end of the lawsuit, the subcontractor or material supplier obtains a judgment of foreclosure.
5. Enforcement of the judgment of foreclosure. The judgment of foreclosure is taken to the county sheriff or other government entity, which will conduct a public sale of the property, collect the proceeds, pay off the lien, and return any overage to the owner.

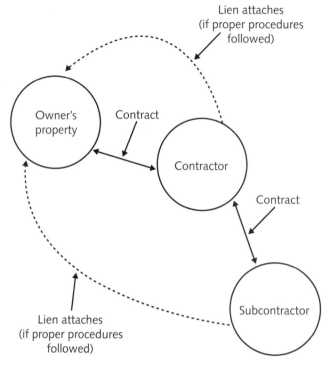

Lien attaches
(if proper procedures
followed)

Owner's property

Contract

Contractor

Contract

Subcontractor

Lien attaches
(if proper procedures
followed)

FIGURE 5-1 The Lien Process

EXAMPLE

Ames Construction is the general contractor and Elias Electrical is the electrical subcontractor. The owner has posted the required information on the site for potential lien claimants. Shortly before beginning the electrical work, Elias Electrical serves the owner with a notice of its intent to use the state's lien law to assure payment from Ames. After the work is completed, the owner pays Ames all sums due on the construction contract. Before paying Elias, Ames goes out of business. Elias Electrical files a lien for $10,000 for the unpaid electrical work on the project and serves a copy of the lien on the owner. Elias Electrical performs all of the steps necessary, and the lien is perfected.

If the owner does not pay Elias, Elias can file a lawsuit to foreclose on the lien. Because owners realize that the lien can be foreclosed and the property sold, the owner usually pays off the lien. Forms are then filed in the office of the county recorder to discharge the lien.

MANAGEMENT TIP

Lien laws vary in every state, and it is necessary to follow the specific requirements of the applicable lien law in order to perfect the lien.

If the owner has not paid the contractor because the owner is disputing the amount owed to the contractor, the owner can bond around the lien: the owner can buy a bond that will pay the contractor should the dispute eventually result in the contractor's favor. The bond will release the lien and allow the owner to sell the property or obtain a mortgage. Any dispute or litigation continues, but the title to the property is clear.

EXAMPLE

Fissel Design bills the owner $10,000 for additional work it claims the owner ordered on a project. The owner claims that the amount is too much for the work done and Fissel Design is overcharging. Fissel Design perfects a lien of $10,000 on the property. The owner wants to sell the property immediately. The owner buys a bond from Surety Company that will pay Fissel Design the $10,000 should Fissel Design prevail in the dispute. The cost of this bond is $11,000, and Surety agrees to refund to the owner $10,000 should the owner prevail in the lawsuit and the lien be removed by the court.

The court determines that the owner owes Fissel Design only $5,000. Surety pays Fissel Design $5,000 and refunds the remaining $5,000 to the owner. The court also extinguishes the lien.

■ Direct Payment and Joint Checks

Owners can protect themselves from liens in a number of ways. One is to buy a payment bond (discussed above) and waivers of liens (discussed below). Other alternatives include directly paying subcontractors or issuing joint checks to the general contractor and the subcontractor.

EXAMPLE

Ames Construction is the contractor and Elias Electrical is the electrical subcontractor. Shortly before beginning the electrical work, Elias Electrical serves the owner with a notice of its intent to use the state's lien law to assure payment from Ames. Later, when Ames Construction submits its monthly submittal containing the electrical work, the owner issues a joint check to Ames Construction and Elias Electrical. This requires the signatures of both Ames and Elias, and Elias will likely demand that it be given the check or it will refuse to endorse it.

■ Waivers of Lien

A **waiver** is a document or act relinquishing or giving up a legal or contractual right. Almost any type of legal or contractual right can be waived, and it is not unusual for parties to sign waivers in a variety of situations.

I release and hold harmless _____, its employees, agents, and officers from liability for any loss, damage, and claims, actions or losses for bodily injury, or property injury, including attorneys' fees, arising directly or directly from _____. Signed.

MANAGEMENT TIP

Waivers are very powerful and should never be signed lightly, as the signer is usually agreeing to give up valuable legal rights, and the law will enforce most waivers.

Waivers are commonly encountered in the construction industry in connection with liens or potential liens. The owner may refuse to pay the contractor unless it has received signed waivers of liens from subcontractors and suppliers. A **waiver of lien rights** is a document signed by a potential lien claimant waiving the potential lien claimant's right to file a lien. These waivers are enforceable.

WAIVER OF LIEN

Description of real property: (fill in legal description of property)
Description of project: (fill in a description of the construction project)
In consideration of the promise of _____ , owner to pay all sums due the undersigned in connection with the undersigned's work or supply of material to the above real property for the above project, and other good and valuable consideration, the receipt of which is acknowledged or promised, the undersigned waives, releases and relinquishes all liens, claims or rights of lien which he, she, or it has or might have on or against the above described real property arising under and by virtue of the laws of this state on account of labor or materials, or both, furnished or which may subsequently be furnished by the undersigned to, or on account of, the owner of the real property, or his, her, or its agents or contractors, including but not limited to _____
(name of general contractor). Signed.

By signing this document, a contractor, subcontractor, or supplier waives its right to file a lien against the property in event it is not paid. A waiver of lien does not waive the right to be paid, only the right the file a lien. Therefore, if a contractor, subcontractor, or supplier signs a waiver of lien, it cannot file a valid lien on the property. The party signing the lien can still pursue its legal rights. The lien waiver only waives the right to file a lien; it does not waive the right to pursue legal claims, though such a breach-of-contract case can take years to wend its way through the legal process. Since most states have expedited processes for lien foreclosure lawsuits, by signing the waiver of lien, the lien claimant is giving up a speedy resolution of its claim.

EXAMPLE

Ames Construction is the general contractor and Elias Electrical is its electrical subcontractor. Shortly before beginning the electrical work, Elias Electrical served the owner with a notice of its intent to use the state's lien law to assure payment from Ames. The owner refuses to pay Ames without a waiver from Elias. Elias signs the waiver. Ames never pays Elias. Elias cannot file a lien on the property and force the owner to pay it. Elias can only pursue litigation or other forms of dispute resolution against Ames.

Because waivers of liens are so powerful and often demanded by owners, subcontractors and suppliers have developed what is called a conditional waiver of lien. A **conditional waiver of lien** does not waive lien rights until payment is received by the signer; the signer maintains all legal rights to file a lien if she is not paid. A savvy owner is unlikely to accept such a waiver.

FORM

Waiver of Lien

Description of real property: (fill in legal description of property)
Description of project: (fill in a description of the construction project)
The undersigned has furnished materials, machinery, fixtures, and/or labor for the construction, erection, improvement, and/or repair to the above described real property for the above described project and has a legal right to a lien upon the property under the laws of this state.
The undersigned waives, releases and relinquishes all liens, claims, or rights of lien which he or she has or might have on or against the above described real property arising under and by virtue of the laws of this state on account of labor or materials, or both, furnished or which may subsequently be furnished by the undersigned to or on account of the owner of the real property, building, premises, or project, or his or her agents or contractors upon receipt of payment of the sum of $_____ to the undersigned. Signed.

■ Endnotes

[1] In Re: *Outer Banks Contractors, Inc.* Debtor, *Outer Banks Contractors, Inc. v. American Casualty*, 1995 Bankr. Lexis 1255 (U.S. Bankr. E.D., N.C. 1995).

[2] 40 U.S.C. § 3131 to 3134.

[3] Example based on *City of Fairfield, Iowa v. Harper Drilling Co.*, 2004 Iowa App. Lexis 740 (unpublished opinion).

[4] Example based on *L. G. Defelice, Inc. v. Fireman's Ins. Co.*, 41 F. Supp. 2d 152 (D.Conn. 1998) (based on Connecticut law).

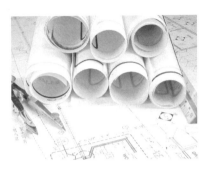

CHAPTER **6**

Legal Issues Relating to Plans and Specifications

■ Introduction

In the construction industry, a **plan** is a pictorial or visual drawing of a particular part of a construction project. A **specification** is a written description of a particular part of a construction project. **Boilerplate** is the term used for standardized specifications or language usually chosen from books and inserted into contracts, often without reconciling the boilerplate to the actual plans, specifications, or special conditions.

Because of the complexity of plans and specifications, the difficulty in reading and understanding them, discrepancies between boilerplate and specially designed plans and specifications, and inadequately prepared plans and specifications, these documents are often a source of controversy. Complicating the problem is the tendency of workers to build a project according to past practices rather than taking the time and effort to decipher the current plans and specifications. Finally, just as with any other human endeavor, plans and specifications may contain errors or workers may commit errors during the construction process.

The plans and specifications are part of the contract, and thus the contractor is required to perform in accordance with them or be in breach of the contract. However, it is not always clear what performance is demanded of the contractor, or it may be that a particular plan or specification is impossible to build. It is not always clear whether a defect in the construction is caused by faulty plans and specifications or faulty construction practices. Each party tends to put responsibility for a defect on other parties and not on themselves. It is costly to take responsibility for defects. In a traditional design-bid-build project, the owner may be in the middle of a fight between a contractor, claiming the fault lies with the designer, and a designer, claiming the fault lies with the contractor. In the real world, the problem may be a combination of both.

The first and most difficult step in a dispute involving plans and specifications is to determine which category the problem fits in. Once the problem has been placed in the correct category, a further and final determination is needed as outlined in the table below. Once this final determination is made, the legal ramifications are usually clear. See Table 6-1 for an overview.

TABLE 6-1 Categories of problems involving plans and specifications

Problem category	Simple rule, if any	Further applicable law
1. Determining whether the defect is faulty construction or faulty design.	Factual issue and is determined by jury (if any) or judge, arbitrator if jury waived. Liability for faulty construction lies with the contractor. Liability for faulty design lies with the owner and rarely with the designer.	Factual issues are discussed in Chapter 1.
2. Determining if the specification is a design or performance specification.	Owner is liable for faulty construction caused by faulty design specifications. The contractor is liable for faulty construction caused by faulty performance specifications. In rare circumstances, the designer is liable to the owner for faulty design specifications.	Discussed in this chapter.
3. Conflicting specifications.	1. Special conditions control over general conditions. 2. Handwritten provisions control over typed or preprinted provisions. 3. Typed provisions control over preprinted provisions.	Discussed in this chapter.
4. Change or addition.	If the owner has changed or added to the contract, the owner owes contractor a reasonable sum to compensate the contractor.	Discussed in Chapter 8.
5. Ambiguous plans and specifications.	Ambiguous specifications: most complicated. The meaning of the specification must first be determined using rules in Chapter 9. Once the meaning is determined, the problem will be a Category 2, 3, or 4 problem. Put the problem into the appropriate category and continue analysis.	Discussed in Chapter 9. Discussed in Chapter 8.

If the problem is a Category 1 problem, a factual issue exists, and the trier of fact must decide whether the defect was caused by faulty construction, faulty design, or some combination of the two. Expert witnesses will be needed to help the trier of fact decide if the error was caused by the contractor or the designer. "It is well settled that in order to prove negligence or malpractice in the design of a structure, the plaintiff must put forth expert testimony that the engineer or architect deviated from accepted industry standards."[1]

An **expert witness** is one who has specialized knowledge in the area such as an engineer, architect, or contractor. The expert is brought in, at least in theory, to render an independent evaluation of the situation and makes a report. Expert witnesses can disagree, and the **battle of the experts** occurs when each side hires experts whose testimony conflicts. In this situation, the trier of fact must decide which expert to believe.

Category 2 problems require determining if the plan or specification is a design or performance specification. Once that is determined, the liability rules outlined below are simple to apply.

Category 3 problems require the application of specific legal rules developed to solve this type of problem. See discussion of Conflicting Specifications below.

Category 4 problems arise when the owner makes a change or addition to the contract. See Chapter 8. The simple rule is that the owner must compensate the contractor of changes and additions that cost the contractor time or money.

A Category 5 problem is the most difficult. Because the plan or specification is ambiguous, it must first be determined what the plan or specification says. To do this, the rules of Chapter 8 must be employed. Once it has been determined what the plan or specification says, the problem will be a Category 2, Category 3, or Category 4 problem.

■ Types of Specifications

This section discusses the types of specifications. The next section discusses the rules. Category 2 problems are resolved by placing the plan or specification causing the problem into one of four specification categories and using the applicable rule to determine liability:

1. Design specification. Liability rests with the owner and, in rare instances, with the designer.
2. Performance specification. Liability rests with contractor.
3. Combination specification. Determine whether the specific section of specification causing the problem is a design or a performance specification. Use rules above.
4. Or-equal specification. More complex. See discussion below.

Historically, liability for defective construction caused by defective plans or specifications always rested with the owner, and liability rested on the designer in only very limited circumstances. Designer liability is discussed below. With the increase in complexity of the building process, liability may rest on the contractor or a specialty subcontractor if the specification causing the problem is a performance specification.

Traditionally, all specifications were what are now called design specifications. A **design specification** is one that tells the contractor exactly what to do and leaves the contractor little latitude except as to means and methods.

EXAMPLE

Provide welded-wire units prefabricated into straight lengths of not less than 10 feet coverage on exterior exposures and ½ inch elsewhere. Wire sizes: Side rod diameter: 0.1483 inch. Cross rod diameter: 0.1483 inch.

    C. [Raceway] Supports shall be mounted to structure with the following:

1. Toggle bolts on hollow masonry.
2. Expansion shields or insets on concrete.
3. Machine screws on metal.
4. Wood screws on wood.
5. Nails, rawl plugs, or wood plugs shall not be permitted.

In recent years, more responsibility for design has been placed on the contractor and the specialty subcontractors such as HVAC subcontractors. The designer gives the contractor a more general specification detailing the desired result and leaving it to the contractor and/or specialty subcontractor to design the system to meet the desired result. Such a specification is called a **performance specification.**

EXAMPLE

Install steel overhead coiling doors designed to withstand a wind load of 20 psf (pounds per square foot).

EXAMPLE

Contractor is to provide the complete electrical design and is free to provide any design that meets all applicable codes, the design build criteria, and the electrical specifications.

In reality, many specifications are a combination of both design and performance characteristics; this type of specification is called a **combination specification.** Where a specification does "not tell a contractor how to perform a specific task that part of the specifications can be performance specifications even if the rest of the specifications are design specifications."[2]

2.1 Sheathing: Shall be strong enough to retain its shape under the weight of concrete and resist unrepairable damage during construction....Provide drain holes if the tendon is to be placed, stressed and grouted in [**9] freezing climate.

 2.1.1 Placing Sheathing: Place sheathing to form ducts for post tensioning tendons. Fasten sheathing securely at close enough intervals to avoid displacement during concreting....After placing of sheathing, reinforcement and forming is complete, perform an inspection to locate possible sheathing damage. Repair all holes or openings in the sheathing prior to concreting.

 2.1.2: Forms shall be well braced and stiffened against deformation and shall be accurately constructed. The forms shall be such as to produce a smooth dense surface. A bond-breaking substance may be applied to the forms before pretensioning steel is placed; if so, it should be done in accordance with PCI MNL-116. Form ties shall be either the threaded or snap-off type, so that no form wires or metal pieces will be within 1½ inches of the surface.

 2.1.3 Strands for precast-prestressed members shall be given an initial stress equal to approximately 10% of the design load, after which the alignment shall be checked for conformity to the drawings...[3]

■ Category 2 Problem: Owner Liability for Defective Construction Caused by Defective Design Specification

Liability for defective construction caused by a defective design specification rests with the owner. If the contractor has constructed the system according to the design specification, the contractor has no fault or liability for failure of the system to perform. This concept is also known as the **_Spearin_ warranty**, **_Spearin_ doctrine**, or the **owner's warranty of the plans and specifications**. The terms _Spearin_ warranty[4] and _Spearin_ doctrine come from a U.S. Supreme Court case, _United States v. Spearin_,[5] which held that a contractor who had built a cofferdam according to specifications had no liability for defects.

The case of _Stuyvesant Dredging Co. v. U.S._[6] contains the following definition of design and performance specifications: "Design specifications explicitly state how the contract is to be performed and permit no deviation....Detailed design specifications contain an implied warranty that if they are followed, an acceptable result will be produced.... Performance specifications, on the other hand, specify the results to be obtained and leave it to the contractor to determine how to achieve those results."

Because both of the above cases, _Spearin_ and _Stuyvesant_, are federal contract law cases, they only apply to federal contracts. However, most states have adopted the _Spearin_ doctrine.[7]

EXAMPLE

Croll, the contractor entered into agreement to construct a post office building. Excavation revealed inadequate soil to support the structure. The contractor stopped work, informed the owner, and an expert was hired to make recommendations. The recommendation was for additional piling beneath exterior walls, and the owner agreed to pay for same. It became obvious to the contractor that additional piling would be needed beneath the floor itself, but the owner refused to authorize the changes and ordered the contractor to build according to the plans and specifications, which the contractor did. After construction, the floor settled, requiring the owner to make substantial repairs. The court held that the owner was responsible for the defects because the contractor had built according to the plans and specifications.[8]

Spearin also states that "general disclaimers requiring the contractor to check plans and determine project requirements do not overcome the implied warranty and thus do not shift the risk of design flaws to contractors who follow the specifications."[9]

EXAMPLE

The contractor entered into an agreement with the government to build a building to house helicopters. The plans contained the following disclaimer:

CANOPY DOOR DETAILS, ARRANGEMENTS, LOADS, ATTACHMENTS, SUPPORTS, BRACKETS, HARDWARE ETC. MUST BE VERIFIED BY THE CONTRACTOR PRIOR TO BIDDING. ANY CONDITIONS THAT WILL REQUIRE CHANGES FROM THE PLANS MUST BE COMMUNICATED TO THE ARCHITECT FOR HIS APPROVAL PRIOR TO BIDDING, AND ALL COST OF THOSE CHANGES MUST BE INCLUDED IN THE BID PRICE.

After construction began, it was determined that the canopy door design supplied by the government was not workable. The contractor redesigned and modified the canopy doors and submitted a claim for additional costs, which the government denied. The government claimed that the above clause required the contractor to verify the workability of the design. The court held that the clause did not alert the contractor that the design might contain substantive flaws requiring correction and approval before bidding. Because the design supplied by the government was impossible to perform, the contractor was entitled to reimbursement.[10]

In addition, if the specification gives alternative methods of performance, each method is warranted.

EXAMPLE

The following specification existed in a contract for the production of two concrete storage tanks consisting of 96 concrete and steel wall panels that later bowed, causing increased costs.

> Forms: Forms shall be well braced and stiffened against deformation and shall be accurately constructed. The forms shall be such as to produce a smooth dense surface. A bond-breaking substance may be applied to the forms before pretensioning steel is placed; if so, it should be done in accordance with PCI MNL-116. Form ties shall be either the threaded or snap-off type, so that no form wires or metal pieces will be within 1½ inches of the surface.

It was difficult to determine whether or not the bowing was caused by a faulty design specification or by faulty construction practices. However, the court rejected the government's argument that the option given to the contractor to use a bond-breaker transferred the risk to the contractor. The court said, "Rather, when the government provides alternate methods by which a project may be completed, there is an implied warranty that either method will achieve the desired result."[11]

Claims between Owner and Designer for Faulty Design Specifications

If the plans and specifications prove to be defective, the owner may attempt to seek reimbursement for any sums it pays to remedy the defects from the design professional. In general, unless the design professional has guaranteed or warranted the plans and specifications, which is rare, the design professional is not liable for defective plans and specifications. The only exception is this: if the owner can show that the designer has breached the contract *or* negligently prepared the design, the designer is liable. The theories used to find designer liability are breach of contract, negligence, and/or professional malpractice.

The breach-of-contract argument is seldom effective because, as a general rule, the contract between the owner and the designer requires the designer to perform according to the "prevailing standard of professional care." While this vague standard is certainly open to interpretation, it most definitely does not mean "to make no mistakes" or "to produce perfect plans and specifications." It then becomes a factual issue to determine whether or not the designer has operated below the prevailing standard of care. The trier of fact will then make this determination based upon the expert testimony. Breach of contract is discussed in Chapter 7.

Negligence or professional malpractice arguments are also seldom effective for the same reasons. **Professional malpractice** is merely a specialized term for negligence committed by a professional. **Negligence** is the failure to act reasonably in a particular situation with resulting injury to some entity. A party is not negligent just because it made a mistake or did not perform perfectly. "It is well settled that in order to prove negligence or malpractice in the design of a structure, the plaintiff must put forth expert testimony that the engineer or architect deviated from accepted industry standards."[12] Negligence and professional malpractice are discussed further in Chapter 12.

EXAMPLE

Columbus was injured when he was attempting to store a salt spreader weighing approximately 2,100 pounds from wooden hangers suspended from the ceiling of a garage. The storage system had been designed by Smith & Mahoney, P.C. The designer specified number two grade lumber in the design of the wooden members of the storage system. No evidence was presented at trial that the use of this grade of lumber for the purpose intended violated generally accepted industry standards or the state building code. The designer was not negligent.[13]

Disputes as to Classification of Plan or Specification

In the event that the parties disagree on whether or not a certain plan or specification is design or performance, the court will determine this by looking at the detail of the specification. If the plan or specification tells the contractor what to do, it is a design specification. If the plan or specification tells the contractor the result to be obtained, it is a performance specification.

EXAMPLE

The government assumed the responsibility of applying for and acquiring all necessary licenses and permits. Included in the permit application were two detailed drawings indicating that a 24-inch pipe was to be used to "dewater" the construction site. A permit subsequently was granted as a result of this application. The contractor followed the drawings, believing them to be design specifications, but the width of the pipe proved to be inadequate to dewater the site. The court looked at the detail of the drawings and the contract provision requiring the contractor to follow the requirements of the permit and determined that the requirements of the permit were design specifications.[14]

Disputes as to the Meaning of Plans or Specifications

If the parties disagree on what a plan or specification requires, an issue of scope or contract interpretation is raised. This is a classic contract interpretation or scope issue: what does the contract require? In this situation, the law first interprets the contract, and if the owner has demanded something not required by the contract, it is treated as a change. See discussion of contract interpretation in Chapter 9 and discussion of changes to the contract in Chapter 8.

EXAMPLE

The contract specifications called for fire-rated walls, but certain panels leading to those walls contained an industry abbreviation for *non*-fire-rated materials. The contractor installed the fire-rated walls but used non-fire-rated materials in the panels. The government demanded that the contractor use fire-rated materials in all of the panels, and the contractor complied but made a claim for increased costs. The court interpreted the contract as requiring fire-rated walls but not fire-rated materials in the panels. Therefore, the owner's insistence on fire-rated panels was a change in the contract, and the contractor was entitled to additional compensation.[15]

■ Category 2 Problem: Contractor Liability for Defective Construction Related to Performance Specifications

If a contractor or specialty subcontractor fails to build a system in accordance with the performance specification, the contractor and/or specialty subcontractor is liable to the owner for the resulting damages.

EXAMPLE

The performance specification instructed the contractor to "install steel overhead coiling doors designed to withstand a wind load of 20 psf." The overhead coiling doors became damaged with a wind load of 15 psf; contractor was liable for damages.

EXAMPLE

The performance specification says, "Contractor is to provide the complete electrical design and is free to provide any design that meets all applicable codes, the design build criteria, and the electrical specifications." After a fire, it was determined that the electrical design was faulty. The contractor was liable for damages.

Because of lack of expertise, the contractor may have a limited ability to monitor the performance of the subcontractor. Because of this, the contractor should insert indemnity language in the specialty subcontract requiring the specialty subcontractor to indemnify the contractor for defects. See discussions of indemnity in Chapter 12 and Chapter 14.

If possible, the contractor should insert language in the contract between the owner and the contractor eliminating the liability of the contractor for the failure of the specialty system and requiring the owner to seek redress for defects directly from the specialty subcontractor, not from the contractor.

EXAMPLE

The contractor entered into a contract with the owner, a government entity, for the design and installation of a waste-water treatment facility. The filtration system was to be designed by Hydro Engineering, a specialty manufacturer and supplier of waste-water filtration systems. The contract between the government and the general contractor contained this language: "Hydro Engineering, manufacturer and supplier of the waste-water filtration system, shall be solely responsible for the performance of the waste-water filtration system as specified and shall modify, add to, or alter the equipment as necessary, without any additional cost to Government, to provide a satisfactory performance."

This clause takes the contractor out of loop. Should the filtration system not work correctly, the government is to go directly to Hydro Engineering for modifications or alterations.

■ Category 2 Problem: Liability for Combination Specification Failure to Perform

In order to determine liability for a combination specification, it is necessary to determine whether the offending part of the specification is a design or performance specification. "One measure of a design, *Spearin*-type specification is the level of detail provided....An even more important measure is the degree of discretion permitted a contractor in following the directions provided in the specifications."[16] A design specification with a lot of detail and little discretion on the part of the contractor is a design specification. Once this is determined, liability follows the above-described rules.

■ Category 2 Problem: Liability for Or-Equal Specification to Perform

An **or-equal specification** is one that allows the contractor to substitute performance or substitute for a specifically called-for product with a different product. The contractor can use either the product specified in the contract or find a different but equal product. These types of specifications are uncommon in private contracts but common in government contracts.

EXAMPLE

The contract contains the following specification for a door: "Floor supported, overhead braced, Sanymetal "Normandie" model, or approved substitution, minimum 58 inches high." The contractor can use the Sanymetal model or substitute another one of the same quality. The contractor must get approval for the substitution.

Problems arise if the contractor substitutes a product for the specified product and it proves defective. If the substituted product is of lesser quality, the contractor is liable. This area of the law is currently under development and not certain, but if the contractor can prove that the originally called-for product would have been ineffective for the job, the contractor may be relieved of liability. Expert testimony would be required.

Frequently the or-equal specification will include a provision requiring the designer to approve the substitution. This will not relieve the contractor from liability should the contractor's choice prove inadequate. The contractor is still liable for the performance of the substitution. Designer approval does not change the liability associated with a type of specification. See discussion below.

■ Designer Approval of a Performance Specification or Or-Equal Substitution

Most contracts require the contractor to obtain designer approval of shop drawings or a substitution in an or-equal specification. Approval by the designer pursuant to the clause does *not* transfer liability to the owner or designer. Liability for a performance specification remains with the contractor. Liability for the failure of the substituted product to perform remains with the contractor, but see exception discussed immediately above.[17] Most written contracts contain language to this effect.

FORM

> The Engineer shall pass upon the shop drawings with reasonable promptness. Checking and/or approval of shop drawings will be general, for conformance with the design concept of the Project and compliance with the information given in the Contract Documents, and will not include quantities, detailed dimensions, nor adjustments of dimensions to actual field conditions. Approval shall not be construed as permitting any departure from contract requirements...nor as relieving the Contractor of the responsibility for any error in details, dimensions, or otherwise that may exist.[18]

■ Category 3 Problem: Conflicting Specifications

It is not the contractor's duty to compare the plans and specifications and find any conflicts. If a conflict arises, the owner is responsible.

MANAGEMENT TIP

Contractors should always inform the owner of conflicts in the plans and specifications and clarify which the owner wants performed. Many contracts contain clauses with wording such as the following: "In the event of inconsistencies between the contract documents, the plans shall control over the specifications." This wording will protect the contractor should the contractor follow the plans, but problems and arguments can be avoided by informing the owner and seeking clarification.

Unless the contract specifies otherwise, the following laws control conflicts in the contract documents:

- Special conditions control over general conditions.
- Handwritten provisions control over typed or preprinted provisions.
- Typed provisions control over preprinted provisions.

EXAMPLE

A printed contract between two experienced companies, Wood River and Willbros, contained detailed language regarding how claims were to be handled. The following handwritten provision was added to the contract, initialed, and dated by representatives of both parties to the contract.

"2.03 Contractor shall not be liable under any circumstances or responsible to company for consequential loss or damages of any kind whatsoever including but not limited to loss of use, loss of product, loss of revenue, or profit."

The handwritten provision was in conflict with detailed provisions of the printed contract, requiring Wilbros to be responsible for all types of damages.

After completion of the project, the pipeline ruptured, and Wood River sued Willbros for various types of damages, including consequential damages, which are discussed in Chapter 7. The court dismissed all claims for consequential damages pursuant to the handwritten provision. The court held that the handwritten provision was clear and unambiguous and prevailed over the preprinted provisions.[19]

MANAGEMENT TIP

Do not be afraid to insert handwritten provisions into contracts.

■ Additional Concepts Relevant to Plans and Specifications

Because plans and specifications are so important to the industry, they are often the subject of disputes. For this reason, many laws have developed around them. The most common legal concepts one can expect to encounter in disputes related to plans and specifications are summarized here.

Application of *Spearin* to Availability of Materials or Fixtures

The *Spearin* warranty does not apply to the availability of fixtures or materials. That is, the owner does not warrant that fixtures or materials will be available to the contractor as needed to complete the project on time. It is the contractor's responsibility to obtain the fixtures and materials as needed to complete the project. Failure to do so is a breach of the contract by the contractor.

In the event that a particular fixture or material is not available, either because it has been discontinued or because a typographical error has occurred, the contractor is relieved of liability under the doctrine of impossibility of performance, discussed below. The contractor must inform the owner of the mistake as soon as it is discovered so that the owner can decide on a suitable alternative.

Clauses Designed to Bypass *Spearin* Liability

Owners may attempt to overcome their liability for faulty plans and specifications by inserting language into the contract designed to transfer the liability onto the contractor. The law is unlikely to uphold such a clause, as it is exculpatory or onerous. An **exculpatory or onerous** clause attempts to transfer liability from a party with control over a situation to one who has no control. Another definition of exculpatory or onerous is a clause that is contrary to fundamental justice or rights. Some courts say that such a clause violates public policy. The law does not favor such clauses, but the line between an exculpatory or onerous clause and a valid clause is not always clear. Parties have little to lose by including exculpatory or onerous clauses in contracts.

The *Spearin* case, in which the owner's implied warranty of the plans and specifications was recognized by the U.S. Supreme Court, involved a clause attempting to transfer risk for performance of the project onto the contractor. The court would not allow this. "This implied warranty is not overcome by the general clauses requiring the contractor to examine the site, to check up the plans, and to assume responsibility for the work until completion and acceptance."[20]

FORM

The owner does not warrant the plans and specifications.

The law will not enforce the above clause.

Strict Compliance/Completion/Performance vs. Substantial Compliance/ Completion/Performance

In order to be entitled to payment, the contractor must substantially comply with the plans and specifications. Strict or exact compliance is not required. On the other hand, the contractor may not be entitled to full payment. The owner can deduct items from a punch list, for example, from the final payment.

The exact point of substantial completion depends on the type of project and the facts and circumstances surrounding the event.

The owner of a casino facility awarded a contract for renovation of the rooms and erection of a new glass façade. The façade, while nonfunctional, was intended to attract customers to the renovated property. The project without the façade was not substantially complete, and the contractor was not entitled to payment. This was true even though the interior work had been finished.[21]

In 1971, an owner contracted with a contractor to design and install a water system for a housing development. Except for a water storage tank, the construction of the system was completed in September 1972. The water storage tank, which was part of the original design, was installed in July 1973 by another contractor. Shortly thereafter, leaks were discovered in the water service lines. On July 21, 1980 the owner sued the original contractor who designed and installed the system minus the water service tank.

The issue in the case was whether the project was substantially complete in September 1972, when there was no water storage tank, or in July 1973, with the water storage tank.

The court determined that as a matter of law the project was not substantially complete until July 1973. Between September 1972 and July 1973, the project was used by the sales office only, not by any owners. The engineers testified that although the system was capable of use, it could not be used for its intended purpose—servicing the development—until the water storage tank was installed in July 1973.

In 1975, the Showalters began construction on their home, doing all of the work themselves. They began occupying the home in 1975 when it was approximately 50% complete. In July 1975, a permit was issued for an electrical panel. In 1977, the Showalters began adding a utility room and a living room. They continued to work on the home. On April 5, 1981, the home was sold to the Smiths and was approximately 90% complete. The utility room was still unfinished at the time of sale, but its electrical wiring was complete. When the Smiths purchased the home, they were not told that the Showalters were the builders, that the Showalters had not received final building or electrical inspections, or that the wiring was substandard.

continued

On May 18, 1984, the home was destroyed by fire. The Smiths' expert witness inspected the fire scene and was of the opinion that the fire began in a substandard outlet and circuit breaker in the utility room. The Smiths sued the Showalters under various contract and tort theories.

An issue in the case was when substantial completion of the project occurred. The Showalters claimed substantial completion occurred in 1977 when the utility room was wired and occupied in 1977. The Smiths contended that substantial completion did not occur until 1981. The appeal court agreed that substantial completion has occurred in 1981 and said, "[t]he phrase "substantial completion of construction" shall mean the state of completion reached when an improvement upon real property may be used or occupied for its intended use."[22]

Causation

In order to be entitled to recovery, the party seeking damages must always prove that the damages were caused by the other party. Though this may sound too fundamental to be mentioned, causation is too often assumed in the construction industry. The party seeking damages must prove by a preponderance of the evidence, usually in the form of expert testimony, that the damage or injury was caused by some act or failure to act of the contractor's.

EXAMPLE

The contractor is hired to put an addition onto a home. Shortly after completion of the work, cracks develop in the walls adjacent to the addition. The owner must prove that the cracks in the walls were *caused* by some act of the contractor in order to recover any damages from the contractor. Just because the cracks appeared after the contractor did the work is not proof of causation.

EXAMPLE

The contractor is hired to put on a new roof. After the next rainstorm, leaks are observed in the roof, and the owner observes that the contractor used a different type of roofing tile than what was called for in the contract. Just because the contractor breached the contract does not mean that the contractor is liable for the damages caused by the leaks. The owner must prove that some act of the contractor's caused the damage.

EXAMPLE

Pickens, a homeowner, sued the general contractor and the electrical subcontractor for damages to his home caused by a fire. Pickens' expert ruled out all other causes of the fire except for a problem with the wiring. The subcontractor's expert testified that the wiring had been inspected and approved. He also testified that "there were a number of tradespersons...such as plumbers, carpenters, and air conditioning installers. After the electrical installation was performed, [the electrical subcontractor] had no control over who was allowed to work around the electrical lines." The court stated, "While numerous witnesses testified in the present case, the record reveals a lack of evidence of the fire's *cause...causation may not be presumed* (emphasis added)."[23]

Objective Impossibility

The law does not require the performance of a plan or specification that is impossible to perform. The inability to perform is measured against the inability of a hypothetical contractor, not the contractor involved in the situation. That is, the plan or specification must be impossible for any reasonable contractor to perform, not just impossible for the contractor involved in the situation.

EXAMPLE

Pursuant to a contract between the government and the contractor, the contractor was to install a boiler furnished by the government. The contract contained a specification detailing a certain level of efficiency, and after installation, the boiler failed to reach this level. The government sued the contractor. Experts testified that the boiler provided by the government could not reach the level of efficiency required by the specification. The contractor was not liable under the theory of impossibility of performance.[24]

It may be that the particular contractor cannot perform due to inexperience or lack of personnel. This type of inability is not excused under this doctrine, and failure to perform a contract due to inexperience or lack of personnel is a breach of contract.

Practical Impossibility/Commercial Senselessness

Much more common than the problem of objective impossibility is the problem of practical impossibility, also referred as commercial senselessness. These concepts are often included in contracts in what is called a *force majeure* clause. They relieve the contractor of liability when the cost of performance is so impractical that the court cannot in good conscious enforce it. "The crucial question in applying that doctrine to any given situation is whether the cost of performance has in fact become so excessive and unreasonable that the failure to excuse performance would result in grave injustice...."[25]

In other words, the law will not require a contractor to perform the contract, or a part of the contract, if performance is impractical due to extreme expense unless the contractor has specifically accepted that risk. Performance will not be relieved merely because a contract becomes unprofitable.[26] "The doctrine of impracticability of performance is not invoked merely because costs have become more expensive than originally contemplated but may be utilized only when the [contractor] has exhausted all its alternatives, when in fact it is determined that all means of performance are commercially senseless; and the burden of proof in this respect is upon [the contractor]." See Figure 6-1 for a graphical depiction of when the contractor may be relieved from performing because the cost is commercially senseless.[27]

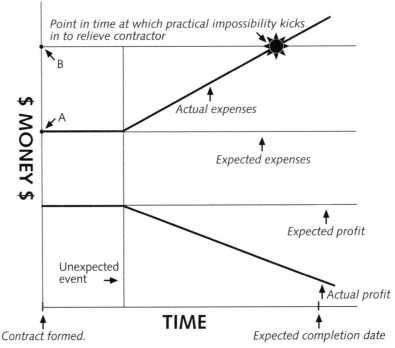

FIGURE 6-1 Privity of Contract

EXAMPLE

Asphalt rented a vessel from Enterprise for one year at a rate of $1,000 per month. After two months, the vessel was destroyed by fire. The vessel could have been repaired, but the costs to do so were extremely high and much higher than the value of the vessel. Enterprise refused to repair the vessel.

Asphalt rented a replacement vessel for $1,500 per month and sued Enterprise for the difference in the rents. Enterprise was excused from fulfilling its contract under the doctrine of impracticability of performance.[28]

> **EXAMPLE**
>
> A new environmental regulation was held to support the application of the doctrine of impracticability of performance.[29]

> **EXAMPLE**
>
> The cost of completing a highway construction contract increased dramatically due to the Arab oil embargo. The contractor attempted to recover the cost increases from the government-owner on a claim of commercial impracticability, but the court denied the claim. The cost to complete the project had certainly increased but not the level of commercial senselessness.[30]

As with the doctrine of impossibility, if the contractor takes a calculated risk as to its ability to perform, it must bear the consequences. In other words, the doctrine is a shield to prevent unfairness but cannot be used by a contractor who knowingly accepts a risk.[31]

> **EXAMPLE**
>
> Natus Corporation, the contractor, entered into a contract with the U.S. government to produce portable steel airplane landing mats. At the time, this was a newly designed and innovative product. Natus was aware of production problems during the development phase that led to high rejection rates. Natus adopted a production method that was "sound in theory [but] proved unattainable in practice." Two different methods of manufacture existed, but neither "could be counted on to operate satisfactorily in a high-speed-production process." The mats could be produced using a slower speed process that was not as profitable to Natus. Natus requested a change order seeking an equitable adjustment in the contract price to increase its profit. It claimed that the specification provided by the government was impossible to perform or commercially senseless.
>
> The court held that the contractor had taken a calculated risk that it could successfully produce the mats using the specific high-speed technology the contractor wanted to use; this proved impossible. The risk stayed with the contractor because it was not impossible to manufacture the mats according to the design, it was just not as easy and cost effective as the contractor had thought.[32]

Open, Obvious, and Apparent Defects in the Plans and Specifications

The contractor is not required to find errors or omissions in the contract documents. It is not uncommon for contracts to contain a provision to this effect.

> AIA 3.2.1. Since the Contract Documents are complementary, before starting each portion of the Work, the Contractor shall carefully study and compare the various Drawings and other Contract Documents....These obligations are for the purpose of facilitating construction by the Contractor and are not for the purpose of discovering errors, omission, or inconsistencies in the Contract Documents..."

However, if the contractor knows, or should have known, that a particular specification or plan would cause a problem or was defective, and does not inform the owner, liability for defects will be transferred to the contractor.

EXAMPLE

The contract contained what the court called a glaring ambiguity. One specification called for weather stripping for doors only and another required weather stripping for windows and doors. The contractor was required to install weather stripping on both the windows and doors at no additional cost. "We do not mean to rule that, under such contract provisions, the contractor must at his peril remove any possible ambiguity prior to bidding; what we do hold is that, when he is presented with an obvious omission, inconsistency, or discrepancy of significance, he must consult the [owner's] representative if he intends to bridge the crevasse in his own favor." This rule does not apply to subtle or latent flaws in the plans and specifications, just those that are glaring and obvious.[33]

EXAMPLE

Terry installed highway sign in accordance with the state's specifications. A driver was injured when he collided with the sign. The contractor had no liability to the driver or the owner because the contractor had no duty to judge the plans and specifications, and the contractor was justified in relying upon the adequacy of the specifications unless they were so obviously dangerous that no competent contractor would follow them.[34]

EXAMPLE

A highway contractor is not liable for wrongful death arising out of a one-car accident on a curve on a highway constructed by a contractor. The contractor followed the plans and specifications provided by the state, and the plans and specifications were not obviously dangerous to a reasonable contractor.[35]

Open, Obvious, and Apparent Defects in the Work

In general, the owner is precluded from collecting for open, obvious, and apparent defects in the work after acceptance of the project. At or near substantial completion of the project, it is common for the owner to prepare a punch list of defective items, and the contractor is required to correct them.

EXAMPLE

Tentinger submitted a bid of $690 to McPheters, a home builder, for touch-up and repainting on a spec home. The parties later agreed to a per-hour fee, not a flat fee. McPheters claimed that there was some overspray on a deck, and Tentinger said it was not enough to make a difference. No further complaint was made. After the work was done, Tentinger billed McPheters $420. McPheter offered $250 in accord and satisfaction, but Tentinger refused it and sued. McPheter countersued for $3,000 in damages.

After testimony, the court held that it was standard practice for a builder to provide subcontractors with a punch list. At the close of testimony, the judge resolved the conflicting testimony in Tentinger's favor and concluded that there was a contract between the parties, and that despite some minor defects, Tentinger had substantially performed the painting job. McPheters had waived any defects in workmanship by failing to request that Tentinger cure such defects. Tentinger was awarded $420 plus $4,000 in attorney fees and $900.02 in costs.[36]

Waiver/Acceptance

As discussed in Chapter 5, a **waiver** is the knowing relinquishment of a known right. Waiver requires proof of facts showing the waiver. This is an issue of fact. A contested waiver is generally proved by evidence of conduct showing that a party has waived its right to receive some benefit to which it is entitled.

Any clause, plan, or specification of a contract can be waived. A provision requiring all waivers and modifications to be in writing can be waived. It is impossible to write a contract that cannot be waived by the acts of the parties. The only way to prevent a contract provision from being waived is to follow that provision. Not following the provision waives it.

Peinado was a supplier to a plumbing contractor. Peinado, the plumber, and the general contractor entered into a joint-check agreement. This agreement required the general contractor to issue only joint checks to Peinado and the plumbing contractor. The parties then totally ignored the agreement. The general contractor issued individual, not joint, checks to the plumbing subcontractor. Peinado never objected. The plumbing subcontractor failed to pay Peinado out of the last check received from the general contractor. Peinado sued the general contractor for breach of the joint-check agreement. The court determined that the supplier had waived the provisions of the joint-check agreement. The supplier's only remedy was to sue the plumbing subcontractor.[37]

If the owner or his representative has had the opportunity to inspect and reject work but fails to do either, the owner is generally deemed to have waived nonconforming work. Acceptance is a form of waiver. It applies to the specific situation where the owner has accepted work with the knowledge of patent or obvious defects. The owner cannot later require the contractor to repair or replace the patently defective work. Occupancy itself is not always acceptance under the AIA A201, and this provision has been upheld. This doctrine does not apply to latent or unobvious defects.

The contract between the supplier, Henley, and the contractor, Universal, required the supplier to install either Slimfold Manufacturing Company or National Industries doors in the "Manufacturer's White Painted finish." National Industries manufactured only one prefinished white metal bifold door. Slimfold Manufacturing Company manufactured prefinished white doors in two colors, "Windsor white" and "Navajo white." In addition, the contract required the supplier to provide shop drawings and manufacturer's product data showing the shop finish. The supplier failed to comply with the requirement to provide the shop drawings and product data.

The contractor chose the National Industries doors, and the first shipment of doors was delivered to the site with the finish visible. The contractor and the owner saw some of the installed doors shortly after delivery and installation. The contractor paid for this first shipment of doors. National Industries was having trouble delivering the next shipment, and the supplier substituted the Slimfold Manufacturing Company "Navajo white" door because it matched the National Industries white.

During a routine inspection, a representative of the owner mentioned to the contractor and the architect, but not to the supplier, that the doors seemed to have a greenish tint. After installation of the doors, the owner decided that the color of all of the doors was unacceptable and refused to pay for them. The owner ordered replacement of all of the doors with another brand with a slightly warmer white color.

continued

Faced with this problem, the contractor notified the supplier that it had failed to submit the shop drawings, thus breaching the contract. The contractor ordered the supplier to remove all the doors and replace them with the new brand. The supplier offered to repaint the doors, but this offer was rejected. The supplier refused to remove and replace the doors. The contractor then removed all the doors, replaced them with doors acceptable to the owner, sold all of the original doors for salvage, and received less than $1,000. The contractor refused to pay Henley, the original supplier. Henley sued for payment for the doors it had supplied.

The court held in favor of Henley and entered a judgment in favor of Henley for the value of the original doors. The court determined that both the owner and contractor had waived any nonperformance by the supplier by failing to object to the original doors when they were delivered to the site with the finish visible. Both had the opportunity to and did in fact inspect the doors. "Inaction in situations where a person would normally be expected to act can be conduct amounting to a waiver....[Parties who fail in] making a timely objection have waived their right to insist on strict performance." In addition, "As a general rule, a buyer who, after a reasonable opportunity to inspect the goods, uses them to construct a building is deemed to have accepted the goods."[38] The fact that the supplier had failed to submit the shop drawings and product data was irrelevant.

Construction Defect Litigation

Many states have begun to implement statutory limitations and other requirements on construction defect litigation. States may limit the amount of recovery or require that certain formal processes be followed before a claim can be litigated. The theory behind these laws is that preventing construction claims litigation will reduce the costs of construction and make housing more affordable.

EXAMPLE

In California, homeowners' associations must first serve builders or developers with a "Notice of Commencement of Legal Proceedings" before filing a lawsuit. The statutory scheme developed in California contains requirements and penalties for all parties who fail to follow the law before filing a lawsuit.[39]

■ Claims by Third Parties

Owners, contractors, and designers [40] are liable to third parties for injuries caused by faulty design under tort theories such as negligence or product liability. Historically, contractors were not liable for injuries to third parties after acceptance of the project by the owner, but the recent trend has been for states to hold contractors liable where it is reasonably foreseeable that third parties will be injured by the contractor's work or by failure to disclose a dangerous condition.[41] See more detailed discussion of owner, designer, and contractor liability to third parties in Chapter 12.

■ Endnotes

[1] *Columbus v. Smith & Mahoney, P.C.*, 59 A.D.2d 857, 686 N.Y.S.2d 235 (1999).

[2] *PCL Constr. Services, Inc. v. U. S.*, 47 Fed. Cl. 745, 794, 796 (2000) (citing *Spearin*, 248 U.S. at 136–37).

[3] Example from *Neal & Co., Inc. v. U.S.*, 19 Cl. Ct. 463, 1990 U.S. Cl. Ct. LEXIS 37, 36 Cont. Cas. Fed. (CCH) P75, 802 (1990).

[4] The word *Spearin* is in italics in these terms because it refers to a case, and names of cases are either underlined or italicized.

[5] *U.S. v. Spearin*, 248 U.S. 132, 39 S.Ct. 59, 63 L.Ed. 166 (1918).

[6] *Stuyvesant Dredging Co. v. U.S.*, 834 F.2d 1576, 1582 (Fed. Cir. 1987).

[7] See *Construction Contractor's Liability to Contractee for Defects or Insufficiency of Work Attributable to the Latter's Plans and Specifications*, A.L.R.3d 1394 (2005).

[8] Example based on *Ridley Invest. Co. v Croll*, 192 A2d 925, 6 ALR3d 1389 (Del. Sup. 1963).

[9] *Spearin v. U.S.*, 248 U.S. at 136, at 137. See also *Al Johnson Constr. Co. v. United States*, 854 F.2d 467, 468 (Fed. Cir. 1988).

[10] Example based on *White v. Edsall Constr. Co., Inc.*, 296 F.3d 1081 (Fed. Cir. 2002).

[11] *Neal & Co., Inc. v. U.S.*, 19 Cl. Ct. 463, 1990 U.S. Cl. Ct. LEXIS 37, 36 Cont. Cas. Fed. (CCH) P75, 802 (Ct. Cl. 1990).

[12] *Columbus v. Smith & Mahoney, P.C.*, 59 A.2d 857, 686 N.Y.S.2d 235 (N.Y. 1999).

[13] Example based on *Columbus v. Smith & Mahoney, P.C.*, 59 A.D.2d 857, 686 N.Y.S.2d 235 (1999).

[14] Example based on *Apollo Sheet Metal, Inc. v. U.S.*, 44 Fed. Cl. 210 (Ct. Cl. 1999).

[15] *Turner Constr. Co., Inc. v. U.S.*, 367 F.3d 1319, 1320 (Fed. Cir. 2004).

[16] *Trans Metro Construction, Inc. v. U.S.*, 2005 U.S. Claims Lexis 72 (Ct.Cl. 2005).

[17] See *A & A Insulation Contractors, Inc.*, 92-2 B.C.A. (CCH) P 24,829, at 123,881, ruling that the government's approval of a submittal does not relieve the contractor of its contractual duties.

[18] Language from contract in *D. C. McClain, Inc. v. Arlington County*, 249 Va. 131, 452 S.E. 2d 659 (Va. 1995)

[19] *Wood River Pipeline Co. v. Willbros Energy Services Co.*, 241 Kan. 580, 738 P.2d 866, 94 Oil & Gas Rep. 228 (Kan. 1987).

[20] *U.S. v. Spearin*, 248 U.S. 132, at 137, 63 L. Ed. 166, 39 S. Ct. 59 (1918).

[21] *Perini Corp. v. Greate, Bay Hotel & Casino, Inc.*, 610 A.2d 364 (N.J. 1992), CCM November 1992, p. 2.

[22] Example based on *Smith v. Showalter*, 47 Wn. App. 245, 734 P.2d 928 (Wash. App. 1987).

[23] Example based on *Doyle Wilson Homebuilder, Inc. v. Pickens*, 996 S.W.2d 876 (Tex.App. 1999).

[24] *American Hydrotherm Corp.* ASBCA 5678, 60-1 BCA ¶2617 (1960), 2. G.C. ¶352.

[25] *International Minerals & Chemical Corp. v. Llano, Inc.*, 770 F.2d 879 (10th Cir. 1985).

[26] *Seaboard Lumber Co. v. U.S.*, 308 F.3d 1283 (Fed. Cir. 2002).

[27] *Piasecki Aircraft Corp. v. U.S.*, 229 Ct. Cl. 208 (Ct. Cl. 1981).

[28] *Asphalt International, Inc. v. Enterprise Shipping Corp., S.A.*, 514 F. Supp. 1111 (D.N.Y. 1981).

[29] Example based on *Tractebel Energy Mktg. v. E.I. du Pont de Nemours & Co.*, 118 S.W.3d 60, (Tex. App. 2003).

[30] Example based on *Helms Constr. & Dev. Co. v. State*, 97 Nev. 500, 1981 Nev. LEXIS 575 (Nev. 1981).

[31] *Natus Corp. v. U.S.*, 371 F.2d 450, 178 Ct. Cl. 1 (Ct. Cl. 1967).

[32] Example based on *Natus Corp. v. U.S.*, 371 F.2d 450, 178 Ct. Cl. 1 (Ct. Cl. 1967).

[33] Example from *Beacon Construction Co. v. U.S.*, 161 Ct. Cl. 1, 314 F.2d 501 (Ct. Cl. 1963).

[34] Example from *Hunt v. Blasius*, 74 Ill. 2d 203, 23 Ill. Dec. 574, 384 N.E.2d 368 (1978).

[35] Example from *Terry v. New Mexico State Highway Com'n*, 98 N.M. 119, 645 P.2d 1375 (1982).

[36] Example based on *Tentinger v. McPheters*, 132 Idaho 620, 977 P.2d 234 (Id. App. 1999).

[37] Example based on *Consolidated Electrical Distributors of El Paso v. Peinado*, 478 S.W.2d 565 (Tex. App. 1972, *no writ*).

[38] Example from *Henley Supply Co., Inc. v. Universal Constructors, Inc.*, 1989 Tenn. App. LEXIS 260 (1989).

[39] Cal. Civ. Code 1375.

[40] Donald M. Zupanec, *Architect's Liability for Personal Injury or Death Allegedly Caused by Improper or Defective Plans or Design*, 97 A.L.R.3d 455 (updated 2004).

[41] Emmanuel S. Tipon, *Modern Status of Rules Regarding Tort Liability of Building or Construction Contractor for Injury or Damage to Third Person Occurring After Completion and Acceptance of Work*, 75 A.L.R.5th 413.

CHAPTER 7

Contract Formation, Breach, and Damages

■ Contract Formation

Several elements must come together in order for a contract to be formed. In the construction industry, the law of contract formation is most applicable in the area of changes to the contract, as discussed in Chapter 8. It is unusual for parties in the construction industry to argue that no contract was ever formed. The major elements required to form a contract are in Table 7-1. The effect of mistakes in contract formation is discussed in this chapter. Finally, even if no contract is formed, a party performing services may be entitled to compensation under the doctrine of detrimental reliance or promissory estoppel. This doctrine actually goes by many different names and is discussed below. See Table 7-1 for a summary of the concepts presented in this chapter. See Chapter 13 for a discussion of the law of contracts as applied to contracts for the sales of goods.

TABLE 7-1 Contract formation and detrimental reliance

Elements needed to form contract	Simplified rule
1. Offer	Expression of willingness to enter into contract. Must contain material terms.
2. Acceptance	Must mirror the offer; if not, it is a counteroffer. No contract formed.
3. Consideration	Both parties must get something, and both must give up something. If only one party gets something, no consideration exists and no contract is formed.

TABLE 7-1 Contract formation and detrimental reliance (continued)

4.	Writing signed by the person to be charged; only required for some contracts	Sale of land Performance will take more than one year Sales of goods over $500 (UCC) or over $5,000 (Revised UCC) Home construction and/or improvement contracts in some states Public contracts
5.	Legal purpose	If the purpose of the contract is to perform some illegal act, the law will not uphold the contract.
6.	Capacity (age, mental competency)	Parties must be over the age of 18 and have sufficient mental capacity to understand what they are doing.
7.	Meeting of the minds	The parties' internal mental processes concerning the material terms of the contract must be the same.
8.	No fraud or unconscionability	No contract is formed if one party lies about something related to the contract. Unconscionability is a difficult concept rarely applied, but if the judge cannot sleep at night because the term is so one-sided or unfair, the term will not be enforced.
Mistakes		
Obvious mistake		Party making mistake may void the contract.
Unilateral mistake		Contract is still formed; the party making the mistake must absorb costs related to the mistake.
Mutual mistake		Either party may void the contract.
Performance has begun, but no contract has been formed because of a problem with one of the elements above		
Detrimental reliance, promissory estoppel		The party performing is entitled to a reasonable sum for the work completed even though no contract has been formed.

Offer

An **offer** is an expression of a willingness to enter into a contract, and it must be definite and certain as to the major elements of the proposed contract. A **preliminary negotiation** on the other hand does not rise to the level of an offer but is merely an attempt to open negotiations. An offer can be terminated or withdrawn at any time prior to acceptance. An offer is terminated by a counteroffer. An offer is terminated by a rejection of the offer or by a reasonable time. The doctrine of detrimental reliance (discussed below) may come into play to pay a party who does work for another without a contract having come into existence for some reason.

In the construction industry, a bid is considered an offer. Upon acceptance of the bid or offer and assuming that no problems with any of the other elements listed above exist, the contract is formed. If the bid is not accepted, no contract is formed. An offer or bid may be withdrawn at any time prior to acceptance unless the doctrine of detrimental reliance (discussed below) is applicable.

EXAMPLE

Charter Township advertised for bids for the construction of a firehouse. The bid documents stated the bids could not be withdrawn for 30 days and that acceptance would be made upon "formal notice of intent...duly served upon the intended awardee by the Owner or Architect."

Reliance was informed it was the low bidder. The architect mailed ten sets of construction documents to Reliance. However, the township board did not meet for more than 30 days after the bid opening, and no formal notice of intent, as required by the bid documents, was sent to the Reliance. Reliance withdrew its bid after the 30-day limit.

Charter Township entered into a contract with another contractor for approximately $32,500 more than Reliance's bid. It then sued Reliance for the $32,500, claiming that the contract had been entered into when the architect sent Reliance the 10 sets of construction documents.

The court held that because the acceptance of the offer (bid) had not been sent in accordance with the bid documents, no acceptance had been made, and no contract formed. Since the contract had not been formed, the contractor was allowed to withdraw its bid.

Charter Township testified that it was the custom and practice in the industry to consider the contract formed when the multiple sets of construction documents were sent to the bidder. However, the court said the bid documents were unambiguous, the acceptance must be "a formal notice of intent" as stated in the bid documents, and the custom and practice in the industry could not modify the clear meaning of the contract.[1]

Acceptance

An acceptance must mirror the offer and if any of the terms vary, a **counteroffer** is formed, not an acceptance. A counteroffer terminates the offer it has countered. Counteroffers can be exchanged until the parties agree on the terms; at that point, the contract is formed. If the subcontractor submits a bid to the contractor, and the contractor sends back a contract to the subcontractor with different terms, the contractor has not accepted the bid but instead has submitted a counteroffer. No contract is formed.

EXAMPLE

Saguara Construction obtained a bid of $50,000 from Jeon for framing on a certain project. This was the offer. Saguara contacted Jeon and asked if Jeon could do the job for $45,000 because O'Brien said he would do it for $46,000. This was a counteroffer, which terminated the offer. Jeon said no. Saguara said, "Okay, I will accept the bid of $50,000." This is another counteroffer. Jeon said he did not want to work for a bid-shopper and would not do the framing for Saguara at any price. Saguara eventually gave the job to Bodren for $55,000. No contract was ever entered into between Saguara and Jeon and Saguara is not entitled to damages from Jeon.

EXAMPLE

Tower, the general contractor, solicited bids for masonry work on a project. After it was awarded the contract, it presented a subcontract to Gomez for signature. However, Gomez refused to sign it, because the subcontract contained a mandatory arbitration clause and a strict timetable for completion, neither of which was reflected in the documents supplied to it to prepare its bid. No contract has been formed between Tower and Gomez.

Use of a subcontractor's bid is not an acceptance of the bid, and no contract is formed at that point or even after award of the prime contract. The contractor can use the subcontractor's bid in its bid to the prime, but the contractor is not obligated to hire the subcontractor once the contractor is awarded the prime contract. This is why bid shopping is legal although considered unethical.

EXAMPLE

Electro submitted an oral subcontract bid to Sharp Construction for the electrical work on a school construction project. Sharp used Electro's subcontract bid in the prime bid and listed Electro in its bid to the owner. Sharp contended that it mistakenly included Electro's bid in its prime bid, because it actually had received a lower bid from another company, Ind-Com. Shortly after being awarded the contract, Sharp discovered the lower bid from Ind-Com and asked Electro if it could do the work for the same amount. Electro agreed and Sharp sent Electro a confirming fax. However, Electro was unable to obtain the performance and payment bonds as required by the subcontract, and Sharp gave the subcontract to Ind-Com, which could obtain the bonds.

Electro sued Sharp for breach of contract. However, the court held that the contractor's use of the subcontractor's bid was not an acceptance of the subcontractor's bid. No contract was ever formed between the parties because Electro could not obtain the bonds as required by the contract.[2]

Consideration

Consideration exists if each party gives up something of value and receives something of value. If only one of the parties in the transaction is to obtain something of value, no consideration exists, and no contract is formed.

EXAMPLE

A homeowner and a contractor enter into a contract for remodeling of a bathroom. While the contractor is working, the homeowner asks the contractor to repaint the kitchen. The contractor agrees. Later, the contractor gives the owner an invoice for the estimated cost of painting the kitchen, but the owner says he should not have to pay because the contractor agreed to paint the kitchen and the owner thought that would be included in the price of the bathroom remodel. The contractor is not obligated to paint the kitchen, because even though there was an offer and an acceptance, the contractor has received nothing of value in exchange for his promise to paint the kitchen. Consideration does not exist, so no contract to paint the kitchen has been formed.

Owners sometimes offer the contractor future work in exchange for no-cost work at the present time. If in fact the owner does give the contractor future work, this would probably be consideration for the no-cost work now—though no case was found to support this statement. In that circumstance the contractor would probably be prevented from recovering anything additional for the original work. What if the owner never gives the contractor future work? In the following case, the court said that the promise of future work, which was never fulfilled, was not consideration. The new contract between the parties never came into existence, because there was no consideration. This result may not occur in all jurisdictions.

EXAMPLE

The contract between the owner and the contractor, Glazetherm, was for the installation of windows in a motel and restaurant complex. Neither the contract nor the plans and specifications specified the color of the windows to be installed. A dispute arose over compensation to the contractor for the removal and replacement of white windows with bronze windows. Glazetherm testified that when the owner paid the first installment due under the contract, the owner verbally specified that white windows should be installed. Subsequently, Glazetherm began installing the white windows over the next two weeks and installed 195 windows. After the 195 windows had been installed, Glazetherm testified that the owner's wife wanted to change the window color to bronze. Glazetherm immediately sent the owner a change order, included in it an estimate of $50,000 as the cost to install and remove the existing 195 white windows, and requested a time extension of three weeks. The owner objected to the $50,000 and contended that he had ordered bronze windows in the first instance. The owner did not object to the time extension, as the delay did not delay the completion of the entire project.

Glazetherm stopped work until the matter could be resolved because of the potential loss to it. A few days later, a new written "agreement" was entered into between the parties. In this document, Glazetherm agreed that it would "remove existing white windows and install bronze windows and bronze exterior trim at no additional cost." The

continued

owner testified that the consideration paid to Glazetherm for the "agreement" to remove the existing windows and to install the bronze windows at no cost was the owner's promise to make up the substantial loss to Glazetherm by awarding Glazetherm future lucrative contracts on other properties controlled by the owner. Glazetherm agreed with this, but the owner never awarded Glazetherm any future contracts. Owner admitted that it did not award Glazetherm any future contracts.[3]

The trial court determined that no consideration existed for Glazetherm's "agreement" to remove the white windows and replacement by the bronze windows at no cost. Because no consideration existed, no contract existed.

Since Glazetherm had in fact removed the white windows and installed the bronze windows, Glazetherm was entitled to payment under the doctrine of detrimental reliance. This is a noncontract theory that requires one who has obtained something of value from another to pay for it. This doctrine is discussed below.

MANAGEMENT TIP

Always be leery of an offer of future work in exchange for no-cost work at the present time. If possible, try to get an agreement or written memorandum detailing the specific future work to be completed. This will obligate the owner to give the future work or be in breach of contract.

Statute of Frauds and Requirement of a Writing

The term **statute of frauds** is an old term used for a particular list of contracts that must be evidenced by a writing signed by the person to be charged in order to be upheld. Other statutes in a jurisdiction may have added additional types of contracts to the historical list. Unless the subject of the contract falls under the statute of frauds or other statutory requirement, no writing or signature is needed to form a contract. Oral, unsigned contracts are just as valid as signed, written ones although the terms of oral contracts are more difficult to prove. In one case discussing oral contracts, the court said, "The primary issue in the instant case is whether the signatures were required for the formation of an enforceable contract. The law of this Commonwealth [Pennsylvania] makes clear that a contract is created where there is mutual assent to the terms of a contract by the parties with the capacity to contract....As a general rule, signatures are not required unless such signing is expressly required by law or by the intent of the parties..."[4]

MANAGEMENT TIP

Avoid oral contracts. Get everything in writing.

The types of contracts that must be evidenced by a writing and signed by the person to be charged include the following:

1. Contracts for the sale of land or for the creation of an interest in real property, including leases longer than a year.
2. Contracts that are not to be performed within a year from the date of contract formation.
3. Contracts to be responsible for the debt of another.
4. Sales of goods over $500 (UCC) or over $5,000 (Revised UCC).
5. Many states require home construction and/or improvement contracts to be in writing.[5]
6. Public contracts may require a formal contract with signatures.[6]

Notably absent from the list are contracts for services, although with fax machines and e-mail, it is common to have an electronic record. Electronic records are admissible forms of evidence under the statute of frauds.

If no writing or record exists, the ability of the contractor to hold a subcontractor to an oral bid depends on whether or not the bid is one for services, not covered by the Uniform Commercial Code, or one for material or supplies, covered by the Uniform Commercial Code.

Bids for materials and supplies over $500 (UCC) must be evidenced by a writing. If the state has adopted the Revised UCC, the amount is $5,000 (Revised UCC). The requirement is that the contract must be evidenced by a record. Many cases exist defining a writing, and the term has been broadly defined to include initials, faxes, letterheads, and electronic records Additional requirements of the UCC are discussed in Chapter 13.

Any bid for a subcontract that is primarily for services is enforceable without a writing unless the work cannot be done within a year.

EXAMPLE

The subcontractor's oral bid is for framing of 20 new homes, starting in one month. This contract can be performed within a year and therefore need not be evidenced by a writing signed by the person to be charged.

EXAMPLE

The subcontractor's bid is for electrical wiring on a major project. The subcontractor outlines work needing to be done over a two-year period. In order for the contractor to hold the subcontractor to the contract, it must be evidenced by a writing signed by the subcontractor, the person to be charged. In order for the subcontractor to hold the contractor to the contract, it must be evidenced by a writing signed by the contractor, the person to be charged. If only one of the parties signs the contract, only that party is bound by it.

> **EXAMPLE**
>
> The subcontractor's bid is for two new air conditioners, and the value of the subcontract is $5,000. This bid is governed by Article 2 of the UCC and must be evidenced by a writing or record.

> **MANAGEMENT TIP**
>
> Try to get a signed fax or writing to support a contract. If nothing else, always send your own signed, written memorandum outlining the agreement. This will support an argument of detrimental reliance (discussed below).

Public contracts may be covered by a specific state or federal law that requires formal execution of the contract before the contract is formed. These laws are upheld, and no contract can be formed just upon an offer and an acceptance. The additional step of a formal, signed contract must be taken. This rule also applies in private contracts if the bid documents or offer specifically state that no contract will be formed unless a formal, signed contract is entered into.

> **EXAMPLE**
>
> A school district solicited bids for the construction of new classrooms. The school district voted to accept Ry-Tan's bid, and the board's executive director signed a Notice to Proceed. A date was set for the contractor to execute the formal contracts. However, before that date, the contractor started work and the board thereafter refused to enter into the contract. The contractor sued, claiming the contract was formed when the board approved the contract. However, based on statute and historical precedent in the state, a contract with a public agency cannot be formed until the formal contracts are signed. Since no contract was formed, the school district could revoke its offer under the rule that an offer may be revoked at any time prior to formation of the contract.[7]

Legal Purpose and Capacity

The law will only uphold contracts for legal acts. In addition, parties must have the capacity to form a contract; that is, they must be over 18 and of sufficient mental capacity to understand what they are doing. These issues seldom arise in the construction industry.

EXAMPLE

Ball and Lewis desired to enter into a joint venture and bid on a contract with the government to construct a highway. However, Lewis had no state license so Ball bid, and eventually won, the contract in his name only. Under state law, Ball was required to perform at least 50% of the work himself. However, to circumvent this requirement Ball entered into two subcontracts with Lewis: one for work and one labeled a "rental" agreement for road building equipment. Ball and Lewis had an outside oral agreement that the "rental" agreement actually required Lewis to perform excavation and grading work. The result of these two subcontracts was that Lewis, who was unlicensed, performed more than 50% of the work on the project, contrary to the law.

Ball did not pay Lewis the sums owed under the "rental" agreement. Lewis sued but did not recover since the "rental" agreement was for an illegal purpose.[8]

In the days when the validity of indemnity agreements was being litigated, many courts refused to uphold broad form indemnity agreements claiming that they were for an illegal purpose.[9] Broad form indemnity clauses are discussed in Chapter 14 in the section entitled "Broad Form."

Meeting of the Minds

The meeting of the minds requirement means that the parties' mental concepts of the major terms of the contract are the same.

EXAMPLE

Quinn Construction and the owner orally agreed that the contractor was to install a pool on owner's property at a price of $10,000. When the contractor arrived at 123 Cedar Dr., the owner told the contractor that she meant the pool to be installed on her lake front property at 456 Oak St. The requirements for putting a pool near a lake were very different, and the contractor refused to put a pool in at 456 Oak St for $10,000. The owner hired D'esposito Construction to install the pool at 456 Oak St. at a cost of $20,000 and sued Quinn for the difference. The owner is not entitled to damages from Quinn because no contract was ever formed because of a failure of a meeting of the minds. Quinn had no obligation to put in the pool.

Of course, mental concepts may be hard to prove, but it is left up to the trier of fact to determine such issues from the surrounding circumstances and circumstantial evidence. "An essential element of any valid contract is a meeting of the minds. When there is no written contract in evidence, and one party attests to a contractual agreement while the other vigorously denies any meeting of the minds, determining the existence of a contract is a question of fact under Texas law....A meeting of the minds can be inferred from the parties' conduct and their course of dealing."

EXAMPLE

Lamb filed a bankruptcy claim for approximately $200,000 for architectural services to The Palms. The Palms objected, stating that no contract was formed because there was no meeting of the minds. The court said, "Here, the evidence shows that there was a meeting of the minds between Lamb and the debtor. Although Eaton [the agent of The Palms] denies the existence of an oral contract, conduct by both parties indicates otherwise. Eaton on behalf of [The Palms] sought out Lamb to produce architectural drawings for a condominium project to be built on Mustang Island. Several meetings [were held].... With the exception of the price terms, the two parties agreed that Lamb would produce architectural work to be used to entice developers and build the project on Mustang Island [and Lamb completed the architectural drawings]." The Palms attempted to say that no contract was formed, because no price had been agreed on. However, the court did not allow The Palms to get out of the contract. The court said the lack of the price term was not fatal, because a reasonable price could be implied into the contract.[10]

MANAGEMENT TIP

A "contract" upon which neither party has started to perform can easily be voided when one of the parties realizes there is no meeting of the minds on a major aspect of the contract. If one or both of the parties starts to perform, a meeting of the minds can be inferred from the acts of the parties. The doctrine of detrimental reliance (discussed below) will often come into play and require a party performing work to be paid a reasonable sum for the work, even if no contract has been formed because of a failure of a meeting of the minds.

No Fraud, Duress, Unconscionability, Onerous

Contracts based upon fraud or duress are unconscionable or onerous and are not enforced by the law. If the contract is based upon a fraud, often called "fraud in the inducement to enter a contract," the victim "has two options: affirm the contract and sue for damages from the fraud or breach, or promptly rescind the contract and sue in tort for fraud.[11]

EXAMPLE

The roof on a certain building leaked into the living areas. The owners got a bid to replace the roof, but since they were selling the house anyway, they decided to just repaint the ceiling and not tell anyone about the leak. The house sold, and the new owners found out about the leak and the bid. They sued the original owners for fraud and were entitled to damages. In this situation, the new owners could void the sale but more likely would just sue for damages.

EXAMPLE

In this case, the buyers were not able to prove fraud. The Ainsworths bought a home from the Perreaults. Prior to the purchase, the Perreaults had had a pool installed by Atlas. The pool contained a limited warranty, and shortly after the pool was installed, there was a rupture in the bottom of the pool. Atlas performed the necessary repairs per the recommendation of a structural engineer. No further problems were noticed with the pool for about two years after the Ainsworths purchased the home. At that time, one corner of the pool began to sink. The Ainsworths sued the Perreaults for fraud in the inducement, alleging they failed to inform them of the problems with the pool.

The trial court granted the Ainsworths a summary judgment motion dismissing the fraud claim, because the agreement contained an entire agreement clause that stated, "No representation, promise, or inducement not included in this Agreement shall be binding upon any party hereto." Since no representation was made in the contract with regard to the pool, there could be no fraud.

In addition, fraud based on active concealment did not exist. This type of fraud occurs when a seller takes steps to conceal a defect. Passive concealment is when a seller keeps quiet about a defect that is not readily discernable. There was no evidence that the sellers knew of the defect.[12]

"Unconscionability is 'a narrow doctrine' that may invalidate the contract at issue upon a showing that the contract is 'one which no reasonable person would enter into, and the inequality must be so gross as to shock the conscience.'"[13] This doctrine is rarely used and only in cases where extreme injustice will result.

EXAMPLE

NYPC was in the business of manufacturing plastic products including plastic sheets 2 millimeters thick and 54 × 54 inches in size. Each sheet is similar to the type of plastic sheet frequently found under desk chairs to allow the chair to easily move. Owens fairly regularly purchased these sheets and regularly paid between 34.5¢ and 45¢ per sheet.

On February 14, 2001, NYPC ordered approximately 8,200 sheets at a price of $172.50 per box of 200 sheets. However, apparently due to an error on the purchase order, the invoice read "$172.50 per sheet." The total purchase price was therefore over $1 million. NYPC shipped the sheets and billed Owens $1 million.

An employee of NYPC testified that the cost per sheet was actually $172.50 per sheet. The company had stopped manufacturing the sheets, and he had to redirect his plant back into manufacturing the product. However, other purchases by Owens into May 2002 had prices from between 34.5¢ per sheet to a high of 51¢ per sheet.

After several attempts to correct the error, Owens refused to pay the $1 million. Eventually, it paid approximately $7,000 for the sheets, a cost that represented 86¢ per sheet, an amount almost double other similar transactions but far below $172.50 per sheet.

continued

The court upheld this payment and said, "I find that the contract price was a good faith error created by Ms. Berry [the Owen employee who inadvertently entered 'per sheet' instead of 'per box'] when she put the information into [Owens'] computer system, and compounded by NYPC when, upon seeing the huge discrepancy in the price listed on the purchase order, a price that was based on a quotation NYPC had provided to Debtors, it failed to correct the situation....I find that the purchase order contained an error in the price based upon a mistake of material fact....Nonetheless, the error is so huge that it shocks the conscience of this court. Therefore, I further find that the contract must be reformed and the price must be modified to avoid an unconscionable result."[14]

The doctrine of unconscionability is most often used in consumer law or employment law where the bargaining power of the parties may be grossly unequal. While the law does not usually look at the bargaining power of either party, if the party with superior power uses their power to gain a grossly unfair contract, one that the court in good conscious cannot uphold, the court can employ this doctrine to void the contract or reform it. Indeed, unconscionability usually focuses on "a grossly unequal bargaining power at the time the contract is formed."

EXAMPLE

In 1960, the Storeys entered into a contract with Hubscher, Inc. for the removal of gravel and sand from their property. The agreement was amended in 1987. In 1995, the Storeys terminated the contract because Hubsher, Inc. did not remove any sand or gravel between 1960 and 1995. Hubsher, Inc. filed a lawsuit to enforce the contract. On appeal, the appeal court said the matter had to be tried to determine if the contract was void for unconscionability.

The appeal court said, "[T] there is a two-pronged test for determining whether a contract is unenforceable as unconscionable, which is stated as follows:

1. What is the relative bargaining power of the parties, their relative economic strength, the alternative sources of supply; in a word, what are their options?
2. Is the challenged term substantively reasonable? Reasonableness is the primary consideration."

The Storeys were unrepresented by counsel during negotiations and execution of the original contract and the modification. Additionally, the Storeys' pleadings contained several allegations of unreasonableness attributed to the length of time since the contract's inception, during which Hubscher, Inc. never extracted sand or gravel from the property.[15]

The applicability of this doctrine is of limited use in the construction industry when the parties are professionals. It might apply in homebuilding contracts if the contractor attempts to take advantage of the homeowner.

EXAMPLE

In a case governed by the UCC, the court held that the following damage clause in the warranty was *not* unconscionable: Envirotech "will replace or repair...any part or parts... which Envirotech's examination shall show to have failed under normal use and service by the original user within one year following initial shipment to the purchaser."[16]

EXAMPLE

In this case with experienced construction professionals, the doctrine of unconscionability was not applied. Reliable entered into a contract to lease a crane and operator to Benco, the customer. The parties had a longstanding relationship, and Reliable had provided cranes to Benco in the past. The normal process was for Reliable to provide the services, and at the end of the day, Benco's foreman would verify the hours worked and sign the agreement to pay Reliable. This agreement contained, on the reverse, a clause stating, "Customer [agrees] to indemnify Reliable Crane and Rigging against all loss, expense, claims and liability of any nature resulting from injury or damage to person(s) or property caused by or arising from performance of such work."

After an employee of another subcontractor was injured in a crane accident, Reliable sought indemnification from Benco. Benco claimed that the above clause was unconscionable. The appeal court declined to so hold. The form clearly stated it was a contract and that the provisions on the back of the contract were included. The parties had used this same form dozens of times, and although Benco's president testified that he had never read it, that was irrelevant.[17]

◾ Mistakes

Mistakes can be divided into three categories: obvious, unilateral, and mutual. Each is treated differently by the law.

The basic rule regarding obvious mistakes is simple: a party making an obvious mistake may void the contract. The law expects mistakes to be fixed, not taken advantage of.

A **unilateral mistake** is a nonobvious mistake made by one party to the contract. The contract with a unilateral mistake is still formed, and the party making the mistake must still perform.

EXAMPLE

After the contract was awarded, the contractor found an error in the calculation of the amount of carpeting needed, and therefore its bid was too low. This reduced the profit on the job. Unless the mistake was obvious, it was a unilateral mistake, and the contractor must still perform the contract for the contract sum.

An exception to the rule of liability for a unilateral mistake does exist. If a party claims that the contract contained a unilateral mistake or that it mistakenly signed a contract, many jurisdictions will allow the party to void the contract if it can prove the mistake goes to the substance of the agreement, the error does not result from an inexcusable lack of due care, and the other party has not relied upon the mistake to his detriment.[18]

EXAMPLE

On a major fast-track construction project, the parties exchanged several contracts for parts of the work. Hardy, Inc., the contractor, accidentally signed a prior draft of a contract. The next day, it discovered its error and informed the owner. The owner attempted to hold the contractor to the signeddraft. Hardy was not obligated. Of major importance was the fact that Hardy informed the owner quickly and the work had not yet begun. Because the matter was fast track and several versions of the contract had been exchanged in a short time period, the mistake was not due to inexcusable neglect.[19]

A mutual mistake is one made by both of the parties. It is similar to the concept of meeting of the minds. If a mutual mistake occurs, either party may void the contract.

EXAMPLE

Both the owner and the contractor mistakenly believe the financing has been approved for a project. The parties sign an agreement. Later, the owner discovers that the financing has not been approved. Either the owner or the contractor can void the contract.

■ Modification

Any common law contracts can be modified as long as the modification is supported by additional consideration. Changes to the contract are discussed in Chapter 8. This rule is applicable even if the contract contains a provision that attempts to prevent modification. UCC or sales contracts do not need consideration to modify the contract as long as the modification is reasonable. Sales contracts are discussed in Chapter 13.

■ Detrimental Reliance, Promissory Estoppel, Quantum Meruit, Part Performance, Implied-in-Law Contracts, Constructive, or Quasi-Contracts

If no contract is formed because of a lack of one of the elements of contract formation discussed above, a party may still be able to obtain relief under the theory of detrimental reliance. This theory goes by different names in different jurisdictions including detrimental reliance, promissory estoppel, quantum meruit, part performance, implied-in-law contract, constructive contract, and quasi-contract. Recovery under any one of these theories "is based upon a promise implied by the law to pay for beneficial services rendered and knowingly accepted."[20] One state court lumped all of these theories together just as done here when it said, "Contractor essentially abandoned the contractual provision of the Subcontract regarding change orders, and in the process, ignored its corporate policy, which mandated written change orders before the extra work was commenced. The absence, therefore, of a written change order is no defense to the claim of the Subcontractor for compensation for the extra work entailed by the oral changes directed by the Contractor (internal citation omitted). Whether this conclusion is propelled by implication, or by promissory estoppel, or by quantum meruit, or common sense, is needless speculation."[21]

For the purposes of this book, only the term detrimental reliance is used. These theories are applicable only if *no* express or implied contract exists.[22] However, a party can present both a contract theory and one of these theories to a jury, and the jury then decides which is applicable.[23] An **express contract** is one that is memorialized in words, either oral or written. An **implied contract** is a contract that is formed by the acts of the parties, rather than by words. An implied contract is not the same as an implied-in-law contract.

EXAMPLE

Walker, an architect, entered into a contract with Gateway for architectural services for the design and construction of a golf course. Walker submitted a proposal, and Gateway signed it. Gateway asked Walker to restructure the payments so that the bulk of the money owed was due at the end of the contract. No written modification was ever entered into by the parties, but Walker did restructure the payments.

A dispute arose regarding the amount of construction management services Walker was to perform. Gateway fired Walker and refused to pay the sums due Walker. Walker sued on both a breach-of-contract theory and a theory of detrimental reliance. Gateway argued that Walker must choose one of the two theories. Gateway was in error: both theories could be presented to the jury, and the jury could decide if Walker was entitled to payment under either theory or not at all.[24]

Detrimental reliance is a legal doctrine giving relief to a party when no contract has been formed *or* there are questions about the validity of the contract. It applies only in situations where at least part of the noncontract has been performed. If performance has not yet begun, the law will likely say that no contract has been formed because of a failure of the meeting of the minds requirement or a mutual mistake. Neither party will be entitled to damages.

Detrimental reliance awards damages to a Party B when the following occur:

1. Party A makes a promise to Party B.
2. Party A should reasonably expect that Party B will take action on the promise in some way, such as by performing work.
3. Party B does actually take some action on the promise, such as by performing work or ordering supplies.
4. Injustice will result if Party A's promise is not enforced.[25]

EXAMPLE

A written agreement existed between an owner and a contractor for a kitchen remodel. The agreement required that all changes and additions to be in writing. The owner asked the contractor to finish out the basement, and the contractor completed this work. Later, a dispute arose between the owner and the contractor about the price to be paid for finishing out the basement. The owner said that she was not required to pay the contractor because the additional work was not in writing. In addition, the owner claimed there was no meeting of the minds on the price to be paid.

The contractor is entitled to payment under a claim of detrimental reliance even if no contract has been performed. The amount paid to the contractor is a reasonable amount for the work done.

EXAMPLE

Stanton, a plumbing subcontractor, quoted the contractor $92,000 for a plumbing subcontract. Using this bid for the plumbing work, the contractor submitted a bid to the owner for the construction of a kitchen and culinary arts building. Shortly after being awarded the contract, the contractor called and informed Stanton that it had been awarded the contract and had accepted Stanton's bid. Stanton refused to sign the subcontract and refused to perform. After several attempts to get Stanton to perform, the contractor hired Biknell, who had submitted the second-lowest bid of $138,900. The contractor was awarded the difference between the two bids as damages under a theory of detrimental reliance.[26] Not all jurisdictions apply this doctrine in the case of a subcontractor's mistake in bid. See more complete discussion in Chapter 2.

EXAMPLE

Insite, the hardscape and landscape subcontractor on Georgia International Plaza for Atlanta's 1996 Summer Olympic Games, accepted a bid from sub-subcontractor SKB for the landscaping. SKB failed to sign an agreement with Insite but did start the work. SKB failed to complete a significant portion of the work. Insite later sued SKB on the theory of detrimental reliance, claiming that it was damaged when SKB did not complete the work as promised. The jury awarded Insite $711,573.42 on its detrimental reliance claim.

On appeal, SKB argued detrimental reliance could not exist because there was no justifiable reliance by Insite when there was no written agreement. The court said that the existence of a written contract was irrelevant in a claim for detrimental reliance and in fact was employed specifically when some element of contract formation, such as the need for a writing, did not exist.[27]

■ Contract Breach

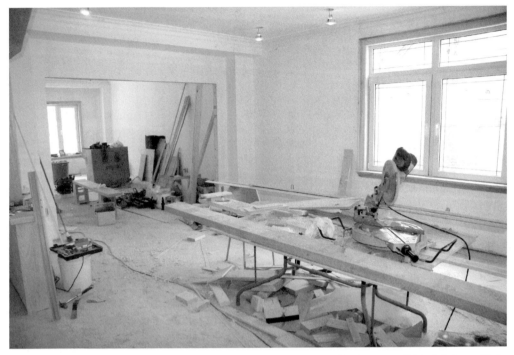

A contract breach is the failure of a party to perform some aspect of the contract. A breach of contract is "an unjustified failure to perform all or any part of what is promised in a contract."[28]

A breach of the contract does not necessarily justify a party in discontinuing performance. In the construction industry, a party must continue to perform under the contract unless the breach is so major that it justifies termination of the entire contract. The only type of breach that allows the termination of a construction contract is a material breach. "[W]hen a party commits a material breach of a contract, the nonbreaching party is discharged or excused from any obligation to perform and may sue the breaching party for the benefit of the bargain.[29]

The definition of material breach, like all such vague standards, is easy: a **material breach** is a significant breach. The difficulty arises in determining if any particular set of circumstances arises to the level of material breach. This is a factual issue.

EXAMPLE

The following matter had to be sent back to the jury to determine whether or not a material breach had occurred. Morganti was awarded the general contract for a 1,000-bed prison facility. It entered into a $4 million subcontract with Petri for the mechanical and plumbing work. The subcontract contained a pay-if-paid clause. It also contained a provision that Morganti would not be liable to Petri for delays or disruptions caused by Morganti, although Petri could receive an extension of time to complete performance under such circumstances.

At some point into the project, the owner terminated the general contractor Morganti. Morganti informed Petri, a subcontractor, that it considered the termination wrongful and that it intended to challenge the termination, but pending a resolution of that challenge, Morganti was suspending Petri's work. A dispute existed as to whether Petri had been paid for all of the work it had completed to the point that the owner terminated the Morganti's contract.

Petri declared that it was of the opinion that Morganti had terminated its subcontract by defaulting on the general contract, making the site unavailable for Petri to perform, and failing to pay Petri.

After several months, the dispute between Morganti and the owner was resolved, and Morganti directed Petri to continue construction. Petri refused. Another subcontractor was hired to complete the subcontract at a higher cost, and Morganti sued Petri. The appeal court said that the matter had to be tried by a jury to decide if Morganti had materially breached the subcontract. "There is also sufficient evidence from which reasonable jurors could conclude that Morganti's suspension of Petri's work constituted a material uncured breach of the subcontract. As a general rule, indefinite suspension of work by an owner or general contractor constitutes material breach of a construction contract...jurors could conclude that Morganti's suspension of Petri's performance, along with preventing Petri from having access to the job site, was a material uncured breach of the subcontract."[30]

Failing to make a progress payment is generally considered a material breach and grounds for terminating a contract. Most contracts contain this as a provision in the contract.[31]

FORM

The Subcontractor may terminate the Subcontract...for nonpayment of amounts due under this Subcontract for 60 days or longer...[32]

Termination of a contract for a *nonmaterial* breach is itself a *material* breach of the contract entitling the party committing the nonmaterial breach to damages. "[A] contractor is deemed to have breached a contract when it is not plainly justified in default terminating a subcontractor."[33]

EXAMPLE

The contractor is tired of what it considers poor-quality work performed by Mispelon, the subcontractor. The contractor terminates Mispelon from the job, pointing out several defects in the subcontractor's work. Later, a jury determines that while there are defects in Mispelon's work, these are curable. The contractor has breached the subcontract and owes the subcontractor damages including lost profit.

MANAGEMENT TIP

A decision to terminate a contract is a major legal step and cannot be taken lightly. Always seek legal advice before terminating a contract. If a contract is terminated for what is later determined to be a nonmaterial breach, the terminating party has in fact breached the contract. The documentation supporting the material breach of contract must be extensive.

Breach of Implied Terms

Implied terms of a contract can be breached just as express terms can be breached. Both lead to damages for their breach. Implied terms are also discussed in Chapter 9 in the section entitled "Implied Terms."

EXAMPLE

Lamb, an architect, claimed that The Palms entered into an oral contract with him for approximately $260,000 for architectural work done on a condominium project. However, no price term was ever agreed upon. The Palms denied the existence of the contract, stating that no meeting of the minds occurred. The evidence showed that The Palms had sought out Lamb to produce the drawings and worked with him during the project. The Palms used the drawings of Lamb to entice developers into financially backing the project.

The court determined that a contract did exist, and it contained an implied term that The Palms would pay Lamb a reasonable fee for the work. The court said, "Courts may supply a reasonable price term if all other elements of the contract are shown (supporting citations omitted). Lamb and his expert witness on architectural fees both testified that the charges of $266,875.00 were reasonable. Lamb established that his expert

continued

witness was an expert on what is reasonable to charge for architectural services. Lamb also testified that he worked 3,000 hours on this project and produced volumes of work product in the court done for this project. Based on this testimony and the amount of time and effort expended, $266,875.00 is the reasonable value of Lamb's services.... The debtor contends that Lamb's work is worth $40,000. However, this number does not appear anywhere in the evidence or in the hearing and seems to have been pulled from thin air. Therefore, the court will ignore it."[34]

Every contract contains an implied term of good faith, and operating in bad faith is a breach of the contract. "To prove a claim for bad faith under Connecticut law, the plaintiffs [are] required to prove that the defendants engaged in conduct designed to mislead or to deceive...or a neglect or refusal to fulfill some duty or some contractual obligation not prompted by an honest mistake as to one's rights or duties...Bad faith is not simply bad judgment or negligence, but rather it implies the *conscious doing of a wrong because of dishonest purpose or moral obliquity*...it contemplates a state of mind affirmatively operating with *furtive design or ill will*....Moreover, bad faith is an indefinite term that contemplates a state of mind affirmatively operating with some design or motive of [self] interest or ill will."[35]

In the construction industry, the implied term of good faith, is often called the **implied covenant to cooperate** or the **implied covenant not to hinder.** "Every contract includes the implied covenant of good faith and fair dealing. Two closely related aspects of the covenant of good faith and fair dealing are 'the duties to cooperate and not hinder the contractor's performance.'"[36] "The Government breaches these duties when it acts unreasonably under the circumstances—specifically, if it unreasonably delays the contractor or unreasonably fails to cooperate."[37]

EXAMPLE

The government breached its duty not to hinder the contractor's performance when it was unreasonably slow in doing necessary survey and staking work on a piece of land. The government's failure to do this work in a timely manner hindered the contractor's performance and rendered the government liable for the contractor's damages related to the delay.[38]

This claim also arises in insurance litigation. If the insurance company fails to fulfill its obligations under the contract, this could be a breach and lead to damages, including punitive damages. Punitive damages for breach of the implied covenant of good faith appear to be limited to insurance litigation and have not been awarded for breach of the implied duty in a construction contract. Insurance is discussed in more detail in Chapter 11.

EXAMPLE

The Ballards purchased a home at a foreclosure sale, and within a few years, some plumbing leaked due to frozen pipes. The Ballards filed claims for plumbing leaks and hardwood floor damage, and testing revealed moisture problems. The insurance company paid for damage to the flooring. Later, Ballard suspected a mold problem and made a claim for mold damages. The family had to move out of the residence. The insurance company did not pay the claim, and the Ballards filed a bad faith claim. After trial, the jury awarded them $2.5 million to replace the home, $1.1 million to remediate the home, $2 million to replace the contents of the home, $350,000 for living expenses, and $176,000 for appraisal fees, plus $5 million in mental anguish, $12 million in punitive damages, and more than $8 million in attorneys' fees.

On appeal, the court carefully reviewed all of the evidence and held that there was sufficient evidence to support the finding of a breach of the duty of good faith and fair dealing. An insurance company breaches its duty of good faith when it fails to pay a claim even though its liability is reasonably clear.[39]

■ Contract Damages

It is not illegal to breach a contract in the United States; it is only illegal not to pay the damages resulting from the breach. The damages awarded are designed to, as far as is practicable, put the injured party in the same position they would be in had the contract been performed.[40] See Table 7-2 for a summary of contract damages, mitigation, and waiver.

TABLE 7-2

Contract damages, mitigation, and waiver	
Basic rule of contract damages	The damaged party is entitled to the damages that were reasonably foreseeable on the date the contract was entered into. Other names for this broad category are compensatory damages or monetary damages.
Specific types of damage	**Simplified rule**
Cost of repair or replacement	Cost rule: actual, reasonable costs to repair are recoverable. Value rule: if the cost to repair materially exceeds the value added by the repair, the damaged party is only entitled to the diminution in value of the project, not the cost to repair.
Lost profit	Reasonably foreseeable of the date the contract entered into.
Home office overhead or *Eichleay* damages	Reasonably foreseeable of the date the contract entered into.
Consequential or special damages	Damages specific to a particular transaction. The party causing the damage must be aware of these damages could result if the contract is breached.

TABLE 7-2 (continued)

Costs	Sums paid in connection with the lawsuit such as filing fees. Does not include attorney fees.
Prejudgment interest	Set by statute and awards the winning party the time value of money from the date the damage was incurred until the date the judgment is entered.
Postjudgment interest	Set by statute and awards the winning party the time value of money from the date the judgment is entered into until the date the judgment is paid. Usually higher than prejudgment interest to encourage payment.
Attorney fees	Only recoverable if allowed by the contract or a specific statute.
Liquidated damages	The parties may agree that instead of the above types of damages, the injured party will be awarded a specific sum enumerated in the contract. The court will enforce such a provision as long as the amount to be paid is not a penalty.
Damages unforeseeable at the time contract entered into	Not a form of compensatory damages and therefore not recoverable.
Mitigation and waiver	
Mitigation of damages	A damaged party must take all reasonable steps to limit their damages. Damages that could have been avoided by reasonable action are not recoverable.
Waiver of damages	Parties can waive damages but cannot waive all damages. It is common for the party with superior bargaining power to force the other party to waive all damages except a time extension. See Chapter 14.

EXAMPLE

Contractor Small enters into a construction contract to build a house for Egan for $200,000, construction to begin immediately. However, shortly after entering into the Egan contract, Small is offered a long-term project to build 20 houses for Lazy Acres Estates, construction to start immediately. Small does not have the personnel to do both jobs. He cancels the Egan job and takes the Lazy Acres Estates job. Egan finds a substitute contractor to build the house for $210,000. Small owes Egan $10,000.

EXAMPLE

Pinewood entered into a contract with the owner of certain land to develop it for commercial and residential purposes. Pinewood sought bids from concrete subcontractors and obtained a bid from Smith Paving for $195,000. The bid said it was good for 30 days only. At that time, the final plans and specifications had not been finalized by the owner, and the permits had not yet been issued by the government.

Pinewood and Smith signed a subcontract for the concrete work for $195,000. No additional contract documents or project drawings were submitted to Smith. nor did Smith provide Pinewood with shop drawings and other documentation provided by the subcontract. Pinewood never called Smith to begin the work. At one point, Smith's project foreman traveled to the job site and discovered the concrete work had been completed by a third party. In explaining why another subcontractor had been hired to complete the job, Pinewood said that the final job plans called for significant additional work and therefore they contracted with another company, which completed the work for $380,000.

Pinewood maintained that the subcontract with Smith was not a contract but merely an "agreement to agree," because it was based on preliminary specifications and drawings and Smith never submitted shop drawings.

Smith sued Pinewood for breach of contract. The court awarded damages of $700 for estimation costs and $38,000 for lost profit. It found that a contract had been entered into between Smith and Pinewood, and Pinewood breached it by hiring another subcontractor.[41]

Contract damages are designed to place the injured party back in the position she would have been in had the breach not occurred. "[T]he general rule in breach of contract cases is that the award of damages is designed to place the injured party, so far as can be done by money, in the same position as that which he would have been in had the contract been performed."[42]

EXAMPLE

Clark entered into a subcontract with Allglass Systems, Inc. for the installation of leak-free windows. Problems resulted with the windows, and Clark sued for breach of contract. The court determined that the contract had been breached based upon extensive testing of the windows by various experts.

Allglass attempted to get out of the breach by claiming that the owner had required a more stringent chamber test rather than the hose test required by the specifications. The court rejected this, because the contract allowed the owner to conduct more stringent testing. Allglass also contended that the testing was done before the sealant had a chance to cure. However, Allglass's own remedial work action plan noted that "several deficiencies have been observed...of the existing installation...and are direct contributing factors to the results of field water testing..."

continued

The following costs and damages were recoverable by Clark:

1. Costs incurred to complete Allglass' work and repair damaged work
2. Payments to Allglass' subcontractors
3. Clark's increased "General Conditions," i.e., supervision and time-related costs
4. A reduction in the final payment to Clark from the Owner, via settlement
5. Attorney's fees as provided for in the contract[43]

Not all damages incurred as the result of a contract breach are recoverable, only those damages reasonably foreseeable at the time the contract is entered into.[44]

EXAMPLE

Soraka, a plumbing subcontractor, entered into a subcontract with Platov but failed to show up on the job. After repeated calls, Platov finally told Soraka it could not perform the work because it was too busy. Soraka spent three days finding a substitute plumber and paid the substitute plumber an additional $5,000 to get the work done and the project back on schedule. In addition, the project manager had to put in additional hours solving this problem and could not take his daughter to the doctor for what appears to be a cold or the flu. The daughter had to be hospitalized for three days, and the doctor said that if the child had been brought in immediately, she would not have required hospitalization.

Soraka is responsible for the $5,000 paid to the substitute plumber and any overtime pay the contractor has to incur to get the project back on schedule. Soraka is not responsible for the child's medical bills or payment of damages for her illness.

Substantial Completion

Upon substantial completion of the project, the contractor is entitled to payment of the full contract price minus the costs of repairs.

EXAMPLE

The general contractor enters into contract to build a new home and substantially completed the home. The homeowner never paid the final payment, claiming that the contractor was not entitled to the payment because of defects. The contractor was entitled to payment minus the costs of repair.

In most areas of contract law, the failure to complete the contract means that the breaching party is not entitled to be paid. However, in the construction industry, some jurisdictions will award a breaching party sums for the work done if the money was not used to pay for repairs or replacement.

An owner hired a contractor to remodel a bathroom and agreed to pay the contractor $15,000. The contractor completed roughly 50% of the job and then walked off the job. The owner had not yet paid the contractor. The owner hired a substitute contractor to complete job, and the substitute contractor charged the owner $10,000 to complete the job. The owner owed the original contractor $5,000.

Lost Anticipated Profit

A party may recover lost anticipated profits when the following criteria are found: (1) there is evidence to establish the damages with reasonable certainty; (2) the losses were the proximate consequence of the wrong; and (3) such losses were reasonably foreseeable at the time that the parties entered into the contract. However, these profits may be difficult to collect, as the parties must always mitigate their damages by taking on additional work.[45]

FSEC entered into a long-term subcontract with McFadden related to FSEC's prime contract with the Navy for construction and relocation services at a Naval shipyard. Problems developed, and FSEC ordered McFadden off the job. McFadden sued for amounts owed to it for additional work and several other categories of damages, including lost anticipated profits.

The court ordered FSEC to pay McFadden damages but not lost anticipated profits. The court said, "Although McFadden is not required to prove the amount of lost profits with mathematical exactitude, McFadden must establish its losses 'with a fair degree of probability' (internal citation removed). Mindful of this standard, I find that McFadden's argument is too speculative to be tenable. At trial, McFadden was unable to point to any specific work that it lost out on as a result of FSEC's actions, and McFadden admitted that it failed to submit any bids for construction projects after it was terminated by FSEC."[46]

Interest

Parties are entitled to **prejudgment interest,** which is interest on the amount of the judgment back to the date of the breach of contract. This amount is set by statute. Prejudgment interest is only recoverable when the amount due is certain as of a specific date.

EXAMPLE

FSEC entered into a long-term subcontract with McFadden related to FSEC's prime contract with the Navy for construction and relocation services at a Naval shipyard. Problems developed and FSEC ordered McFadden off the job without paying McFadden for extra work it had completed at FSEC's request.

The court said, "Furthermore, under Pennsylvania law, the fact that McFadden's claim relates to extra work does not affect McFadden's entitlement to prejudgment interest....I award prejudgment interest at the rate of 6% from November 1, 1997 [to the date of the judgment] on McFadden's claims, totaling $67,096 (internal citations omitted)."[47]

The prevailing party is also entitled to postjudgment interest. This amount is set by statute. If the losing party does not pay the judgment on the date it is entered, interest accrues at the rate set by law until it is paid. Postjudgment interest may be at a higher rate than prejudgment interest to encourage payment.

Consequential or Special Damages

Consequential or special damages are damages suffered by a party due to some unique characteristic of the injured party or because of some unique situation of the injured party *known* to the breaching party. "Consequential damages or special damages may not naturally flow from the breach, but may be awarded by the court upon proof of certain additional elements. It might be said that consequential damages are peculiar to the particular plaintiff and would not necessarily arise [with another plaintiff] in a similar circumstance."[48]

EXAMPLE

Discovery entered into a subcontract with Charter for the installation of drywall on a home. At the time the contract was entered into, Charter knew that Discovery had obtained the money for the project from various sponsors and that Discovery intended to enter the house into a homebuilders show to showcase the sponsors products and work. Charter knew that its work had to be completed in time for Discovery to complete the project for the show. Charter failed to complete the work, and Discovery had to hire another subcontractor. Discovery had to show the house in an unfinished condition, which did not make its sponsors happy. Discovery attempted to salvage its relationship with its sponsors by entering the house in a different show at a later date. Discovery had to pay an additional $10,000 entry fee for this second show.

Discovery did not pay Charter for its work but offered to settle with Charter when Charter began threatening Discovery with a lawsuit. Charter sent Discovery threatening e-mails including statements that the homeowner was going to be mad at Discovery and the litigation costs could get expensive. Charter threatened to pursue the matter with an open checkbook. Charter eventually sued, and Discovery counterclaimed for breach of contract and damages. The trial court awarded Discovery consequential damages including the $10,000 entry fee for the second show and attorney fees.[49]

When a contractor is terminated from a job, it loses its bonding capacity. What if the termination is later determined to have been wrongful? Is the contractor entitled to damages for its lost bonding capacity? The matter is not well settled in the law. One court has held no.[50]

Cost Rule, Value Rule, and Economic Waste

Image courtesy © istockphoto.com/leventince

Two measures of damages exist in construction contracts: the cost rule and the value rule. The **cost rule** states that the damages are the costs to repair the work. This is the preferred measure of damages in construction cases.

E X A M P L E

The DelGrecos entered into a written contract with Shewmake to remodel their home. The remodeling consisted of adding a playroom and a hobby room underneath an existing wooden deck, replacing the existing wooden deck with a concrete deck, adding a screened-in porch, and adding a third garage. The total amount of the remodel was $119,000.

continued

Problems developed with the new concrete deck. It allowed rainwater to leak into the new playroom, and several other less-significant problems existed, but the major problem was with the concrete deck. The DelGrecos complained to Shewmake regarding all of these problems, saying they would not pay Shewmake any further money until the leak in the deck was corrected. After this conversation, Shewmake did not perform any further work.

The DelGrecos paid a substitute contractor $18,470.36 to demolish the defective deck and to install a new deck. In addition, the DelGrecos paid the substitute contractor approximately $57,000 to complete the remodeling of their house. The DelGrecos sued Shewmake for a return of all of the funds they had paid Shewmake.

The DelGrecos damages were

1. the cost to remove and replace the deck, *plus*
2. the cost to complete the remodel, *minus*
3. the amount they would have paid Shewmake to complete the work.

For some reason, the DelGrecos waived the second item but sued for the return of all of the money they had paid Shewmake, a sum to which they were not entitled. The damages they were legally entitled to were

1. the cost to remove and replace the deck, *plus*
2. zero because this amount was waived, *minus*
3. the amount they would have paid Shewmake to complete the work.

The above calculation was $1,000.[51]

The **value rule** says the damages are the diminution in the value of the project due to the defects. The value rule is related to the concept of economic waste. **Economic waste** exists when the costs of repair materially exceed the value of the project and the project should not be repaired.

EXAMPLE

After substantial completion of a project, defects in the slabs are discovered. It would cost $1 million to correct the defects. The value of the project is only reduced by $50,000 due to the defective slabs. The owner is entitled to $50,000, not $1 million.

EXAMPLE

Garrett, the general contractor, entered into a subcontract with Canty for the masonry work on a new restaurant. After Canty failed to make corrections to its faulty work, failed to return to the site, and generally ignored Garrett, Garrett sued Canty for damages. At trial, Garrett was awarded the cost of removal of defective masonry and replacement, including costs for overhead.

On appeal, Canty argued that "the measure of damages [for] real property is the diminution of the fair market value of the property and/or the cost of repair or restoration but limited by the fair market value at the time of the breach or tort." The appeal court disagreed and stated that damages for breach of a construction contract could be measured by the costs to correct. The value rule is only applied when the costs of repair are material as compared to the increase in value due to the repairs.[52]

Eichleay or Overhead Damages

Recoverable damages include the amount of overhead that must be transferred to or born by other projects when a construction contract is wrongfully terminated or delayed.[53] This type of damage is often called *Eichleay* damages, because they were first enumerated in a case of that name.[54] Home office overhead costs may include the following:

1. Rent and depreciation costs
2. Licenses and fees
3. Property taxes
4. Utilities and telephone expenses
5. Salaries or fees of directors, officers, management, and clerical personnel
6. Automobile costs
7. Travel costs
8. Data processing costs
9. Insurance and bond premiums
10. Legal and accounting expenses
11. Interest costs
12. Office supply costs
13. Photocopying
14. Other general administrative expenses[55]

These damages are commonly associated with federal government projects and are to compensate the contractor for overhead costs during government caused delays. Many states have recognized and awarded *Eichleay* damages in appropriate circumstances.[56]

In order to recover overhead damages, the contractor must show owner-caused delay or suspension during which time the contractor was required to remain ready to perform *and* the contractor was unable to take on outside work during that time. This is the concept of mitigation of damages, discussed in more detail below.

Other damages recoverable for delay include increased payroll, increased labor, increased materials costs, costs resulting from the loss of efficiency of the use of labor or equipment, increased costs for extended bonding and insurance coverage, increased storage costs, and other items of overhead that are reasonably attributable to the delay.

Attorney Fees

Attorney fees are not recoverable in breach-of-contract actions unless recovery is provided for in the contract or by statute; this is called the **American Rule**. A sample form allowing attorney fees to be recovered is below. The **British Rule** is that attorney fees are an element of damages and payable to the winning party.

FORM

The prevailing party in any legal proceeding related to this Agreement shall be entitled to payment of reasonable attorney's fees, costs, and expenses, including expenses relating to arbitration.

MANAGEMENT TIP

All agreements should have a clause awarding attorney fees to the prevailing party.

Liquidated Damages

Liquidated damages are precalculated damages included in the contract. Usually calculated per day of delay, these damages compensate the injured party for delay and obviate the need for the parties to actually calculate their damages for delay. As long as the amount of liquidated damages is a reasonable reflection of the actual damages and is not a penalty, the law will uphold a liquidated damages provision.

EXAMPLE

The contract between the owner and the contractor was for the building of a casino in Las Vegas. It contained a liquidated damage clause that stated, "Contractor shall pay to the owner $5,000 per day for each day beyond the contract term for liquidated damages. Owner waives any other forms and types of damages for delay."

The contractor was 30 days late and paid the owner $150,000 in damages. The owner claimed that it could prove by competent experts that had the casino been open for 30 days, the owner would have received $1 million in profit. The contractor is required to pay only the $150,000 in liquidated damages.

If the liquidated damage provision is designed to penalize the contractor, it will not be upheld. If a liquidated damage clause is not upheld, it is as if it were stricken from the contract. The breaching party then owes the reasonably foreseeable damages.

EXAMPLE

The contract between the owner and the contractor was for the building of a casino in Las Vegas. It contained a liquidated damage clause that stated, "Contractor shall pay to the owner $1 million per day for each day beyond the contract term for liquidated damages."

The contractor was late by 30 days. Under the liquidated damage clause, the contractor owes the owner $30 million. The contractor can prove by competent experts that had the casino been open for 30 days, it would have made $1 million in profit. The contractor is required to pay the owner $1 million in damages, not $30 million. The liquidated damage provision is unenforceable, because it is a penalty and does not reasonably reflect the actual damages.

MANAGEMENT TIP

A liquidated damage clause can be beneficial to both sides and should be seriously considered. It must always be reasonable and not a penalty. See the sample form below.

FORM

In the event that the project is not completed on the date so specified in this contract, the Contractor shall pay to the owner a reasonable sum to the owner for substitute lodging, not to exceed $75.00 per day. Owner waives any other forms and types of damages for delay.

Total Cost Calculation for Extra Work

The total cost method of calculating damages is used by a contractor when making a claim for extra work. The total cost method takes the difference between the original contract price and the contractor's costs and awards that difference to the contractor as damages. This method is *not* favored by the courts because of the possibility of the contractor charging an excessive amount.[57] The total costs theory is not favored and has been "tolerated only when no other mode of recovery was available and when the reliability of the supporting evidence [is] fully substantiated."[58]

The party attempting to use this method must convince the court of the following: (1) it is impracticable to prove the cost of the extra work by other means, (2) the contract price was reasonable, (3) the actual costs for the extra work are reasonable, and (4) the contractor is not responsible for the increased costs.[59]

EXAMPLE

Mergentime, the prime contractor, entered into a contract to build a subway complex. The contractor was entitled to an adjustment of the contract price due to multiple changes and acceleration of the job. However, the contractor's request to use the total cost method of calculating damages was rejected by the court. The court said, "There is a more reliable means of proving damages: direct or 'actual costs' based upon expert testimony, contemporaneous records, and [the owner's] cost audits....The 'actual cost' method...requires the contractor to submit...detailed documentation regarding the 'extra' costs it incurred due to modification in performance....The method requires cumbersome segregation of those costs incurred due to the original contractual obligations from those associated with the modification."[60]

MANAGEMENT TIP

To the extent possible, parties should always keep costs for extra work segregated from contract costs for the project. It will be extremely time consuming and difficult to win on a total cost theory of damages.

Specific Performance

Specific performance is a court order to actually perform a contract. It is seldom applied in the construction industry, because courts prefer to award monetary, or compensatory, damages in the event of a breach. However, specific performance is a common remedy in the enforcement of real estate sales contracts. "[S]pecific performance is an exceptional remedy, to be granted only if damages would not be adequate, and not always even then."[61]

EXAMPLE

Takamara entered into a contract with Emmet to purchase two acres zoned as residential on a lake. Emmet then decided that he did not want to sell. Takamara is entitled to an order of specific performance requiring Emmet to sell.

Mitigation of Damages and Curing

The law requires parties to take reasonable steps to limit their damages. "The amount of loss that [the nonbreaching party] could reasonably have avoided by...making substitute arrangements or otherwise is simply subtracted from the amount that would otherwise have been recoverable as damages."[62]

EXAMPLE

A contractor entered into contract to build a house in a small town in Michigan and with a completion date of June 1. The contractor was aware that the project needed to be completed by this date, because the owners had to be out of their old house by that date. However, the house was not completed until June 15. The owners took a vacation to New York City and claimed that the contractor owed them $300 per night of lodging expenses in New York City. Lodging in their small town would have been $75 per night. The contractor is required to pay only $75 per night for lodging, not $300.

EXAMPLE

The contractor enters into a contract to remodel three buildings for the owner. The work is the same for each of the three buildings. The schedule calls for the contractor to remodel Building 1, then Building 2, then Building 3. However, after completing the remodel on Building 1, Building 2 is not yet available. Building 3 is available. The contractor must begin work on Building 3 even though it is not yet scheduled. This will reduce or mitigate the damages that the owner will incur.

EXAMPLE

The architects had breached their contract with the developer by failing to reveal the extent of the bricked in or "toothed-in" condition of windows on an historic building. The owners were therefore entitled to compensatory damages. For some reason, the developers did not employ the fastest and most efficient method of removing the brick from the windows, but the court found "the credible evidence discloses that the means of the demolition of affected window openings, i.e., by sledgehammer or air hammer from the interior of the building, was unreasonably expensive and impractical. Moreover, this means of demolition in these circumstances was, largely due to the cost factor, not only impractical but rarely heard of in the business, especially considering the object to be accomplished. To persist in this method...unreasonably aggravated the damages..." The architect did not have to pay for the method used to remedy the breach, only for a reasonable method to remedy the breach.[63]

EXAMPLE

Cizek entered into a subcontract with A-1 for the installation of a tennis court. A-1 in turn subcontracted the asphalt layer to Asphalt Maintenance. The cost of this asphalt layer was $13,000.

After Asphalt Maintenance completed the work, it was found to be unacceptable due to unevenness of the surface. Asphalt Maintenance installed an additional layer of asphalt, but the problems persisted. Asphalt Maintenance then hired American Asphalt to install another layer at a cost of $3,000, but the problems persisted.

Finally, Parking Area Maintenance was hired to remedy the problem, and it applied two additional layers of asphalt for $20,000. Expert testimony at trial stated that this cost was excessive, and the amount for the two layers installed by Parking Area Maintenance should have been about $6,000.

A-1 also incurred extra costs for fencing, because the thickness of the court was about 12 to 14 inches, rather than the normal 4 to 6 inches. This required the fencing subcontractor to jackhammer through the additional layers of asphalt, and cost for the extra work by the fencing subcontractor was $700.

A-1 sued Asphalt Maintenance for the $20,000 paid to Parking Area Maintenance and the $700 paid to the fencing subcontractor. Asphalt Maintenance was only responsible for the $700 paid to the fencing contractor and $6,000, the reasonable cost, for the repair of the surface. The $20,000 paid to Parking Area Maintenance was excessive, and only $6,000 was allowed. The court said, "A repair that cost nearly 50 percent more than the original cost is unreasonable. We therefore find that the amount for which A-1 contracted with Parking Area Maintenance was unreasonable and inordinate."[64]

Some states require the owner to give the contractor a chance to cure; some do not. The UCC, which covers suppliers but not contractors, does require a chance to cure. Because so many concepts from the UCC have found their way into the common law, it can be expected that more states will require the owner give the contractor the right to cure.

EXAMPLE

In this case, the court said that based on the facts, the owner was *not* required to give the roofing contractor a chance to cure. Harrington entered into a roofing contract with Grillo. After several months but still within the warranty period, the roof began to leak, and Harrington tried to find Grillo, who had changed his address and phone. Harrington had a county inspector out to look at the roof. The inspector warned Harrington that the work was substandard. Harrington hired a different contractor to replace the roof and sued Grillo, who had by this time been found, for damages. At trial, it was determined that Grillo had failed to install the roofs in a good and workmanlike manner and therefore had breached the contract. Harrington was entitled to the amount paid to the substitute contractor to make the repairs.

continued

Grillo argued that he could have done the repairs cheaper and therefore the owner failed to mitigate damages. Grillo argued that he was required to pay the owner only the amount that Grillo would have charged to do the repairs. "Under the circumstances of this case, where [owner was] warned by the Building Inspector of the inadequate installation of the [roof], this court finds that it would be unreasonable to require or even to expect [the owner] to rely upon [Grillo], who already had two strikes against him, to come back and provide replacement roofs for them."[65]

EXAMPLE

In this state, the law requires that a party be given the change to cure an error. In this case, the court found that the subcontractor had been given the chance to cure and did not cure. Therefore, the contractor was entitled to hire another to perform the work.

Byrd, the contractor, entered into a plumbing subcontract with Herring for the installation of plumbing in 20 condos. Problems arose, and eventually Herring walked off the job but sued Byrd for amounts due. Byrd cross-complained for the cost of completing Herring's work and repairing Herring's shoddy work. The amount Byrd paid the substitute contractor, over the amount it would have paid Herrington to do the job, was approximately $4,600. After a trial on the merits, the court determined that Herring had performed substandard work, thus breaching the contract. Herring claimed that Byrd should have allowed it to repair the work. The court determined that Herring *had* been given the opportunity to cure, because the contractor had pointed out the shoddy work several times while Herring was still on the job. Herring never repaired the deficiencies.[66]

The court said, "A party who has breached or failed to properly perform a contract has a responsibility and a right to cure the breach. The nonbreaching party must give him a reasonable opportunity to cure the breach. However, the right to cure is not unlimited. Where the breach is a material one, the nonbreaching party has a right to end the contract, but in doing so he is also obligated to minimize his damages. Likewise, when the conduct of the breaching party has been of such a nature as to cause a loss of confidence or 'shaken faith,' the offended party is entitled to end the contract, but he remains responsible for mitigating damages." The contractor did not owe the subcontractor anything (internal citations removed).

■ Conditions Precedent to the Receipt of Damages

Contracts frequently list "things" one party must do before it is entitled to payment by another party, and in the law, these "things" are called **conditions precedent.** Conditions precedent are upheld unless the party who is obligated to pay deliberately prevents performance of the condition precedent.

EXAMPLE

Aleda entered into a contract with the Winns for the construction of a house. The contract, written by Aleda, contained a provision that draws would be paid "in consideration of the covenants and agreements being *strictly* performed...The final draws was due when the house was 'fully complete and [Aleda] supplied [the Winns] with a final survey and executed Releases of Liens by the subcontractors.'"

The Winns never paid the final draw, because about $1,000 of punch list items remained unrepaired, the final survey was never given to them, and they never received the lien releases. Aleda sued for the amount of the final draw. Aleda was not entitled to the final draw, because it had not fulfilled the conditions precedent of (1) *strictly* performing the construction, (2) fully completing the construction, (3) supplying a final survey, and (4) supplying executed lien releases. Until Aleda did those "things" and fulfilled the conditions precedent, Aleda was not entitled to payment.[67]

■ Accord and Satisfaction

The doctrine of accord and satisfaction comes into play when (1) the parties have a dispute; (2) one party pays an amount in an attempt to settle the dispute and includes the words "full payment of all sums due" (or similar words) on the check; (3) the party receiving the check cashes it; but (4) the party who received the check tries to get more money from the party who wrote the check. An **accord and satisfaction** is the acceptance of something of value—usually money, but that is not required—in settlement of a claim. The doctrine of accord and satisfaction says that when party accepts something of value in settlement of a claim, the claim is settled and the party is not entitled to more money.

EXAMPLE

Hawthorne entered into a contract to grade and clear 12 acres. The developer claimed that Hawthorne had not completed the work. Hawthorne billed the developer $26,400 for the work, and the developer sent Hawthorne a check for $10,000 with the words "full payment of all sums owed to Hawthorne." Hawthorne endorsed the check "with reservations" and deposited it. The court held that cashing the check was an accord and satisfaction and that Hawthorne was not entitled to any additional sums. Not all states would have reached the same result."[68]

MANAGEMENT TIP

Depositing a check with the words "full payment of all sums due" may prevent a party from being entitled to any additional sums, because this is an accord and satisfaction. Some states will not find an accord and satisfaction if the check is endorsed "with reservations." Parties have been known to take advantage of this doctrine, always send a check for less than the actual amount due, and include the appropriate words on the check.

■ Endnotes

1. *Charter Township of Independence v. Reliance Building Co.*, 437 N.W.2d 22, 175 Mich. App. 48 (Mi. App. 1989).

2. Example based on *Electro-Lab of Aiken, Inc. v. Sharp Construction Co. of Sumter, Inc.*, 357 S.C. 363, 593 S.E.2d 170 (SC App. 2004).

3. Example based on *Thermoglaze Inc. v. Morningside Gardens*, 23 Conn. App. 741 (Conn. App. 1981), app.den. 217 Conn. 811 (Conn. App. 1991) and used almost verbatim in White, *Principles and Practices of Construction Law* (Prentice Hall 2002).

4. *Shovel Transfer and Storage v. Pennsylvania Liquor Control Board*, 559 Pa. 56, 739 A.2d 133 (Penn. 1999).

5. *California Business & Professions Code section 7159.* This requirement is not part of the statute of frauds but is found in a separate code. Most states have such a law as part of code regulating building contracts.

6. These statutes are usually in some type of public contracting code.

7. Example based on *Ry-Tan Construction, Inc. v. Washington Elementary School District*, 210 Ariz. 419, 111 P.3d 1019 (Az. 2005).

8. Example based on *Lewis and Queen v. N. M. Ball Sons*, 48 Cal. 2d 141, 308 P.2d 713 (Cal. 1957). See also Corbin on Contracts. §580 *Oral Proof of Fraud, Illegality, Accident, or Mistake.* Matthew Bender & Co., Inc. (2006).

9. See *Redfern v. R.E. Dailey & Co.*, 146 Mich. App. 8, 379 N.W.2d 451 (Mich. App. 1985).

10. Example based on *In re: The Palms at Water's Edge, L.P.*, 334 B.R. 853 (Bankr. D. Tex. 2005).

11. *Paden v. Murray*, 240 Ga. App. 487 (Ga. Ct. App. 1999), *Estate of Sam Farkas, Inc. v. Clark*, 238 Ga. App. 115, 117 (Ga. Ct. App. 1999), *Ben Farmer Realty Co. v. Woodard*, 212 Ga. App. 74 (Ga. Ct. App. 1994).

12. *Ainsworth v. Perreault*, 254 Ga. App. 470, 563 S.E.2d 135 (Ga.App. 2002).

13. *Sydnor v. Conseco Fin. Servicing Corp.*, 252 F.3d 302, 305 (4th Cir. 2001) (internal citations omitted).

14. Example based on *In Re: Owens Corning*, 291 B.R. 329 (Del. 2003).

15. Example based on *Hubscher & Son, Inc. v. Storey*, 228 Mich. App. 478, 578 N.W.2d 701 Mi.App. 1998).

16. Example based on *Envirotech Corp. v. Halco Engineering*, 234 Va. 583 (Va. 1988).

17. *Marin Storage & Trucking, Inc. v. Benco Contracting and Engineering, Inc.*, 89 Cal. App. 4th 1042, 107 Cal. Rptr. 2d 645 (Cal. App. 2001).

18. *Langbein v. Comerford*, 215 So. 2d 630, 631 (Fla.Dist.Ct.App.1968) (citing *Maryland Cas. Co. v. Krasnek*, 174 So. 2d 541 (Fla.1965)), *Pennsylvania Nat'l Mut. Cas. Ins. Co. v. Anderson*, 445 So. 2d 612, 613 (Fla.Dist.Ct.App.1984).

19. Loosely based on *Roberts & Schaefer Co. v. Hardaway Co.*, 152 F.3d 1283 (11th Cir. 1998).

[20] *Vortt Exploration Co. v. Chevron U.S.A., Inc.*, 787 S.W.2d 942, 944, 33 Tex. Sup. Ct. J. 409 (Tex. 1990), *Truly v. Austin*, 744 S.W.2d 934, 936, 31 Tex. Sup. Ct. J. 228 (Tex. 1988).

[21] *Amprite Electric Co. v. Tennessee Stadium Group*, LLP, 2003 Tenn. App. Lexis 686 (Tenn. App. 2003).

[22] *Albrecht v. Comm. on Employee Benefits of Fed. Reserve Employee Benefits Sys.*, 360 U.S. App. D.C. 47, 357 F.3d 62, 69 (D.C. Cir. 2004).

[23] *Wingate Land & Development, LLC v. Robert C. Walker, Inc.*, 252 Ga. App. 818, 558 S.E. 2d 13 (Ga. App. 2001).

[24] Example based on *Wingate Land & Development, LLC v. Robert C. Walker, Inc.*, 252 Ga. App. 818, 558 S.E.2d 13 (Ga. App. 2001).

[25] See *Restatement (Second) of Contracts* (1981).

[26] Example based on *Hinson v. N&W Constr. Co.*, 890 So.2d 65 (Miss. App. 2004).

[27] Example based on *SKB Industries, Inc. v. Insite*, 250 Ga. App. 574, 551 S.E.2d 380 (Ga. App. 2001). Court's ruling slightly varied from law as explained herein.

[28] *Bembery v. District of Columbia*, 758 A.2d 518, 520 (D.C. 2000) (quoting *Fowler v. A & A Co.*, 262 A.2d 344, 347 (D.C. 1970)).

[29] *Hernandez v. Gulf Group Lloyds*, 875 S.W.2d 691, 692 (Tex. 1994), *Graco Robotics, Inc.*, 914 S.W.2d at 641.

[30] Example from *Morganti National, Inc. v. Petri Mechanical Co., Inc.*, 2004 U.S. Dist. Lexis 8659 (Dist. Conn. 2004).

[31] Corbin on Contracts §692 at 269-271, n.34 & 36. *U.S. ex rel. Endicott Enters. v. Star Bright Constr. Co.*, 848 F. Supp. 1161 (D. Del. 1994).

[32] Example from *U.S. ex rel. Endicott Enters. v. Star Bright Constr. Co.*, 848 F. Supp. 1161 (D. Del. 1994).

[33] See *Snyder v. Reading Sch. Dist.*, 311 Pa. 326, 166 A. 875, 877 (Pa. 1933), see also *Accu-Weather, Inc. v. Prospect Communications*, 435 Pa. Super. 93, 644 A.2d 1251, 1254 (Pa. Super. Ct. 1994) ("conditions precedent to a contract termination must be strictly fulfilled").

[34] Example based on *In Re: The Palms at Water's Edge, L.P.*, 334 B.R. 853 (WD Tx. 2005).

[35] *Chapman v. Norfolk & Dedham Mutual Fire Ins. Co.*, 39 Conn. App. 306, 320, 665 A.2d 112, cert. denied, 235 Conn. 925, 666 A.2d 1185 (1995).

[36] *Centex Corp. v. U.S.*, 395 F.3d 1283, 1304 (Fed. Cir. 2005). *C. Sanchez & Son, Inc. v. U.S.*, 6 F.3d 1539, 1542 (Fed. Cir. 1993).

[37] *Orlosky Inc. v. U.S.*, 68 Fed. Cl. 296 (Ct.Cl. 2005).

[38] Example based on *Lewis-Nicholson v. U.S.*, 550 F.2d 26, 213 Ct. Cl. 192 (1977).

[39] *Allison v. Fire Ins. Exch.*, 98 S.W.3d 227 (Tex. App.—Austin 2002, pet. withdrawn). This is the famous mold litigation case. However, it was not a mold case; it was a bad faith insurance case.

[40] *Radecki v. Mutual of Omaha Ins. Co.*, 255 Neb. 224, 583 N.W.2d 320 (Neb. 1998), *Larsen v. First Bank*, 245 Neb. 950, 515 N.W.2d 804 (Neb. 1994).

41 *Smith Paving & Excavating v. Vermillion Shores Development Group, Inc.*, 2005 Ohio 3196 (Ohio App. 2005).

42 *Harrington v. Grillo*, 2000 Conn. Super. LEXIS 3344 (Conn. Sup. 2000) quoting *Rejouis v. Greenwich Taxi, Inc.*, 57 Conn. App. 778, 784, 750 A.2d 501, (Conn. App. 2000).

43 Example based on *The Clark Construction Group Inc. v. Allglass Systems, Inc.*, 2004 U.S. Dist. Lexis 15459 (D. Md. 2004).

44 *Indiana Mich. Power Co. v. U.S.*, 422 F.3d 1369, 1373 (Fed. Cir. 2005).

45 See *Advent Sys. v. Unisys Corp.*, 925 F.2d 670, 680 (3d. Cir. 1991) (citing *Delahanty v. First Pennsylvania Bank, N.A.*, 318 Pa. Super. 90, 464 A.2d 1243, 1258 (Pa. Super. Ct. 1983)).

46 Example based on *U.S. for the use and benefit of McFadden Mechanical, Inc. v. FSEC, Inc.*, 2001 U.S. Dist. LEXIS 24201 (D. Penn. 2001).

47 Example based on *U.S. for the use and benefit of McFadden Mechanical, Inc. v. FSEC, Inc.*, 2001 U.S. Dist. LEXIS 24201.

48 Construction Law §11.02 *Losses for Which Recovery Is Available on a Breach of Contract Theory*, Matthew Bender (2006).

49 Example based on *Charter Drywall Atlanta, Inc. v. Discovery Technology, Inc.*, 271 Ga. App. 514, 610 S.E.2d 147 (Ga. App. 2005).

50 *Lewis Jorge Construction Management, Inc. v. Pomona Unified School District*, 34 Cal. 4th 960, 102 P.3d 257, 22 Cal. Rptr. 3d 340 (Ca. 2004).

51 *Shewmake v. DelGreco*, 926 So. 2d 348 (Ala. 2005).

52 Example based on *Robert E. Canty Building Contractors, Inc. v. Garrett Machine & Construction, Inc.*, 270 Ga. App. 871, 608 S.E.2d 280 (Ga. App. 2004).

53 For a discussion of the weaknesses of *Eichleay,* see Lisa Liberman Thatch, *Eichleay Jurisprudence: Its Strengths, Weaknesses and Inappropriately Broad Application*, 12 Fed. Cir. B.J. 107.

54 *Eichleay Corp.*, ASBCA No. 5183, 60-2 BCA 2688 (1960), aff'd on reh'g, 61-1 BCA 2894 (1960).

55 Braude & Kovars, *Extended Home Office Overhead. Basic Principles and Guidelines,* Construction Briefings (June 1984).

56 *Broward County v. Russell, Inc.*, 589 So. 2d 983 (Fla. 4th DCA 1991), *Moore Constr. Co. v. Clarksville Dep't of Elec.*, 707 S.W.2d 1, 14 (Tenn. Ct. App. 1985).

57 For a more in-depth legal analysis of this method, see Karl Silverberg, *Construction Contract Damages: A Critical Analysis of the "Total Cost" Method of Valuing Damages for "Extra Work,"* 17 St. John's J.L. Comm. 623 (2003).

58 *G. M. Shupe, Inc. v. U.S.*, 5 Cl. Ct. 662, 676 (1984).

59 *Servidone Constr. Corp. v.U.S.*, 931 F.2d 860, 861 (Fed.Cir. 1991).

60 Example based on *Mergentime Corp. v. Wash. Metro. Area Transit Auth.*, No. 89–1055 (HHG), 1997 U.S. Dist. LEXIS 23408, at 16 (D.D.C. 1997).

61 Melvin A. Eisenberg, *Actual and Virtual Specific Performance, the Theory of Efficient Breach, and the Indifference Principle in Contract Law*, 93 Calif. L. Rev. 975 (2005).

[62] Restatement (Second) of Contracts § 350 cmt. b.

[63] *Center Court Associates LP v. Maitland/Strauss & Behr* (1994 Conn. Super. unreported case).

[64] *A-1 Track & Tennis, Inc. v. Asphalt Maintenance, Inc.*, 2000 Neb. App. LEXIS 187 (Neb. App. 2000).

[65] *Harrington v. Grillo*, 2000 Conn. Super. Lexis 3344 (Conn. App. Ct. 2000).

[66] Example based on *Byrd Brothers, LLC v. Herring*, 861 So. 2d 1070 (Miss. App. 2003).

[67] Example based on *Winn v. Aleda Constr. Co., Inc.*, 227 Va. 304, 315 S.E.2d 193 (Va. 1984).

[68] *Hawthorne Grading & Hauling v. Rampley*, 252 Ga. App. 771, 556 S.E.2d 912 (Ga. App. 2001).

186

CHAPTER 8

Changes, Additions, and Delays

■ Changes to the Contract

The construction industry differs from other industries in that the owner is allowed to make unilateral changes to the contract. A **unilateral** change is a change made by one of the parties to the contract, which under traditional contract law is not allowed. However, the law recognizes the unique characteristics of the industry and allows the owner to make unilateral changes as long as the contractor is reasonably compensated. Most contracts have a provision to the effect.

AIA A201-1997, Article 7 reads:

"Article 7 CHANGES IN THE WORK

7.1 GENERAL

7.1.1 Changes in the Work may be accomplished after execution of the Contract, and without invalidating the Contract, by Change Order, Construction Change Directive or order for a minor change in the Work, subject to the limitations stated in this Article 7 and elsewhere in the Contract Documents."

The Contracting Officer may at any time, by written order, and without notice to the sureties, if any, make changes within the general scope of this contract…[49 CFR §52.243(2)].

Often the contract will have specific procedures that the contractor must follow in order to be entitled to payment. The law will enforce the processes and procedures outlined in the contract. These are some of the most common clauses waived, and if the contractor can prove the requirements were waived, then the contractor is entitled to payment without following the prescribed procedures. As with most civil matters, the contractor must prove waiver by a preponderance of the evidence.

EXAMPLE

In this case, the contractor failed to prove a waiver of the change procedures had occurred and so was not entitled to payment, because it had failed to follow the procedures. Hall, the contractor, claimed that Entergy, the owner, had waived the contract's requirements for written change orders. However, in order to support a waiver, the party must submit evidence of waiver. Absent evidence of a waiver, the written terms of the contract are enforced. The strongest evidence of waiver is evidence of a series of oral changes that were approved. In this case, all change orders, except the one the contractor was attempting to collect on, had been submitted as required by the contract. There was no evidence to support the contractor's waiver theory.[1]

Field Changes

A **field change** is a change order made on the site without following the procedures outlined in the contract. Contractors are generally entitled to compensation for field changes if they immediately document the field change to the owner and no objection is raised. If the owner pays the contractor for field changes, the law will assume that the owner approved the change.

EXAMPLE

The U.S. Navy awarded a construction to FSEC. FSEC entered into a subcontract with McFadden for plumbing work. FSEC terminated McFadden prior to the completion of the subcontract, and McFadden sued for the unpaid balance on the plumbing subcontract. The unpaid balance was all related to additional work pursuant to the contractor's field changes. Prior to the dispute, McFadden submitted invoices to FSEC for work done on the subcontract and for the extra work pursuant to the field changes. No formal change orders were ever issued despite a contract provision requiring them. FSEC made periodic payments for both work on the subcontract and the extra work and never objected to any irregularity in the procedure. McFadden is entitled to payment for the field changes.[2]

MANAGEMENT TIP

Contractors should always follow up on field changes with written confirmation. Owners should always follow up with a notice that costs for the field changes are subject to review.

Deletion Clause

A **deletion clause** allows the owner to delete a portion of the work. Again, as long as the deletion is made in good faith, the clause is upheld.[3]

FORM

The Owner may make Changes and/or Deletions to the work without invalidating the Contract or incurring any liability to the Contractor for lost profit, consequential damages, or any other damage.

■ Construction Change Directive

AIA Document A201-1997 §7.3 and other contracts have redefined the term "change" to mean only a modification of the contract *agreed* to by the owner and the contractor. If the contractor disagrees with some aspect of the change order, the owner or architect issues a construction change directive. The **construction change directive** requires the contractor to perform the work as requested and follow procedures in the contract for making a claim. This allows the work to proceed. The law already requires this result. See Chapter 7 in the section entitled "Contract Breach."

■ Legal Avenues for Payment Available to the Contractor or Subcontractor without a Change Order

Image courtesy Getty Images.

Several legal theories are available to the contractor or subcontractor seeking payment for the changed work but without a change order. These theories include the additional work doctrine, promissory estoppel, waiver, and the constructive change doctrine. In addition, if the change is major, the contract may be destroyed, leaving the contractor or subcontractor to recover under the legal theory of promissory estoppel for the reasonable value of the work, rather than the contract price. See Table 8-1.

TABLE 8-1 Avenues for payment available to the contractor or subcontractor without a change order

Name of legal theory	Simplified rule
Additional work doctrine	Additional work is not covered by the original contract but by a new, usually oral, contract between the parties. If the party seeking payment can prove the existence of the oral contract, the party is entitled to a reasonable sum for the work completed.
Promissory estoppel also known as detrimental reliance	If no contract has been formed, a party performing work under circumstances where a reasonable person would expect to be paid for the work, the entity doing the work is entitled to a reasonable sum for the work completed.
Waiver	A waiver is the knowing relinquishment of a known right. Any clause to a contract can be waived by acts of the parties contrary to the requirements of the written contract. If the party seeking payment has evidence that the changes clause has been waived, the party is entitled to a reasonable sum for the work complete.
Constructive change doctrine	A constructive change is owner conduct that amounts to a change in the contract, but the owner refuses to recognize the change. The contractor is entitled to compensation for a constructive change that requires additional work.
Destruction of the contract, a.k.a. theories of cardinal change or abandonment	The owner destroys the original contract when it makes so many changes that the project completed is materially different from the project as contracted. The contractor is paid a reasonable sum for the work completed; the work is not paid according to the destroyed contract's terms.

Additional or Extra Work

Additional or extra work is not covered by a changes clause in a contract. **Additional work** is work outside the scope of the original contract and not covered by the contract. A new contract, usually oral in nature, is easily formed to support additional work, and this is despite the fact that the original contract may have a provision requiring additional work to be approved in writing or even barring additional work. Not all courts would go as far as the Louisiana court did in saying, "Modification of a written agreement can be presumed by silence, inaction, or implication."[4] The contractor is likely to be paid for any additional work of which the owner is aware.

EXAMPLE

Petroccitto entered into an agreement with Zach for Zach to construct a retaining wall adjacent to an existing pool deck at a cost of $16,400. The contract required all changes and additional work to be in writing. About halfway through the project, Petroccitto requested an additional wall in front of a large gazebo, a terrace to accommodate a children's play structure, and steps to the gazebo. Zach claimed that the cost for the additional work was $3,800.00. The court said, "It is clear in the trial record that a contract was duly performed; Zack completed the extra work as requested by the defendant, totaling $3,800, and should therefore be paid for the work. The fact that [the owner] did not issue a formal change order does not cause the Court to conclude there was not a binding contract between the parties. The Court finds by a preponderance of testimony produced at trial that such a binding contract was entered into between the parties for the additional $3,800 work, and Zack fully performed said additional work."[5] Judgment was in favor of Zach.

EXAMPLE

Aqua Pool entered into a contract to perform pool repairs at Paradise Manor for the price of $50,000. The contract contained a requirement that all changes be in writing. Breaux was a member of Paradise Manor's board of directors and had been authorized to sign the contract and oversee the project. Several unanticipated issues and problems arose during the project, including unanticipated repiping of certain buried worn and outdated pipes and a new manifold pumping system. The president of Paradise Manor testified that Paradise Manor obtained a loan for only $50,000, the amount of Aqua Pool's original bid, and there was no money in the budget for additional work. He also testified that although Breaux had been designated to run this project, the board had to approve any additional work or changes. He also testified that the board probably would have approved some of the changes but was never asked.

The court awarded Aqua Pool compensation for the additional work, amounting to about $1,000. Breaux knew about the extra work and had asked Aqua Pool how much it would cost. Aqua Pool had followed up with an e-mail confirming the extra work.[6]

MANAGEMENT TIP

Owners may want to insert a clause covering additional work into the contract in an attempt to have the existing contract cover the additional work at the prices outlined in the contract.

During the course of the project, Owner may order additional work outside of the scope of this contract. The cost of this additional work will be the Contractor's actual costs plus profit and overhead at a rate of 3%.

Detrimental Reliance, Promissory Estoppel, Quantum Meruit, Contract Implied in Fact, and Unjust Enrichment

Each of these legal doctrines is slightly different, but they all require the same result: someone who receives a benefit from another must pay for it. A more complete discussion of this topic is in Chapter 7 in the section entitled "Detrimental Reliance, Promissory Estoppel, Quantum Meruit, Part Performance, Implied in Law Contracts, Constructive, or Quasi-contracts."

"Under Maine law, a quantum meruit claimant must show that (1) services were rendered to the defendant by the plaintiff; (2) with the knowledge and consent of the defendant; and (3) under circumstances that make it reasonable for the plaintiff to expect payment.[7] While the formalities of an express contract are not a prerequisite to recovery in quantum meruit, there must be a reasonable expectation on the part of the claimant to receive compensation for his services and a 'concurrent intention' of the other party to compensate him (internal citations omitted)."[8]

"In order to sustain a claim of unjust enrichment, plaintiff must establish (1) the receipt of a benefit by defendant from plaintiff, and (2) an inequity resulting to plaintiff because of the retention of the benefit by defendant. If this is established, the law will imply a contract in order to prevent unjust enrichment....However, a contract will be implied only if there is no express contract covering the same subject matter (internal citations omitted)."[9]

These theories are not related to the existence of a contract and in fact are used specifically when no contract has been formed but the circumstances indicate that the person performing the work should be paid for it.

EXAMPLE

Madison was the owner and operator of Anson Dam, a hydroelectric facility on the Kennebec River in central Maine. It awarded to Abington, the contractor, a contract to reconstruct a waste gate and repair a power station. The repairs required the construction of a cofferdam to allow work to be done below the water level.

Contract documents stated that a bidder could rely on "the general accuracy of the 'technical data'" in the materials but authorized only "limited reliance" on drawings. Contract documents stated the average level of the tail pond to be 222.65 feet above sea

continued

level, although it was actually 223.65. The drawings did not reveal several deep contours in the area where the cofferdam was to be constructed. These errors required Abington to increase the number of bulk sandbags from 180 to 300 and the number of small sandbags from 250 to 5,000. The route of the cofferdam also had to be changed. Floods also caused delay to the project.

Abington informed Madison of the problems shortly after beginning the work and requested a change order. Madison did not respond for several months but eventually "a series of interactions [conversations, written memos, minutes of meetings] between the parties occurred in which Madison indicated that it would cover some, if not all, of the additional costs caused by the faulty specifications and high water flows." Later, Madison refused to pay the contractor the additional sums. The court held that the contractor was entitled to an award of $334,175, the reasonable value of its services for extra work.[10]

The U.S. Court of Federal Claims does not have jurisdiction to hear claims of promissory estoppel or contract implied-in-law,[11] therefore, these will only be claims based on state law. On the other hand, theories such as implied terms are often used by the federal courts to produce the same result. Implied terms are discussed in Chapter 9 in the section entitled "Implied Terms."

Waiver

The concept of waiver is discussed several times in this book, because it is a broad doctrine applicable to any situation involving a legal right. A **waiver** is the knowing relinquishment of a known right. Almost all legal rights can be waived. Most contracts contain a provision attempting to prevent the waiver doctrine from applying. This is impossible to do, because the provision preventing a waiver can itself be waived. In other words, it is impossible to write a contract or provision that cannot be waived by acts of the parties.[12]

EXAMPLE

Beneco entered into a contract with the National Park Service for maintenance work to be done at the Grand Canyon. Beneco entered into a subcontract with EPC to perform excavation and related work for utility line installation. The price of the subcontract was about $3.8 million.

Problems developed, and Beneco terminated EPC. EPC filed a proof of claim against Beneco seeking $2.8 million for work performed by EPC prior to termination of the contract. This amount included $1.3 million for work performed under the subcontract and $1.5 million for additional or changed work.

Beneco claimed that EPC was not entitled to the $1.5 million for additional or changed work because EPC did not follow the contract requirements for changed work. The contract contained the following clauses: "Beneco may order changes in the Work. No alteration, addition, omission or change shall be made in the Work or the method or

continued

manner of performance of the Work except upon the *written change order* of Beneco....
The contract may not be changed in any way...and no term or provision hereof may be
waived by Beneco except in writing signed by its duly authorized officer or agent."

The court said, "Utah has long recognized that 'parties to a written agreement may
not only enter into separate, subsequent agreements, but they may also modify a written agreement through verbal negotiations subsequent to entering into the initial written agreement, even if the agreement being modified unambiguously indicates that any
modifications must be in writing.' Thus, it is at least possible that the parties orally modified their agreement at some point despite the requirement that all modifications be in
writing. If the parties did modify their agreement regarding changed work, then EPC's
claims for that work are not automatically barred."

Sufficient admissible evidence in the form of letters and oral conversations existed
to require the matter to be resolved by a jury. It "is a question of fact whether the parties modified the requirements regarding changed work during the course of their
relationship."[13]

Constructive Change

A **constructive change** is owner conduct that amounts to a change though the owner refuses
to recognize it as such. Contractors are entitled to compensation for constructive changes
to the contract. "A constructive change occurs where a contractor performs work beyond the
contract requirements, without a formal order under the Changes Clause, either due to an
informal order from, or through the fault of, the government."[14] A constructive change has
the effect of requiring the contractor to perform more work than outlined in the original contract. This term developed in federal contract law but has bled over into state
contract law. It is applied in the following cases:

- The plans or specifications are defective and require the contractor to perform extra work to correct the defects.
- The owner's representative misinterprets the contract and requires additional work.
- The owner's representative rejects work that is actually in compliance with
  the plans and specifications.

Defective Plans and Specifications

If the owner provides the contractor with defective plans or specifications, and the contractor incurs a cost to repair the work or do additional work to remedy the defects, the
contractor is entitled to compensation.

Stonebrook entered into a contract with Matthews for construction of a new residential neighborhood. Matthews performed what it claimed was additional work at the Stonebrook residential neighborhood in order to repair some portions of the roads due to Stonebrook's faulty road design. Stonebrook argued that the work was "warranty work" related to the original contract. After review of the road design and other evidence, the court agreed with Matthews: the design of the road was faulty and required additional work to correct the faulty designs. The contractor was entitled to additional compensation.[15]

Misinterpretation of Contract or Rejection of Conforming Work

If the owner misinterprets the contract in such a way as to require additional work or rejects conforming work, the owner has made a constructive change to the contract. The contractor is entitled to damages.

Ace entered into a contract with the U.S. Army Corps of Engineers to construct a runway and related improvements at an Army airfield. A specification in the plan provided for two different testing methods to determine whether or not the finished concrete was sufficiently smooth and level: straightedge testing and profilograph testing. Profilograph testing is normally done on road construction and is more expensive than straightedge testing. Ace planned on using the straightedge testing method. However, the Army *required* Ace to use the profilograph testing method. Forcing the contractor to use the more-expensive method was a constructive change, and the contractor was entitled to the increased costs associated with the profilograph testing method.[16]

■ Destruction of the Contract Theory

If the owner makes material changes to the original contract, the original contract is destroyed, and the contractor is paid a reasonable sum for the work done; the contractor is not paid the sum from the destroyed contract. This rule is a summary of the federal law of cardinal change and the state law of abandonment of the contract.

The **destruction of the contract theory**[17] is an outgrowth of other legal theories. If the owner prevents the contractor from performing the contract, the owner breaches the contract. Insisting on substantial changes in the work has been considered a form of owner prevention of the contractor's performance and thus a breach of the contract by the owner.

Federal Law

A **cardinal change** "occurs when the government effects an alteration in the work so drastic that it effectively requires the contactor to perform duties materially different from those originally bargained for."[18] A cardinal change is considered a material breach of the contract[19] and relieves the nonbreaching party from performance.[20] In effect, the contract no longer exists; it has been destroyed.[21] When this happens, the contractor is still paid, just not under the original contract. The contractor is paid a reasonable sum. Contractors never sue using this theory if the original sum is greater than the reasonable sum. Contractors use this theory when the profit realized on the job is greatly reduced due to inefficiencies caused by the changes.

EXAMPLE

Alliant entered into a contract with the United States Army to demilitarize 24,000 bombs. The contract contained an option clause allowing the Army to increase the number of bombs by another 24,000. The contract also contained a monthly schedule or rate at which the bombs were to be demilitarized.

The contracting officer exercised the option to demilitarize the additional 24,000 bombs but at a rate far in excess of the rate called for in the contract. Alliant refused to demilitarize the bombs. The government sued Alliant for breach of contract, but the court held that no contract existed between the parties because it had, in effect, been destroyed. "If the government orders a 'drastic modification' in the performance required by the contract, the order is considered a 'cardinal change' that constitutes a material breach of the contract....The consequences of such a deviation from the proper terms of the option exercise are that the option clause imposed no obligations on Alliant and that its refusal to perform the option did not constitute a breach of the option clause."[22]

EXAMPLE

This doctrine is rarely employed as shown by this case. The contractor, PCL, entered into a contract to construct a visitor's center and parking lot at Hoover Dan in Nevada. Because of the geography of the site and unknown subsurface conditions at the time of the bid, the contract contained provisions for modification of the plans and specifications as excavation was performed. This case details over 30 pages of design and construction problems encountered on this extremely difficult project, but the original contract was never destroyed. "The contract itself explicitly provided that discrepancies, omissions, conflicts, and design changes would or likely would arise, and that the parties would address such issues during contract performance. This concept was part of the fundamental nature of this contract and generated part of PCL's duties under the contract. The fact that there were discrepancies, omissions, and incomplete contract drawings requiring additional work is consistent with, and not a deviation from, the 'nature' of PCL's contract." Given the nature of the contract, 356 changes was not "an inordinate number of changes or accumulatively constituted a cardinal change to a contract of this magnitude." The contractor was denied relief under the cardinal change doctrine. This decision was upheld on appeal.[23]

State Law

The term "abandonment of the contract" is more commonly used in state law to cover this situation, although it is also used in a variety of other situations, such as abandoned property, which can cause confusion.[24] Another theory used to achieve destruction of the contract is the material breach theory. "If a breach constitutes a material failure of performance, then the nonbreaching party is discharged from all liability under the contract...a party who has materially breached a contract may not complain if the other party refuses to perform its obligations under the contract."[25] A recent Nevada Supreme Court case called this concept a "cardinal change/abandonment/quantum meruit claim."[26]

Some courts come to the *effect* of destroying the contract but do not use any of these theories when weighing the merits of a case and deciding in favor of a contractor or subcontractor who has been harmed by numerous changes to the contract.

EXAMPLE

Havens, the general contractor, subcontracted the installation of ductwork on a large project to the subcontractor, Randolph. Havens was to manufacture and provide the ductwork, and Randolph was to install it. The contract contained the schedule for the installation and the price to be paid to Randolph. Havens was not able to manufacture the ductwork per the schedule but provided it to Randolph much later than scheduled. Partway through the performance of the contract, Havens sent Randolph a change order, changing the schedule of installation and including a "no damage except time extension for delay" clause into the contract. This "no damage" clause stated that Havens did not have to pay Randolph any money for the delay caused by late delivery of the ductwork, only grant Randolph additional time. Randolph objected.

Because of the late delivery the ductwork, Randolph needed additional time to install it, because the building was in a much later stage of completion. This necessitated cutting through and repairing walls, ceilings, rerouting ductwork, and other additional work to get the ductwork properly installed. Thus, Randolph incurred greatly increased costs to install the ductwork. The court ruled that Randolph was entitled to all its reasonable costs in installing the ductwork and to a reasonable profit.[27]

All of these theories have the same or very similar result of destroying the construction contract for excessive changes. When the contract is destroyed, the contractor is entitled to the reasonable value of the work, not the contract price.

■ Delays and Acceleration

In the law, delays are always defined from the contractor's perspective. An **inexcusable delay** is one for which the contractor has no excuse, and the contractor is responsible for paying the owner's damages. An **excusable delay** is one for which the contractor has a legal excuse and is excused from completing the project by the due date. If an excusable delay occurs, the contractor must absorb its own damages, and the owner must absorb its own damages. A **compensable delay** is a delay caused by the owner and for which the contractor is entitled to damages.

Inexcusable Delay

As a general statement, a contractor is required to fulfill its contract according to the terms including the date for completion. Failure to complete the project by the completion date is a breach of the contract, and the owner is entitled to any damages incurred.

EXAMPLE

The contract required the completion of a commercial building by February 1. The contractor was delayed because of personnel problems, subcontractors not performing as scheduled, and rain. The contractor breached the contract, and the owner was entitled to damages.

EXAMPLE

Murdock, a masonry subcontractor, was awarded the masonry subcontract on a prison project. It had no experience with the amount of reinforcing needed in a prison masonry project as compared with the types of projects it normally handled. It took much longer to complete the job than Murdock had anticipated. This is an inexcusable delay. The court said, "The prison was more difficult to build than the usual building with four straight walls....Murdock's estimate grossly understated the amount of time and effort required to perform the masonry work for a unique project, a risk that falls squarely on the shoulders of [Murdock]."[28]

MANAGEMENT TIP

Contractors should try and include a clause in the contract limiting their damages for delay. Limited damages for delay and liquidated damage clauses are discussed in Chapter 14.

Excusable Delay

An excusable delay is a delay for which neither the contractor nor the owner is required to pay the other damages. When an excusable delay occurs, each party must absorb its own damages. A list of excusable delays may be detailed in the contract. In addition, the law in many jurisdictions declares the following delays excusable: acts of God, strikes, acts of government, or acts of pubic enemies. Different jurisdictions may have additional excusable delays recognized by the law.

If both parties contribute to a delay, "neither can recover damage, unless there is in the proof a clear apportionment of the delay and the expense attributable to each day."[29] In other words, if both parties contribute to a delay, and it cannot be determined how much of the delay is attributable to each party, neither may recover. This is a form of excusable delay, although courts do not label it as such.

Acts of God include earthquakes, tornadoes, hurricanes, floods, unusual rain or weather conditions, and often fire. Acts of government would include changes in zoning, regulations, or other laws applicable to the project. Acts of public enemies include acts related to war and, in some jurisdictions, criminal acts.

Compensable Delay

A list of compensable delays may be detailed in the contract or implied into the contract. For example, if the contract has a Type 2 or unforeseen site condition clause, the risk of delay caused by unforeseen site conditions rests on the owner. Unforeseen site conditions are discussed in Chapter 2.

EXAMPLE

Kostmayer, a marine contractor, was awarded a contract by the government to build improvements and refurbishments to a drainage pumping station. At one point, Kostmayer was driving steel sheet piles with a large crane located atop a barge. The crew experienced electrical current or shock, a safety hazard. Work had to be stopped to find the source of the electrical current. It was determined that the problem was the existence of some radio towers about 2.5 miles from the site. The metal in the 180-foot crane acted as a giant radio antenna, collecting and focusing the electrical energy from three radio station transmission towers in the distance. Kostmayer immediately hired several experts to aid it in ameliorating the problem, and eventually a complex arrangement of nylon straps and slings was used to insulate the crane and break the electrical circuit in the boom antenna. However, the project was closed for 27 days. The contractor submitted a change order for approximately $100,000 for repair costs to the crane, costs of the nylon pulley system, and costs of 27 days of delay. The government's engineers recommended approval of the change order, but the government refused.

The contract contained the following two provisions:

"However, the contractor shall not be charged with liquidated damages or *any excess cost for delay in* starting or *completing the work* or in making deliveries of materials *when the said delay is due to unforeseeable causes beyond the control of the Contractor and without fault or negligence on his part* [emphasis added]."

"When, for the proper prosecution of the contract, work becomes necessary that has not been provided for in any clause of the contract, the Engineer will issue an order, and the Contractor shall perform the work stated in the order. Such work, frequently called 'Extra Work' may be paid for in any or all of the following ways as determined by the Engineer in each case..."

The court held that the contractor was entitled to damages, because the contract clauses placed the risk of the events on the owner.[30]

In addition, the laws in many jurisdictions have found the following to be compensable delays: delay in turning over the site, restricted site access, changes to the contract, constructive changes, delay in approval of shop drawings, failure to coordinate several prime contractors, failure to divulge information, unreasonable delay in shop drawing approvals, refusal to make timely payments, owner breach of a contract term, including implied terms,[31] and any unreasonable act.[32] The contractor is not entitled to damages for all owner-caused delays, just unreasonable owner-caused delays. Construction projects may be delayed by the owner and the owner is only liable to the contractor for *unreasonable* delays.

The following is an example of a jury instruction regarding this issue:

"You are instructed that in order to recover for delay damages you must find that [contractor's] performance was delayed for an unreasonable period of time due to the act or omission of [the owner].

You are instructed that simply showing that the contract took longer to finish than the allotted time is insufficient to show that [the owner] caused some delay or that [the contractor] suffered damages."[33]

Owner's Attempts to Avoid Delay Damages

Owners have developed the following contract clauses to relieve them of liability for at least some of their compensable delays: no damages for delay clauses, limited damages for delay clauses, clause requiring contractor to give notice of owner-caused delay, and waivers of delays. No damages for delay and limited damages for delay clauses are discussed in Chapter 14.

The law generally requires parties to give notice of acts causing damage as soon as possible in order to mitigate damages. Many contracts have provisions requiring the contractor to give the owner notice of delays, and a failure to do so will prevent the contractor from collecting its damages.[34] Failure of the contractor to give the owner notice could be construed as a waiver of the contractor's right to damages.

All of these clauses or theories will aid the owner in limiting its delay damages; however, it must always be remembered that the law will not enforce an onerous clause or any clause that has an onerous result.

Acceleration

An acceleration of the contract occurs when the owner decreases the time allotted for completion of the project or increases the work but *not* the time allotted for the completion of the project. The contractor is entitled to compensation for damages related to an acceleration of the project. In the second type of acceleration, increase in work but not increase in time, it may first be necessary to determine if the scope of the work has changed. See Chapter 9 for the rules used to determine if the scope of the contract has been changed.

Cases involving acceleration are rare. One court did describe that a "constructive acceleration claim has five elements: (1) the contractor experienced an excusable delay entitling it to a time extension, (2) the contractor properly requested the extension, (3) the project owner failed or refused to grant the requested extension, (4) the project owner demanded that the project be completed by the original completion date despite the excusable delay, and (5) the contractor actually accelerated the work in order to complete the project by the original completion date and incurred added costs as a result."

James was awarded a $10 million highway construction project by the state. The contract contained a completion date of October 30 with an incentive payment plan for payments to James of $15,000 per day for each day, not to exceed 30 days, that the work was completed on or before October 30. The roadway was opened to traffic on October 27.

The Department of Transportation agreed that it owed James $45,000 under the incentive plan. James argued that it was entitled to an additional 10 days of incentive payments, because the Department of Transportation had refused to grant James additional time to complete the project, despite adding work to the project and an Act of God: abnormally severe weather. Both of these added 10 days to the completion time. The evidence supported James, and he was entitled to a total of 13 days of incentive payments, because the owner had accelerated the project by demanding that same amount of work be done despite contract clauses granting the contractor a time extension.[35]

■ Endnotes

1. Example based on *Hall Contracting Corp. v. Entergy Services, Inc.*, 309 F.3d 468 (8th Cir. 2002).

2. Example based on *U.S. (for McFadden) v. FSEC, Inc.*, 2001 U.S. Dist. LEXIS 24201 (2001).

3. *Wright Lining & Constr. Co., Inc. v. Tully Constr. Co., Inc.*, No. 00-7436, 2001 WL 121863 (2nd Cir. 2001).

4. *Professional Const. Services, Inc. v. Lee M. Marcello Contractor, Inc.*, 550 So.2d 968, 971 (La. App. 1989), writ denied, 556 So. 2d 36 (La. 1990).

5. Example based on *Zack Constr. Co., Inc. v. Petroccitto*, 2000 Del. C.P. Lexis 100 (Del. Trial 2000).

6. Example based on *Aqua Pool Renovations, Inc. v. Paradise Manor Community Club, Inc.*, 880 So. 2d 875 (La. App. 2004).

7. *Bowden v. Grindle,* 651 A.2d 347, 351 (Me. 1994).

8. *Abington Constr., Inc. v. Madison Paper Industries*, 2000 U.S. App. Lexis 4454 (1st Cir. 2000).

9. Quoted in *The Lasalle Group, Inc. v. Crowell*, 2006 U.S. Dist. Lexis 90016 (D.E.Mi. 2006).

10. Abington Constr., Inc. v. Madison Paper Industries, 2000 U.S. App. Lexis 4454 (1st Cir. 2000).

11. Conner Brothers Constr. Co., Inc. v. U.S., 65 Fed. Cl. 657 (Ct. Cl. 2005).

12. Jhong C. P. *Effect of Stipulation, In Public Building or Construction Contract, That Alterations or Extras Must Be Ordered In Writing*, 1 A.L.R.3d 1273. *Effect of stipulation, in private building or construction contract, that alterations or extras must be ordered in writing*, 2 A.L.R.3d 620. (2004).

13. *U.S. v. Travelers Casualty & Surety Company of America*, 423 F. Supp. 2d 1016 (D.C. Az. 2006) internally quoting *R.T. Nielson Co. v. Cook,* 2002 UT 11, 40 P.3d 1119, 1124 n.4 (Utah 2002).

14. *Navcom Def. Elecs., Inc. v. England*, 53 Fed. Appx. 897, 900 (Fed. Cir. 2002) (citing *Ets-Hokin Corp. v. U.S.*, 190 Ct. Cl. 668, 420 F.2d 716, 720 (Ct. Cl. 1970)), *Len Co. & Assocs. v. U.S.*, 181 Ct. Cl. 29, 385 F.2d 438, 443 (Ct. Cl. 1967).

15. Example based on *Ex parte Stonebrook Development, L.L.C.*, 854 So. 2d 584 (Ala. 2003).

16. Example based on *ACE Constructors, Inc. v. U.S.*, 70 Fed. Cl. 253 (Ct. Cl. 2006).

17. Material in this section is a summary of the author's article entitled, *Destruction of the Contract through Material Changes*, 2 International Journal of Construction Education and Research, 43–51 (2006).

18. *Ebenisterie Beaubois Ltee v. Marous Bros. Constr., Inc.,* No, 02-CV-985, 2002 U.S. Dist. LEXIS 26625 (N.D. Ohio, 2002).

19. *Air-A-Plane Corp. v. U.S.*, 187 Ct. Cl. 269, 408 F.2d 1030, 1033 (Ct. Cl. 1969). *General Dynamics Corp. v. U.S.*, 218 Ct. Cl. 40, 585 F.2d 457, 462 (Ct. Cl. 1978). *Allied Materials & Equip. Co. v. U.S.*, 215 Ct. Cl. 406, 569 F.2d 562, 563-64 (Ct. Cl. 1978).

[20] *Stone Forest Indus., Inc. v. U.S.,* 973 F.2d 1548, 1552 (Fed. Cir. 1992).

[21] *PCL Constr. Servs. v. U.S.,* 47 Fed. Cl. 745, 806 (2000). *PCL Constr. Servs. v. U.S.,* 96 Fed.App. 672, 2004 U.S.App. Lexis 6706 (Fed.Cir. 2004). *AT&T Communications, Inc. v. WilTel, Inc.,* 1 F.3d 1201, 1205 (Fed. Cir. 1993).

[22] *Alliant Techsystems, Inc. v. U.S.,* 178 F.3d 1260 (Fed. Cir. 1999).

[23] *PCL Constr. Servs. v. U.S.,* 47 Fed. Cl. 745, 806 (2000). *PCL Constr. Servs. v. U.S.,* 96 Fed.App. 672, 2004 U.S.App. Lexis 6706 (Fed.Cir. 2004).

[24] *O'Brien & Gere Technical Services, Inc. v. Fru-Con/Fluor Daniel Joint Venture,* 380 F.3d 447; 2004 U.S. App. LEXIS 18100 (8th Cir. 2004).

[25] *In Re: Cornell & Company, Inc.,* 2000 Bankr. Lexis 357, p. 23 (E.Dis.Penn.2000).

[26] *J. A. Jones Construction Co. v. Lehrer Mcgovern Bovis, Inc.,* 89 P.3d 1009, 120 Nev. Adv. Rep. 32, 2004 Nev. Lexis 39 (Nev. 2004).

[27] *Havens vs. Randolph,* 613 F.Supp. 514 (D.C. Mo. 1985).

[28] Example based on *Murdock & Sons Constr., Inc. v. Goheen General Construction, Inc.,* 461 F.3d 837 (7th Cir. 2006).

[29] *Blinderman Constr. Co., Inc. v. U.S.,* 695 F.2d 552, 559 (Fed. Cir. 1982) (quoting *Coath & Goss, Inc. v. U.S.,* 101 Ct. Cl. 702, 714-15) (1944).

[30] Example based on *Kostmayer Constr., Inc. v. Sewerage & Water Board of New Orleans,* 943 So. 2d 1240 (La.App. 2006).

[31] The case of *Renda Marine, Inc. v. U.S.,* 66 Fed. Cl. 639 (2005) contains detailed discussion and citation of the implied terms concept.

[32] *L. L. Hall Constr. Co. v. United States,* 379 F.2d 559 (Ct. Cl. 1966). *L. L. Hall Constr. Co. v. U.S.,,* 379 F.2d 559 (Ct. Cl. 1966).

[33] *Metropolitan Transit Authority, v. Pyramid Constr. Inc.,* 2001 Tex. App. Lexis 7507 (Tx.App. 2001) (unpublished opinion).

[34] *Seaboard Lumber Co. v. U.S.,* 45 Fed. Cl. 404 (Ct. Cl. 1999).

[35] Example based on *James Cape & Sons Co. v. Illinois,* 52 Ill. Ct. Cl. 322 (2000).

CHAPTER **9**

Scope of the Contract

■ Introduction

Construction contracts are some of the most complex in the legal world. "[T]he process of interpreting a series of written documents as complex as the intricate assortment of plans, drawings, and specifications covering a major industrial construction project is not merely the simple task of picking out and then applying words literally in their normal connotation..."[1]

MANAGEMENT TIP

Because it is extremely difficult to predict the outcome of a scope case, these types of cases should be settled.

Issues of scope arise whenever the parties to a contract disagree as to the meaning of the words, oral or written but usually written, of the contract. Judges and lawyers usually refer to this type of issue as one of "contract ambiguity" or "contract interpretation." However, in the construction industry, this type of issue is referred to as a "scope" issue.

EXAMPLE

The contractor entered into a contract with the U.S. Navy to remove and replace a condensate and steam system. The contractor claimed that the language of the contract did not require the installation of conduit sleeves in existing manholes, only in new manholes installed per the contract. The Navy claimed that the contract language required installation of conduit sleeves in both existing and new manholes. This is a scope issue, because the parties disagree about what the contract says.[2]

A simplified approach to solving scope issues is presented. First, it must be determined if the contract is ambiguous (issue 1). Second, if the contract is ambiguous, one of the two meanings must be chosen as the correct meaning of the contract (issue 2).

Scope issues are extremely common as all languages are ambiguous to some extent, and any word or phrase might have different meanings to different people. It is difficult to predict the outcome of a scope issue, and parties involved in a scope issue should think seriously of trying to settle these issues out of court or arbitration, as the likelihood of winning is always problematic. In one complex construction insurance case, the court said, "It is uncertain whether this tangled dispute, with its idiosyncratic documentation, has a 'right' answer in any meaningful sense, and given the litigation expenses, it is a mystery why it was not settled. Still, it must now be decided."[3]

If the words of the contract are clear, avoid scope arguments, and look for other areas of the law to help.

EXAMPLE

Crosswaite Construction submitted a bid on a project. The bidding documents stated, "The bid may not be withdrawn..." and also required the contractor to purchase a bid bond. Prior to the opening of the bids, Crosswaite noticed a large error in the bid and asked that its bid be withdrawn. The owner refused. At the bid opening, Crosswaite's bid was the lowest, and the owner awarded the contract to Crosswaite.

continued

If Crosswaite attempted to argue the meaning of the bid documents, it would lose. The bid documents were clear that the bid could not be withdrawn. The contract language was not ambiguous, and under a scope analysis, the contractor would lose. The contractor would have to find another area of the law if it had any hope of winning. Using the law of mistakes in bids, the contractor had a good argument that would give it relief despite what the contract said. This legal argument is discussed in detail in Chapter 2.

■ Sidebar: Oral Contracts

Scope issues do not usually arise in a disagreement involving an oral contract. The most common issues are issues of fact, which are determined by the trier of fact based on the credibility of the evidence and not by the law. In oral contracts, if the parties do not agree on *what* terms exist in their oral agreement, the issue is one of fact, and the trier of fact will decide what terms exist in the contract. Issues of fact are discussed in more detail in Chapter 1 in the section entitled "Issues of Law and Issues of Fact."

EXAMPLE

Fallon Painting and Tarif, the owner, had an oral agreement for the painting of Tarif's home. Fallon painted the home. Tarif said the contract was for two coats of paint and that Fallon was also supposed to paint the steps. Fallon said that the contract was only for one coat, and the steps were not included. This is a factual issue, and the jury had to decide what the contract said based on the credibility of the evidence.

If the parties disagree about what the terms of their oral contract *mean*, then an issue of scope arises, and the rules in this chapter are used to determine the meaning of the words.

EXAMPLE

Fallon Painting and Tarif, the owner, had an oral agreement for the painting of Tarif's home. The parties both testified that the Fallon agreed to do a "top-notch job." Fallon painted the home with one coat of paint. Tarif claimed that a "top-notch job" required two coats of paint. Fallon claimed that a "top-notch job" could be done with only one coat of paint. This is an issue of contract ambiguity, and depending on how a particular jurisdiction handles this type of issue, the rules discussed in this chapter help the judge and jury decide what the words "top-notch job" mean.

■ Issue 1: Is the Contract Ambiguous?

The first issue in a scope problem is to determine if the contract is ambiguous. This is always a legal issue decided by the judge. A contract is **ambiguous** if it is capable of having two or more relatively reasonable meanings.[4] The rule is *not* that the contract is ambiguous if it is capable of having two meanings. The rule is *not* that if the parties disagree about the meaning, the contract is ambiguous. Each of the proposed meanings must be *reasonable*. If reasonable persons, who are not usually the persons involved in the dispute, may fairly and honestly differ as to the meaning, then the contract is ambiguous.[5] "However, contracts are not necessarily rendered ambiguous by the mere fact that the parties disagree as to the meaning of their provisions...That the parties disagree with a specification, or that a contractor's interpretation thereof is *conceivable*, does not necessarily render that specification ambiguous....A contract is ambiguous if it is susceptible of two different and reasonable interpretations, each of which is found to be consistent with the contract language..."[6] The emphasis on the word *conceivable* is important because *reasonable* does not mean *conceivable*.

EXAMPLE

At a prebid meeting, the contractor mentioned to the architect that she intended to use plastic vent covers. The architect did not dispute this, probably because the architect did not recall what the specification for vent covers called for.

The actual specification called for "metal vent covers," which are more costly than plastic covers. The contractor installed plastic covers. The owner kept an amount of the retainage to cover the cost of replacing the vent covers with metal covers, and the contractor filed a claim.

The contractor claimed that she was entitled to install plastic vent covers because of the conversation that was not disputed by the architect; the owner claims the contract required metal vent covers. The contract is *not* ambiguous, and the contractor cannot win because the term "metal vent covers" cannot reasonably mean "plastic vent covers." The conversation is irrelevant.

Ordinary Meaning Rule

This rule is so obvious it should not have to be mentioned; however, since it comes up so often in the case law, it must be included as one of the most important rules of contract interpretation. The rule is that the ordinary meaning of words prevails. Bizarre, twisted, odd, perverse, or imaginative but highly unlikely meanings of words are not going to be adopted by the law.

EXAMPLE

"We find that the plain and ordinary meaning of 'liens arising out of the Work' includes mechanic's and materialmen's liens..."[7]

"Further, the policy defines 'your work' to include 'any work that you're performing or others are performing for you.' (Policy at 20). 'Your work' means 'your work' or work you are doing."[8]

EXAMPLE

Dement contracted with the U.S. Department of Transportation to construct approximately 3.6 miles of the Natchez Trace Parkway. Dement and Ladner entered into a subcontract for "ALL TESTING, FIELD AND LAB WORK EXCLUDING CONCRETE TESTING FOR NINETY-FIVE THOUSAND DOLLARS ($95,000.00)." The subcontract contained more detail on the types of testing and scheduling not relevant to the dispute.

At the end of the project, Ladner requested an additional $40,000 for testing. The contractor refused to pay. Ladner sued. The court said, "Giving the term 'all' its usual, natural, and ordinary meaning as we are obligated to do under Tennessee law, we are compelled to rule in [the contractor's] favor [and dismiss Ladner's claim]. The word 'all' means 'all.' The phrase 'all the testing for $95,000' means the subcontractor will do all the testing for $95,000. The subcontractor is not entitled to an additional sum for testing."[9]

It is not illegal to have bizarre, twisted, odd, perverse, or imaginative but highly unlikely meanings for words in a contract, but if the parties want those meanings, they need to spell them out in the contract.

EXAMPLE

The government and the contractor entered into a contract for the demolition of an old commissary and construction of a new one. The contractor had the salvage rights to the old commissary. At the time the contract was entered into, the old commissary was still in operation and contained large quantities of goods being sold to military personnel and their families. The government was supposed to turn over the old commissary to the contractor on June 1, but the old commissary was still in operation on that date because of delays, all of which were caused by the government. However, the contractor was able to do other work on the contract during that time and was not delayed in the performance of the entire contract.

The contractor however made a claim to the all of the contents of the old commissary as of June 1, saying that it had the salvage rights as of that date, and since the commissary was full of goods, it was entitled to those goods or the value of those goods.

continued

The contractor's claim was rejected. "But we can find no language in the [contract] before us that can be reasonably construed to mean that [the government], in the event of a failure timely to deliver the building, agreed to vest in the [contractor] the title to all the merchandise in the building. Such a penalty for late delivery is so bizarre that there would have to be a very clear and express provision to that effect to permit recovery as [contractor] asserts."[10]

MANAGEMENT TIP

Courts are too often faced with such ludicrous arguments. However, since the court system is open to all for all of their claims, even stupid ones, these types of cases are too often found in the records.

Actual Agreement of the Parties

Courts always approach scope issues with the desire to uphold the actual agreement of the parties and not necessarily what is written down on paper. Parties to contracts tend to be more enamored of the written words in their contracts than judges. Judges may toss out the written words of a contract. "The purpose of interpreting a contract is, of course, to accomplish the intentions of the parties" at the time they entered into the contract.[11] The words written down on paper are not necessarily the agreement of the parties; they may just be words on paper. "But paper and ink possess no magic power to cause statements of fact to be true when they are actually untrue."[12] However, the difficulty in actually determining the intentions of the parties may induce many courts to impose a "just result" in these types of cases.[13] This tendency to impose a just result on the parties may account for at least some of the inconsistency in this area of the law.

EXAMPLE

Owner sent Hakan, the low bidder on a sewer construction project, a written contract for signature. Hakan signed and returned the contract but with a letter attached saying that his acceptance was conditional on an extension of the time for completion. The owner removed the letter, returned a signed copy of the writing to the contractor, and claimed a binding contract existed. Since there was no intent on the part of the contractor to enter into the contract returned by the owner, no contract was formed. The contractor had made a counteroffer, one the owner did not accept.[14]

Northern entered into a contract with Alaska Housing to completely demolish 15 buildings. The following provisions existed in the contract:

Scope of the Work: The removal and satisfactory disposal of all buildings.

General Description of the Work: The buildings and foundations will be completely razed.

Section 02055 (salvage provision): Removed items will become the property of the Contractor....The disposal of the building materials is at the contractor's discretion.

At a prebid conference, it was emphasized that all of the buildings were to be razed. Northern Construction was the low bidder, but prior to being awarded the project, the Procurement Officer said that if Northern did *not* intend to demolish the buildings, it would *not* be awarded the contract. Northern agreed that it intended to completely demolish the buildings.

Northern admitted that at the time it bid on the contract, it intended to completely demolish the buildings. However, at some point after it was awarded the contract, Northern found out that other bidders had intended to sell parts of the buildings. Northern looked into this and found it a profitable course of action. However, when Alaska Housing found out that Northern was not going to demolish the buildings but in fact sell them, it terminated the contract and took back the buildings. Eventually, Alaska Housing decided to do exactly what it had said Northern could not: sell the buildings. This turned out to be a profitable course of action. Rather than paying Northern or some other contractor $150,000 to demolish the buildings, Alaska Housing made $200,000 by structuring the contract so that the new contractor could do whatever it wanted with the buildings as long as the contractor turned over a clean site to the owner.

The court held that Northern had breached the contract with Alaska Housing to destroy the buildings, thus allowing Alaska to terminate the contract and take back control of the buildings. Once the buildings were the property of Alaska Housing again, it could sell them if it wanted.

The clause requiring razing the buildings, coupled with the clause that said the contractor could dispose of the buildings at its discretion, was ambiguous but "[a]s a matter of law, 'when, at the time of formation, the parties attach the same meaning to a contract term and each party is aware of the other's intended meaning, or has reason to be so aware, the contract is enforceable in accordance with that meaning [and not in accord with the written words of the contract].' Because it is the intent of the parties that governs the court's interpretation of a contract, a party will thus be "bound not by the outer limits of an ambiguous document, but by the terms agreed upon by the parties." The key question in this case was this: what meaning did the parties attach to the terms of the demolition contract at the time the contract was made? It was actually clear from the evidence what the parties understood the contract to mean at the time it was entered into.[15]

Complete, Integrated Contracts

If a contract is final, complete, integrated, and not ambiguous, parol evidence cannot be used to modify the contract. "An integrated agreement supersedes contrary prior statements..."[16] "[The] terms of a final and integrated written expression may not be contradicted by parol evidence of previous understandings and negotiations for the purpose of varying or contradicting the writing."[17]

> ### EXAMPLE
>
> Draft 1 of a contract said the contract price was $200,000 and contained details of the project. Draft 2 of the contract had a price of $220,000 and contained slightly different details of the project. Draft 3 of the contract had a price of $210,000 and contained a third set of details and a clause that the contract is complete and integrated. The parties signed Draft 3, and it became the contract. Later, the owner said that certain items in Draft 2 should have been included in Draft 3 but were mistakenly left off. Only the contract based on Draft 3 controlled, and Draft 2 could not be used to vary or modify the contract.

MANAGEMENT TIP

Although not always legally effective, every contract should contain a clause stating that it is the entire agreement of the parties and no outside terms or representations exist. See the form below.

FORM

This Contract represents the entire and integrated agreement of the parties, and no terms of the Contract exist except those contained herein, and this Contract supersedes all prior and contemporaneous negotiations and representations, either written or oral.

If a contract contains such a clause and a party believes a term has not been included in the contract, the contract should not be signed until the term is added. The term can be handwritten and initialed but should be included before signing.

EXAMPLE

A contract between the general contractor and the owner contained a provision for mandatory, binding arbitration. The contracts between the general contractor and the subcontractor did not. The subcontract contained a provision stating, "This Contract represents the entire and integrated agreement of the General Contractor and the Subcontractor, and no terms of the Contract exist except those contained herein and this Contract supersedes all prior and contemporaneous negotiations and representations, either written or oral."

The general contractor told the subcontractor that any disputes regarding the subcontract were also subject to binding arbitration. The general contractor's oral statements were not admissible to vary the terms of the subcontract; the subcontract was not subject to arbitration. If the contractor wanted that clause in the contract, the contractor needed to write it in.

Issue 2: Deciding Which of the Two Meanings Prevails

Once the first issue, "is the contract ambiguous?" has been resolved with the answer "yes," the next step is to determine which of the two reasonable meanings prevails. The process for resolving this issue is one of the most complicated and varied. A judge may decide, the jury may decide, or a combination of judge and jury may decide. Different states have different procedures.[18] If the matter is in arbitration, the arbitrator will decide if the contract is ambiguous and also which of the two competing meanings prevails.

The rules outlined in Table 9-1 are used to determine which of the two reasonable meanings prevails. It is not uncommon for different rules to produce conflicting results. In this situation, the decider weighs the results of the different rules and comes to a conclusion. It is unusual for a court to apply only one of the rules, although this is done in here to illustrate a particular concept. See Table 9-1 for a summary of the major rules used in interpreting contracts. The list of rules in this table is not exhaustive, and other rules exist. Jurisdictions may have different names for the rules listed.

TABLE 9-1

Resolving a scope issue	
Issue 1: Is the contract ambiguous?	
Contract ambiguity	A contract is ambiguous if it is capable of having two or more reasonable meanings.
Ordinary meaning	The ordinary meaning of words prevails.
Actual agreement of the parties	The actual agreement of the parties at the time they entered into the contract controls, not necessarily the written words of the contract.
Complete, integrated contracts	If a contract is final, complete, integrated, and not ambiguous, parol evidence (defined below) cannot be used to modify the contract. It is enforced as written.

TABLE 9-1 (continued)

If the answer to issue #1 is yes, proceed to issue 2: **Which of the two reasonable meanings prevails?** Use the rules below and come to decision.

Name of rule	Simplified rule
Parol evidence	Evidence outside of the contract such as prior drafts, conversations, faxes, emails.
Parol evidence rule	Parol evidence can be used to clarify the meaning of an ambiguous contract. It cannot be used to vary the meaning of an unambiguous contract.
Incomplete agreements	Incomplete written agreements may be supplemented with parol evidence to prove provisions of the contract not included in the writings. Parol evidence cannot vary the written terms but can add terms not included in the writing.
Course of dealing, conduct of the parties, or past practice	The conduct of contracting parties is a strong indication of what the writing means.
Fraud, mistake, or illegality	Parol evidence of fraud, mistake, or illegality is always admissible even if the contract is complete and integrated or unambiguous. Such contracts may be void or voidable. See Chapter 7.
Patent ambiguity rule	A patent (obvious) ambiguity is resolved in favor of the party who drafted the contract. If both parties drafted the contract, this rule is inapplicable.
Custom and usage	Custom and usage in the trade can be used to clarify the meaning of a term.
Implied terms	Terms not expressly stated in the contract, but needed to fulfill the purpose of the contract, are implied in the contract.
Good faith	Every contract contains an implied covenant of good faith and fair dealing; that is, the parties must cooperate and not hinder the other party's performance.
Whole agreement	Contracts must be read as a whole, that is provisions cannot be taken out of context.
Order of precedence	Absent a contrary term in the contract, special conditions prevail over general conditions: handwritten terms prevail over typewritten provisions; typewritten provisions prevail over preprinted terms; words prevail over figures.
Latent ambiguity	A latent or hidden ambiguity is resolved in favor of the party who did not draft the document. If both parties drafted the document, this rule is inapplicable.
Clear expression of intention	The party drafting the contract has the duty to clearly express its intent.

TABLE 9-1 (continued)

Loopholes	
Onerous or uncon-scionable clauses	The law will not enforce an onerous or unconscionable clause even if it is unambiguous. An onerous or unconscionable clause is one the judge can-not enforce because to do so prevents her from sleeping at night.
No contract formed	All of the scope rules are bypassed if no contract exists. The party perform-ing work is paid under the doctrine of detrimental reliance. See Chapter 7.
Waiver	Any provision of a contract can be waived. Waiver is discussed in Chapter 7.

Parol Evidence Rule

Parol evidence is evidence outside of or extraneous to the words of the contract. Parol evidence, as shown in Figure 9-1, is any evidence outside of the "four corners of the contract." Examples of parol evidence include conversations, drafts of contracts, let-ters, industry standards, e-mails, and any evidence other than the actual words in the contract.

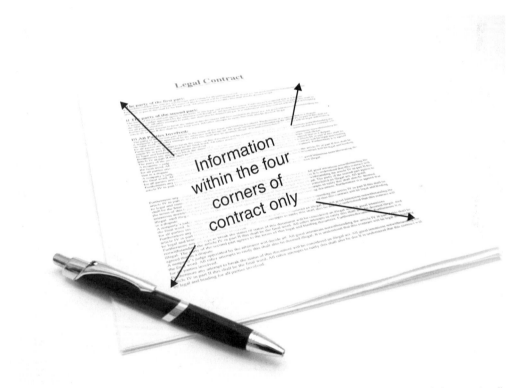

Image courtesy © istockphoto.com/swalls

FIGURE 9-1 Parol evidence is any evidence outside the four corners of the contract.

The word *parol* should not be confused with the word "parole," which is pronounced the same. **Parole** is a criminal law term used to describe a time period after a criminal has been released from jail but is still under the scrutiny of the criminal justice system. It is meant to be a time when a criminal is reintroduced into society as a contributing citizen. Crimes committed during this period are punished more harshly than crimes committed while not on parole.

The parol evidence rule is actually two rules:[19]

- Parol evidence *cannot* be used to *vary* the terms of a *complete and unambiguous* contract. This rule prevents the use of any parol evidence to determine if the contract is ambiguous; parol evidence cannot be used until the contract has already been determined to be ambiguous. "Where the language of the agreement is unambiguous, the intent must be ascertained solely from that language; extrinsic [parol] evidence may be only considered where the language is ambiguous."[20]
- Parol evidence *can* be used to *clarify* the terms of an *ambiguous* contract or to add terms to an *incomplete* contract. Once the contract has been determined to be ambiguous and/or incomplete, parol evidence can be used to aid in determining what the contract means.

If the contract is not ambiguous, the law does not allow parol evidence to create an ambiguity, show obligations not expressed in the contract, or contradict the express terms of the contract.

EXAMPLE

Diversified entered into a contract with the School District to construct several schools. The date on the contract required completion by August 31, 2000, but Diversified testified that during a prebid conference, the School District said it was acceptable for Diversified to complete construction up to 11 months later, and that this was the purpose behind a contract provision that imposed only nominal liquidated damages for the first 11 months after August 31, 2000. The School District pushed the contractor to finish by August 31, 2000, and the contractor sued for acceleration damages.

The court refused to allow the parol evidence, the conversation during the prebid conference, to modify the terms of the contract; the completion date was August 31, 2000 as clearly stated in the contract. No parol evidence was admissible to vary the express term of the contract.[21]

While the judge may prevent parol evidence from being presented to the jury or may refuse to give the parol evidence any credence when presented to the judge, the parol evidence will be reviewed by the judge. In *Still v. Cunningham*,[22] the court said extrinsic evidence may always be received by the judge on the interpretation of contract terms. Not all state courts would agree with the statement, but courts routinely review parol evidence and then decide how to treat it. Parol evidence is seldom ignored by the judge.

TEG entered into a contract with the U.S. Department of Housing and Urban Development (HUD) to perform asbestos abatement work on an old apartment complex. The contract between the parties contained the following provision:

Asbestos-containing materials applied to concrete, masonry, wood and nonporous surfaces, including, but not limited to, steel structural members (decks, beams, and columns), pipes and tanks, shall be cleaned to a degree that no traces of debris or residue are visible by the Observation Services Contractor.

TEG claimed that HUD breached the contract by requiring extraordinary and unnecessary cleaning of pores and cracks and requested an additional $4 million in damages. TEG argued that the term "surfaces" as defined in common usage dictionaries meant only the outer area and not pores and cracks. TEG claimed that the HUD inspector was fanatical about the removal of asbestos.

The court looked at the original draft of the contract, which contained the following language (emphasis added):

Friable materials applied to concrete, masonry, wood and nonporous surfaces, including but not limited to, steel structural members (decks, beams, and columns), pipes and tanks, shall be cleaned to a degree that no traces of debris or residue are visible. **Nonfriable materials** applied to concrete, masonry, [or] wood shall be cleaned until no residue is visible other than that which is embedded in the pores, cracks, or other small voids below the surface of the material.

The court said, "Although extrinsic evidence may not be used to interpret an unambiguous contract provision, we have looked to it to confirm that the parties intended for the term to have its plain and ordinary meaning." The plain meaning of the clause required removal of asbestos from pores and cracks, and the contractor was not entitled to any additional sums.[23]

In this example, the court reviewed the prior drafts of the parties' agreement and came to the conclusion that the final agreement was *not* ambiguous and the owner owed the contractor for changes.

Batzer, the contractor, and Boyer, the developer, had a partnership developing properties for construction and then lease to the U.S. Postal Service. After several years, they decided to terminate their partnership and divide the partnership assets. At the time the assets were divided, three of the buildings to be received by Boyer were incomplete. Batzer agreed to complete the construction of those three buildings as the contractor but was no longer the owner.

Boyer agreed to pay Batzer "a price equal to Batzer's cost including liability insurance costs" to complete the construction of the buildings. A handwritten note in the margin of the contract indicated that liability insurance costs were to be the sole item of overhead and profit paid to Batzer for work on the buildings. Both parties initialed the note.

continued

At the end of the construction, Batzer sent Boyer a final invoice for $95,000 for changes ordered by the U.S. Postal Service. Boyer refused to pay for the changes. Boyer claimed that the term "a price equal to Batzer's cost" was ambiguous and meant the original bid price only and he did not have to pay for changes. Batzer said the term "a price equal to Batzer's costs" meant his actual expenses to construct the buildings.

Boyer argued that he made the statement at the time the contract was signed that "Batzer's cost refers to the bid price," and Batzer remained silent. Later, after the evidence was presented, the judge stated that Boyer was not a credible witness and doubted that this statement had in fact been made.

The court reviewed three prior drafts of the Batzer and Boyer's partnership termination agreement, which were, in order of negotiation: (1) Batzer would be paid "for costs plus 8%"; (2) Batzer would be paid "for costs not to exceed the bid amount"; and (3) Batzer would be paid "per price to be agreed upon." The prior drafts supported Batzer's contention that the terms in the final agreement, "a price equal to Batzer's cost" plus the handwritten note that liability insurance costs were to be the sole item of overhead and profit, was unambiguous and Batzer should be paid for his costs to complete the project, including the change orders. The court agreed.[24]

It is common for parties to attempt to introduce parol evidence to vary the terms of an otherwise unambiguous contract. This is not usually allowed.

EXAMPLE

Utility Contractor entered into a contract with the government for the installation of improvements to the "Joe Creek" Canal Lining Project. The purpose of the project was to eliminate flood damage caused by severe rainstorms. The project was intended to collect all rainwater in the area and channel it through the city and into the Arkansas River. The contractor sought reimbursement for damages and repairs to part of the work caused by severe rain.

The Permits and Responsibility Clause of the General Provisions, paragraph 12, and the Damage to Work Clause of the Construction Special Conditions, paragraph 8, placed the risk of loss for all precompletion damage to permanent work on the contractor with the exception of damage caused by *flood* or earthquake. In addition, the contract contained a definition of **flood**. Technical provisions required the contractor "to prevent damage to the work during construction." When damage to the work was caused by high water levels but not amounting to floods as defined in the contract, the contractor tried to recover additional sums from the owner anyway.

The parol evidence the contractor requested the court to consider included Procurement Regulations (ASPR) 1-109 and the ECI 1-109, which contained a definition of **flood** different from that in the contract.

The court said "there is no need for the court to consider parol evidence as to [contractor's] intentions or interpretations concerning a contractual clause when it is clear and lucid....There is no ambiguity here. Consequently, parol evidence will not be

continued

considered." The contractor was not entitled to any additional sums, because the contract clearly placed the risk of damage for waters below the contract definition of flood on the contractor.

As is common, the contractor raised several other legal theories to try and get the parol evidence admitted. These included the following: (1) the government's design specifications relating to the concrete lining were defective and caused the damage; (2) the amount of rain constituted a differing site condition; (3) the government's nondisclosure of superior knowledge was a breach of contract; (4) mutual mistake because neither party expected any damage if the level of the waters remained below the definition of flood in the contract; and (5) the contractor's unilateral mistake coupled with the government's inadvertent misrepresentation regarding the possibility of flood damage at water levels below the definition of "flood" in the contract. All of these theories are valid, and the contractor obviously had an experienced construction lawyer working on the case. Though these theories work in certain situations, they were irrelevant here, as the contract clearly placed the risk of damage on the contractor.

Incomplete Agreements

Incomplete written agreements may be supplemented with parol evidence to prove provisions of the contract not included in the writings. Parol evidence cannot vary the written terms but can add terms not included in the writing(s).

EXAMPLE

The written contract between the engineers and the owner did not spell out in detail the services to be rendered. Parol evidence is admissible to prove the required services.[25]

EXAMPLE

Iber entered into a subcontract with J&B Steel for work related to the construction of a parking garage. Iber failed to pay J&B Steel certain sums, and J&B Steel filed a construction lien on the property. The only written communication between the parties was a purchase order dated December 4, 1989, which stated the price, $220,000, to be paid to J&B Steel. No completion date was indicated in the purchase order, but the purchase order did reference July 31, 1990 as the date through which J&B Steel was to supply a foreman experienced in the nature of the subcontract.

Three days prior to the date of the purchase order, the parties had had a telephone conversation. J&B Steel testified that it was the party's understanding from that conversation that at the stated price, J&B Steel was not obligated to perform the work beyond July 31, 1990.

continued

The issue was whether or not the purchase order was the entire and integrated agreement of the parties. If it was, the parol evidence rule would prevent the admission of the telephone conversation. If the purchase order was not the entire and integrated agreement, the telephone conversation could be admitted.

The purchase order consisted of five pages. The following sentence appeared at the top of the purchase order:

"ENTER THE FOLLOWING ORDER IN ACCORDANCE WITH THE TERMS AND CONDITIONS OF YOUR telephone proposal on December 4, 1989 by Terry Estes with Jerry Leander. Installation to begin as required by job progress."

It is true that the absence of a completion date does not necessarily mean the purchase order is incomplete. However, the reference to "telephone proposal" indicated that elements in that telephone proposal were part of the contract. Since those elements had not been included in the purchase order but were only referenced in it, the purchase order was not the complete, integrated agreement of the parties. The parol evidence was admissible.[26]

"The California courts usually take the traditional position that finality is determinable from the document itself. If on its face it purports to be a complete expression of the agreement, it is conclusively presumed to contain all of the agreed terms, and extrinsic evidence is excluded."[27] Incomplete or nonintegrated contracts can be supplemented by parol evidence. Because incomplete and nonintegrated contracts can be supplemented by parol evidence, it has become common for parties to include in their contracts a provision stating the written contract is the complete and integrated agreement of the parties and no outside agreements or terms exist. The court will still look to outside evidence to determine if the document is integrated.

MANAGEMENT TIP

If your contract contains the above clause and there is some additional point not in the contract, be sure to handwrite the term into the contract. Terms handwritten into the contract should be initialed by both parties. Handwritten terms are given more credence by the law than preprinted terms.

Letters of confirmation are unlikely to be considered complete integrated agreements and are subject to augmentation by parol evidence.

EXAMPLE

The general contractor and the subcontractor had several telephone conversations regarding the subcontract. The general contractor sent the subcontractor three letters on three different occasions, each of which stated, "This letter confirms our agreements by telephone." The court said that the letters, individually and collectively, were not the complete integrated agreement of the parties. Oral testimony concerning contract terms not contained in the letters, such as time of completion, insurance, and method of payment was admissible.[28]

EXAMPLE

Levy, the engineer, sent a letter agreement to Leaseway, the owner of the proposed project. The letter included a statement that his fee was to be 7% of the "net final cost of construction....The return of one signed copy, as noted below, will be a confirmation of your verbal authorization to proceed." A copy was signed and returned by Leaseway, but Leaseway did not approve the sets of plans and specifications drawn by the engineer. The engineer sued for compensation for his work. Leaseway stated that in the telephone conversation, the engineer had said the total cost of the construction would not exceed $300,000, but Leaseway could not get a construction bid lower than $400,000. Leaseway refused to pay for the plans, claiming that the engineer had breached a warranty. The owner's testimony was admissible to prove that the engineer had given a warranty that the project as designed could be built for $300,000."[29]

EXAMPLE

The government entered into a lumber contract with the West, the lumber supplier on a certain project. A contract provision said, "The contractor shall furnish said products when and as the government may make calls..." Later, the government terminated the contract without ever calling West. West asserted that the contract was a requirements contract and that the government had agreed to purchase all of its requirements for lumber for the project from West. The lumber supplier was permitted to show this by parol evidence of verbal conversations. "The key language [quoted above]...does not exclude the interpretation sought by [the lumber supplier]....The proper inquiry, then, is whether the writing reflects the entire agreement of the parties. The answer depends wholly upon the intent of the parties to the agreement, which must be determined, first, from the expressions in the contract, and if not there expressed, then from the conduct and conversations of the parties and the surrounding circumstances....Here the plain fact is that the question of whether the government is obligated to place calls with the appellant is simply not covered or dealt with in any way in the contract." The court held in favor of West.[30]

The court is not likely to use parol evidence to add terms to a contract that would normally have been included in a contract of that type if the parties had in fact agreed to such terms.

EXAMPLE

Fisher was employed by Jones to build homes in a planned subdivision. The parties had a written contract whereby Jones would pay Fisher $100 per week for three months and $25 per house sold. The subdivision originally planned by Jones contained 500 homes; however, after building 100 homes, Jones discontinued the project. Jones paid Fisher the $100 per week and the $25 per house sold, but Fisher sued, claiming that Jones breached the contract because Jones had told Fisher that he would employ Fisher for the building of 500 homes. The court refused to allow the parol addition to the contract that Fisher was to be employed for a minimum of 500 homes, stating this is the type of provision that, if it existed, would have been included in the contract.[31]

EXAMPLE

Angles entered into a demolition contract with Transit Authority for demolition of a certain building and removal of all materials on the site so that another building could be erected. No clause existed in the contract giving Angles the salvage rights but neither did the owner want any of the materials. Prior to turning over the site to Angles, some salvageable materials had been removed by either the Transit Authority, theft, or both. This reduced the profit that Angles expected to make.

Angles claimed that Transit Authority breached its contract with Angles by failing to safeguard the salvageable materials and/or removing some salvageable materials before Angles could do so. Angles claimed that the custom and usage of the trade required Transit Authority do the above, as it knew that part of a salvage company's profit was from salvageable materials.

The court dismissed Angles' claim against the Transit Authority and would not allow Angles' claim that the custom and usage of the industry required Transit Authority to protect the site from vandals or prevented Transit Authority from removing some material. Had that obligation been intended in the contract, it would have been in there.[32]

MANAGEMENT TIP

At the time a contract is being signed, if a party makes an oral representation, addition to the contract, or statement regarding what she believes is in the contract, these statements should be written into the contract or, at the least, followed up by letter and clarified. The follow-up letter may not be binding, but at least the parties will know to expect a problem and can plan accordingly.

Fraud, Mistake, or Illegality

Under the law of fraud, mistake, or illegality, contracts may be void or voidable. (see Chapter 7). Parol evidence of such fraud, mistake, or illegality is always admissible, even if the contract is complete and integrated or unambiguous.[33] The parol evidence is not seeking to vary the terms of the contract but to prove the fraud, mistake, or illegality. and therefore the parol evidence rule does not prevent its introduction.

Ball and Lewis wanted to form a joint venture and bid on road construction contract with the government however Lewis was not licensed in that state. Ball bid on and was awarded the contract, individually. Under state law, Ball was required to perform at least 50% of the work himself. However, to circumvent this requirement and the fact Lewis had no license, the parties entered into two "subcontracts": one for services and one labeled a "rental" agreement for road building equipment. The rental agreement had a statement that it was the final, integrated contract of the parties. In fact, the parties had an outside oral agreement that the "rental agreement" actually required Lewis to perform excavation and grading. The result of the two subcontracts was that Lewis, who was unlicensed, would be doing more than 50% of the work on the project contrary to the law.

Ball did not pay Lewis the sums owed under the "rental" agreement. Lewis sued. Ball attempted to introduce parol evidence of the "rental" agreement's illegality. Lewis attempted to block the parol evidence of the "rental" agreements illegal purpose by claiming that the contract was the complete, integrated agreement. However, the court said, "The parol evidence rule does not exclude evidence showing that a contract lawful on its face is in fact part of an illegal transaction." Lewis was denied recovery.[34]

Course of Dealing, Conduct of the Parties, or Past Practice

Some of the strongest evidence in support of a particular interpretation is how the parties actually interpreted the contract. The conduct of contracting parties following the formation of a written contract is "perhaps the strongest indication of what the writing means."[35]

The prime contractor, LBL, entered into a subcontract with APG for installation of a metal wall system as part of a new airline terminal. The applicable provision in the subcontract was as follows:

Include all necessary structural steel...to support the work. The structural steel **sized and indicated** on the structural steel drawings will be furnished and installed by the steel fabricator [Cives, another subcontractor]. Those items **not shown, not sized**, or indicated not to be by the steel fabricator are to be designed, engineered, fabricated, furnished, and installed as the support system for the curtainwall, metal panel, skylight, and louver by Contractor [APG].

Partway through the project, APG claimed that the certain support steel for APG's insulated metal panel system was not within APG's scope of work under the subcontract, because it was "sized and indicated on the structural steel drawings..." and was therefore part of Cives' (a different subcontractor) subcontract.

continued

The court held that the term "not shown, not sized" was ambiguous and admitted parol evidence to explain the term. The parol evidence consisted of two engineers and one architect, all of whom testified that the support steel in the metal panel system was not shown and was not sized and was therefore within the scope of APG's subcontract.

However, the court held that the strongest evidence in support of its interpretation that the APG was responsible for the support steel in its metal panel system was APG's own actions. APG failed to object to the inclusion of support steel at issue in shop drawings for APG's work; APG's work schedules showed that APG planned to install the support steel at issue; APG designed a structure to aid it in installing the support steel at issue, and; APG took responsibility for engineering the support steel at issue prior to its claim.[36]

EXAMPLE

Blinderman entered into a contract with the U.S. Navy for certain permanent improvements in military housing. The contract contained the following clauses:

SCHEDULING OF WORK: The contractor shall notify the occupants of the housing unit at least three days prior to commencing any work in a housing unit. The contractor shall perform his work between the hours of 8:00 a.m. and 5:00 p.m. and, having once started work in a housing unit, shall work to completion in consecutive work days.

METHODS AND SCHEDULES OF PROCEDURES: The work shall be executed in a manner and at such times that will cause the least practicable disturbance to the occupants of the buildings and normal activities of the station. Before starting any work, the sequence of operations and the methods of conducting the work shall have been approved by the Contracting Officer.

Contractor shall give a notice to the occupants at least three days before work is commenced.

The contractor experienced considerable difficulty and delays in gaining access to approximately 60 apartments due to various reasons including the occupants being on vacation, on active duty on ship, refusing the contractor access, and locking the contractor out partway through the work.

Notices to the occupants were given in the morning, during the noon hour, or in the afternoon. If the contractor could not reach the occupants during the day, she tried to see them in the evening. If all of these efforts failed, the contractor would, in accordance with a suggestion made by the Navy's project manager, leave a yellow card on the doorknob of the apartment, indicating when the work in that unit would begin.

Whenever the contractor or the subcontractors were unable to gain access to an apartment for any of the reasons mentioned above, they would call on the Navy's project manager who would help them provide the access they needed by calling the occupants and asking if he could use the base master key. The contractor had done similar work for the Navy before, and the Navy had helped the contractor gain access in various ways.

continued

After the project was concluded, the contractor submitted a claim for unreasonable delays. At this point, the Navy's project manager took a position that was inconsistent with his conduct during the performance of the contract. He stated that the claimed delays were due to the failure of the contractor to notify the occupants as required by the specifications, and that placing cards on doorknobs of the housing units did not constitute notification. He also denied that the Navy had any responsibility for assisting the contractor to obtain access to the apartments.

The court held that the Navy had an implied obligation to provide access so that the contractor could complete the contract within the time required by its terms.[37]

EXAMPLE

In the following example, the past practice of the parties was held not to govern in a different contract situation. Batzer, the contractor, and Boyer, the developer, had a partnership developing properties for lease to the U.S. Postal Service. In the partnership's agreement with the Postal Service, the Postal Service would not accept change order requests from the contractor but only from the owner. As the construction end of the partnership, Batzer would submit the change orders. Batzer and Boyer were the owners. After several years, they decided to terminate their partnership and divide the assets. At the time the assets were divided, three of the buildings to be received by Boyer were incomplete. Batzer agreed to complete the construction, but Boyer became the owner of the buildings upon the signing of the partnership termination.

During the course of construction, $95,000 in change orders was incurred. Batzer submitted all to Boyer, but Boyer failed to submit them to the Postal Service and therefore was never paid. At the end of the project, Batzer presented Boyer with a final invoice for $95,000, but Boyer refused to pay.

Boyer claimed that the past practice of the parties showed that Batzer was obligated to submit the change orders to the Postal Service. The court refused to accept this argument, because the undisputed evidence established that Batzer was responsible for submitting the change orders to the Postal Service.[38]

Patent Ambiguity Rule

Patent means open or obvious, and a **patent ambiguity** in a contract is one that is obvious to a reasonable contractor or owner. Patent ambiguities are usually the result of a mistake by the party drafting the contract. A patent ambiguity can be something left out of the contract or a typographical error. A patent ambiguity is one that blatantly jumps out at the reader: a direct numerical conflict exists or the contract contains an obvious internal inconsistency.[39]

The patent ambiguity rule states that a patent ambiguity is resolved in favor of the party who drafted the contract. In other words, the patent ambiguity is resolved in favor of the party who made the error. This rule may seem odd because it favors the party who made the mistake, but it encourages mistakes to be fixed rather than taken advantage of. A basic maxim of the law is that mistakes are to be fixed, not taken advantage of. A contractor's failure to seek clarification of a patent ambiguity prevents the court from accepting the contractor's interpretation of the ambiguity. In the construction industry, the patent ambiguity rule usually means that the patent ambiguity is resolved in favor of the owner versus the contractor and the contractor versus the subcontractor.

> ## EXAMPLE
>
> Triax entered into a contract with the Army to renovate military housing and build lanais. The contract drawings did not require Triax to paint the lanais, but the specifications contained a painting schedule requiring it to paint the lanais. Triax filed a claim for additional compensation for painting the lanais, claiming that it was extra work. The court said that the drawings and specifications, taken together, were patently ambiguous and that Triax should have sought clarification. Failure to do so required the ambiguity to be resolved in favor of the Army. Triax was not entitled to an additional payment to paint the lanais.[40]

> ## EXAMPLE
>
> The contract drawings, but not the specifications, called for the installation of a specific brand-named radio system. The special conditions contained a provision that if the equipment was included in the drawings, it was part of the specifications. The contractor billed the owner for the cost of the specific brand-named system over and above another less-expensive system. The contractor's failure to seek clarification before submitting its bid subjected the bid to the owner's interpretation.[41]

Custom and Usage

Custom and usage in the trade can be used to clarify the meaning of a term—for example, to explain that "2 x 4s" are not actually 2 inches by 4 inches. "It is well settled that where words or expressions are used in a written contract, which have in particular trades or vocations a known technical meaning, parol evidence is competent to inform the court and jury as to the exact meaning of such expression in that particular trade or vocation, and it is for the jury to hear the evidence and give effect to such expressions as they may find their meaning to be."[42]

EXAMPLE

Pinney and Arb entered into a contract instructing Arb to "reclaim the parking lot." Arb claimed that the term "reclaim the parking lot" meant digging up the existing asphalt surface, grinding it up, and depositing it evenly over the surface area in preparation for the next step called "paving." The surface was not suitable for use as a parking lot until the paving was done. Pinney claimed that the term "reclaim the parking lot" meant what Arb said it did but also included the paving and that Arb should have made the surface suitable for parking.

After hearing the testimony, the court determined that in the construction industry, the term "reclaim the parking lot" meant what Arb said it did and was separate from the paving of the parking lot.[43]

EXAMPLE

Glover Construction Company entered into a contract with Hampton Roads Sanitation District, the owner, to construct approximately 10,000 feet of sewer line in the city. During construction, traffic was detoured around the site. The contract clearly required a uniformed police officer at the intersection near the site. A dispute arose over to who must pay for the police officer.

The owner argued Section 2(a) of the contract stated, "Contractor shall provide and pay for all...labor,...and other services and facilities of every nature whatsoever...," that the contract was not ambiguous, and that this section required the contractor to pay for the police officer. The contractor argued that this section meant the contractor was to pay for all of the construction work but that the custom and usage of the trade was that the owner paid for the police officer to divert traffic.

Custom and usage of the industry—that the owner normally paid for the police officer—was supplied by the testimony of Matt Glover, the owner of the construction firm, and two witnesses employed by the owner. The owner was required to pay for the police officer.[44]

Custom and usage will not be used to change an express term of a contract, because parties are free to contract for different specifications than custom and usage. It may be the custom in the industry to allow for a ½-inch tolerance in a certain situation, but the parties are free to contract for a ¼-inch tolerance. The contractor must bid the ¼-inch tolerance even though the custom in the industry is a ½-inch tolerance.

EXAMPLE

General Plumbing entered into an agreement with American Air Filter to deliver certain parts. The contract stated, "Any shipping date stated in this quotation or any acknowledgement is American's best estimate, but American makes no guarantee of shipment by any such date and shall have no liability or other obligation for failure to ship on such date, regardless of cause, unless expressly stated otherwise herein." American Air Filter was late in the shipment of the parts. This caused General Plumbing to be late, and General Plumbing attempted to hold American Air Filter liable for its damages. General Plumbing attempted to introduce that the custom and usage of the industry was for the supplier to deliver the supplies in time for the buyer, General Plumbing in this case, to meet its own deadlines. The custom of the industry is not admissible at it is contrary to the clear language of the contract.[45]

EXAMPLE

The contract between Jowett and the government contained the following two provisions:
  Section 3.3.1, Insulation and Vapor Barrier for Cold Air Duct: "The following shall be insulated: a. **Supply ducts,** b. Return air ducts,...c. Plenums...."
  Section 3.3, Duct Insulation Installation: "Duct installation shall be **omitted** on the following:
  f. **Return ducts** in ceiling spaces. Ceiling spaces shall be defined as those spaces between the ceiling and bottom of floor deck or roof deck inside the air-conditioned space insulated envelope.
  j. **Ceilings** which form plenums.
  The contractor did not insulate the cold-air supply ducts inside the ceiling spaces. After being required to insulate these cold-air supply ducts, he submitted a claim for extra work that was denied. The contractor supplied two affidavits that attested that it was not "standard practice in the greater Baltimore/Washington area" to insulate supply ducts in ceilings. The court held the affidavits irrelevant, saying, "However, affidavits describing a supposed common industry practice of not insulating air supply ducts in ceilings are simply irrelevant where the language of the contract is unambiguous on its face. It is well-established that the government can vary from the norm in the trade when contracting for goods and services."[46]

Implied Terms

Implied terms come in two broad categories: terms needed to complete the project and good faith. The law adds both types of terms to contracts. In other words, every contract contains terms that are invisible to reader but are there because the law puts them there.

EXAMPLE

Micro entered into a detailed contract with C.S. that said, in part,

    2. You [C.S.] agree to purchase _____ (fill in blank with appropriate term)...

    No provision in the contract actually said, "Micro agrees to provide _____ ."

    Later, a dispute arose between the parties, because Micro could not supply the product that C.S. had agreed to buy. C.S. had to purchase it elsewhere, and it sued Micro for the difference. Micro contended that while the contract obligated C.S. to buy, it did not obligate Micro to sell. If Micro chose not to provide the product, C.S. was out of luck. C.S. won on a summary judgment motion. Micro breached the contract. It is implied in the contract that Micro would provide the product its buyer is agreeing to buy.[47]

Terms Necessary to Complete the Project

Terms not expressly stated in the contract, but needed to fulfill the purpose of the contract, are implied in the contract.

EXAMPLE

The contract between the parties did not contain a time for performance, therefore a reasonable time was implied.[48]

EXAMPLE

A subcontract provided for the removal of certain highway dump material but did not list the quantity to be removed. The subcontract did refer to the principal contract, which specified the amount to be removed. The specific amount to be removed was implied into the subcontract and equaled the amount specified in the principal contract.[49]

EXAMPLE

S&S entered into a heat-treating contract with Quality Metal. The contract stated that Quality Metal "will have indefinite and exclusive rights to heat treat all of the metal treating process required of all goods manufactured by S&S coming under Quality Metal Treatment,

continued

Inc.'s capabilities." The evidence supported the contention that both parties expected the relationship to be a long-term one. The agreement was part of a complex relationship among the parties involving the selling of Quality Metal, which had formerly been part of S&S.

After several months and several problems, some personal, cracks appeared in some lots of S&S Part # K233, a shock absorber, and seven lots were recalled. A small number were determined to be defective. Later, investigation revealed that the problem was apparently with the steel, not the heat treating.

However, shortly thereafter, S&S terminated its agreement to use Quality Metal for all its heat-treating business without actual notice. S&S claimed that it could terminate the contract because of quality problems and also at will because there was no specific termination date.

The law implied a term in the contract that it could not be terminated except for cause. S&S was therefore in breach of an implied contract term and owed Quality Metal damages.[50]

Good Faith

The law implies into every contract the implied covenant of good faith and fair dealing.[51] This term requires the parties to cooperate and not hinder the other party's performance.

EXAMPLE

The government was unnecessarily slow in doing necessary survey and staking work on a piece of land. This hindered the contractor's ability to do the work in a timely manner. The government breached the contract, because it violated the implied covenant of good faith and fair dealing. The government owed the contractor damages.[52]

EXAMPLE

Coakley entered into a subcontract with Blake for spray fireproofing on a hospital project. The subcontract contained a provision that Coakley could not collect damages for any delays.

Blake failed to adhere to the schedule it gave Coakley and allowed other subcontractors to attach pipes, ducts, and conduits to the structural steel. Therefore, Coakley was hampered in applying the fireproofing. Blake had to build scaffolding to apply the fireproofing instead of standing on the structural steel itself. The court awarded damages to Coakley, despite the "no damage for delay" clause, because Blake failed to uphold the implied duty to schedule work reasonably and the implied duty not to interfere.[53]

Whole Agreement

Contracts must be read as a whole; that is, provisions cannot be taken out of context. A contract must be construed as a whole and not by consideration of isolated provisions.[54]

EXAMPLE

Seabury entered into a contract with Jeffrey for industrial chain to be used in a major wastewater treatment plant. The purchase order contained detailed specifications, including a specification for the hardness of the chain. Another provision required testing of the chain. The chain was tested and passed the test, but within the warranty period, the chain developed cracks and was shown to be too hard and so failed the hardness specification.

Seabury sued for breach of the warranty and won. Just because the chain passed the test did not mean that the warranty was null; the warranty still existed. Seabury purchased chain that was required to meet the specifications, and the failure of Jeffrey to meet the specifications was a breach of the contract.[55]

EXAMPLE

Baltimore Contractors entered into a complex project for the construction of an extension to the Senate office building on Capitol Hill. It entered into a subcontract with Vermont Marble. The contract contained a standard "time is of the essence clause" but also many other clauses dealing with how delays were to be handled. Originally, the subcontract contained a "no damage for delay" clause, but it was eliminated and replaced by a clause allowing the subcontractor to receive damages. It also contained the sentence, "(2) Contractor for just cause shall have the right at any time to delay or suspend the commencement or execution of the whole or any part of the Work without compensation or obligation to Subcontractor other than to extend the time for completing the Work for a period equal to that of such delay or suspension."

After 20 months of delays and problems, Vermont Marble walked off the job and refused to perform. Vermont Marble said that the "time is of the essence" clause allowed it to terminate the contract when it was delayed in its performance.

The court rejected this argument and determined that Vermont Marble's termination of the subcontract was a breach entitling the contractor to damages. When read as a whole, the contract contemplated delays and provided remedies for such delays. The "time is of the essence" clause could not be taken out of the context of the entire contract but had to be read in conjunction with the entire contract.[56]

Order of Precedence

Absent a contrary term in the contract, special conditions prevail over general conditions. Handwritten terms prevail over typewritten provisions, which prevail over preprinted terms of the contract. Words prevail over figures. The reason for this rule is that, as a general statement, such terms more accurately reflect the true intent of the parties.

In an interesting case showing how the law adapts to changed circumstances, the court said the rule that "typewritten terms control over preprinted terms" applied to word-processed terms added to a standard form agreement even though the computer-generated form produced the entire document at the same time.

A painting contractor faxed a document captioned "Estimate" to the homeowner. The document contained the words "This is a guesstimate" at the bottom. The terms "This is a guesstimate" had the same effect as a typewritten addition to the standard printed form and controlled over the term "estimate." The painting contractor was not obligated to the guesstimate.[57]

Latent Ambiguity

A **latent ambiguity** is a hidden or unobvious ambiguity. It can be something left out of a contract, but the omission must not be obvious. The latent ambiguity rule states that latent ambiguities are resolved in favor of the party who did *not* draft the contract. This rule is also called the rule of *contra proferentem*, which is that the party with the least bargaining power is preferred when the ambiguity is latent. "[T]he rule of 'construction against the draftsman' [*contra proferentem*]...applies with particular force in cases where the drafting party has the stronger bargaining position."[58] In construction contracts, this rule resolves the ambiguity in favor of the contractor against the owner and the subcontractor against the contractor.[59]

This rule is usually only applied if other rules do not resolve the ambiguity. "We recognize that this rule should be applied only 'where, after examining the entire contract, the relation of the parties, their intentions and the circumstances under which they executed the contract, the ambiguity remains unresolved.'"[60]

The latent ambiguity rule is consistent with the open and obvious errors rule in the law specifically relating to plans and specifications. See Chapter 6.

The contractor, Washtenaw, entered into a contract with the state to repave a particular road. A dispute arose as to the amount to be paid to the contractor.

Relevant provisions of the contract favorable to the contractor's position were as follows:

1. The plans and specifications stated, "The roads requiring work will be pointed out by the inspector..."
2. The estimated base sum for the cost of the project was $16,607.70.
3. "Payment for all work will be on the basis of the contract unit prices applied to the actual quantities installed."

continued

4. "Quantities as listed have been carefully estimated but are not guaranteed. The state reserves the right to increase or decrease the quantity of work to be performed at the unit prices by amounts up to 25 per cent of the quantities stated."

   The contractor repaved the areas pointed out by the inspector, and at the end of the job, the contractor billed the state $23,506.28 or $6,898.58 more than the estimate. There was no evidence that the contractor used more material than necessary to do the job, as pointed out by the inspector.

   The state claimed that it did not have to pay the contractor the sum over the original contract price, because the contractor had an obligation to stop the project once the quantities in the bid had been reached. The state pointed to the following provision in support of this position:

5. "Quantities: It is the responsibility of the Contractor to keep a running account of the quantities. No overrun in excess of 5 per cent will be permitted without authorization from the Director."

However, the court said that since the ambiguity in the contract was latent, it was resolved in favor of the contractor. The contractor was entitled to the full payment.[61]

EXAMPLE

Crowley was awarded a contract to install roofing and repair exterior walls on two government buildings. The contract required the "actual installed thickness of insulation shall be such as to provide a coefficient of heat transmission or U-value, through completed roof construction air-to-air, not in excess of .030 Btu per hour." The written specifications further provided that "insulation shall be laid in two or more layers."

A drawing contained a depiction of each section and containing two layers of insulation with an arrow pointing to the two layers and labeled "Insulation W/R-value min. 12.5." The subcontractor interpreted the drawings as requiring a minimum of two layers of insulation, which together would provide an R-value of at least 12.5. The contractor realized that a total R-value of 12.5 insulation only at this point would not meet the specified U-value of .030. However, the contractor believed the specification could be met through the "completed roof construction air-to-air" by the existing insulation and plaster under a suspended ceiling.

The architect demanded the contractor install two layers of 12.5 R-value insulation. The cost for the additional insulation was $90,000. This would still not meet the .030 U-value, and that value would have to be met by contribution from some other element.

Because the ambiguity was latent, and the subcontractor's and contractor's interpretations were reasonable, the ambiguity was resolved in favor of the subcontractor and contractor. The subcontractor was entitled to $90,000, because the owner had constructively changed the contract by requiring two layers of 12.5 R-value insulation.[62] The concept of constructive change is discussed in Chapter 8 in the section entitled "Constructive Change."

Clear Expression of Intention

This rule does not stand alone—if it did, the drafter of the document would always lose. Since this does not always happen, the importance of this rule is less than other rules. It is used to bolster the conclusion reached by some other rule. It is often coupled with the latent ambiguity rule. The clear expression of intention rule states that the drafting party has the duty to express its intent clearly. It is often used when the interpretation being promoted by the drafter is odd or against the custom in the industry.

■ Application of Several Rules to Resolve a Scope Issue

Courts usually apply several of the above rules when coming to a decision in a scope case.

EXAMPLE

WDC managed four off-base military housing contracts for the U.S. Army. Under the contracts, it was required to maintain the housing and replace carpeting and appliances as needed. Carpeting in one of the units was damaged beyond ordinary wear and tear and was replaced by WDC. An invoice for the total cost to replace the carpeting was submitted to the Army per the following provision:

Damages Caused by Occupants: Damages to a housing unit or to other improvements within the project which are beyond normal wear and tear and are caused by the Government or an occupant, his dependents, or invited guests, which are not corrected by Government or occupant, shall be repaired by the Developer. The cost of such repairs shall be billed to the Government...

The Army only paid WDC the replacement cost minus the depreciated value of the carpet. When this situation had occurred in the past, the Army paid the entire replacement cost and did not subtract the depreciated value.

The change in approach was due to a new officer who claimed that the following clause in the contracts required the Army to reimburse the contractor only the replacement cost minus depreciation.

"The Developer shall, with the approval of the Government, establish a list of cleaning and repair costs for dwelling unit components which will establish the normal maximum amounts to be charged in the event of damage to property and equipment installed within a living unit over and above normal wear and tear."

The court held that the Army had to pay the entire replacement cost because: (1) using the ordinary meaning rule, the lease provision plainly stated the government would replace the carpeting or reimburse WDC the full costs of repairs without deprecation, (2) using the conduct of the parties rule, the parties' past conduct demonstrated that WDC's argument was consistent with the parties' understanding when it signed this contract, (3) using the whole agreement rule, the Army's argument was weak because, if accepted, it would create a latent ambiguity by placing the two lease provisions in conflict and the contract could not be read as a whole, (4) the rule of *contra proferentum* requires the court to construe the lease against the government because the government drafted the lease, and (5) despite the fact this ruling gave WDC an economic windfall, the court was not in the business of rewriting a lease to which the parties have agreed to the terms.[63]

237

■ Loophole: Onerous or Unconscionable Clauses

The law will not enforce an onerous or unconscionable clause even if it is unambiguous. This rule is similar to the rule of illegality, which holds that a clause requiring an illegal act is void. However, the rule is rarely used.

EXAMPLE

The contract between Mellon Stuart and the Metropolitan Water Reclamation District was extremely complicated from an engineering standpoint. Several disagreements arose between the contractor and the engineer regarding delays and compensation to the contractor. The contract between the parties gave the engineer the power to "decide all questions which may arise in relation to the work and the engineer's decisions shall be final and binding, and not subject to appeal." The engineer always decided in favor of the owner, no matter what the facts were. However, if the contractor could show that the engineer's decisions were unreasonable and made in bad faith, the provision would not be upheld. The jury would make that decision.[64]

■ Loophole: No Contract Formed

Another alternative available to a person attempting to prevent enforcement of an otherwise unambiguous clause is to claim that no contract was ever formed; if no contract was formed, the parties have no duties upon which to sue.

However, if one of the parties has actually received some benefit, that party must pay for the benefit received. This result is not required by contract law but by the doctrine of promissory estoppel (see Chapter 7).

■ Loophole: Waiver and Modification

None of the rules in this chapter prevent the modification or waiver of a contract *after* it has been entered into. Parties are always free to modify their contracts. Common law contracts, including construction contracts, require consideration to support the modification. Consideration is discussed in Chapter 7 in the section entitled "Consideration." UCC or sales contracts do not need consideration to modify the contract as long as the modification is reasonable. Sales contracts are discussed in Chapter 13. A waiver is the knowing relinquishment of a known right. Waiver is discussed in Chapter 6 in the section entitled "Waiver and Acceptance."

Endnotes

1. *Rea Const. Co. v. B.B. McCormick & Sons, Inc.*, 255 F.2d 257 (5th Cir. 1958).

2. Example based on *Community Heating & Plumbing Co. v. Kelso*, 987 F.2d 1575 (Fed. Cir. 1993).

3. *Reed & Reed, Inc. v. Weeks Marine, Inc.*, 431 F.3d 384 (1st Cir. 2005).

4. *Edward R. Marden Corp. v. U.S.*, 803 F.2d 701, 705 (Fed. Cir. 1986), *Highway Prods., Inc. v. U.S.*, 530 F.2d 911, 917 (Ct. Cl. 1976), *Sun Shipbuilding & Dry Dock Co. v. U.S.*, 183 Ct. Cl. 358, 393 F.2d 807, 815-16 (Ct. Cl. 1968).

5. *PlaNet Productions, Inc. v. Shank*, 119 F.3d 729 (8th Cir. 1997) quoting *Angoff v. Mersman*, 917 S.W.2d 207, 210 (Mo. Ct. App. 1996).

6. *Community Heating & Plumbing Co. v. Kelso*, 987 F.2d 1575, 1582 (Fed. Cir. 1993).

7. *Hall Contracting Corp. v. Entergy Services, Inc.*, 309 F.3d 468 (D.C. Ark. 2002).

8. *Metric Construction Co. v. St. Paul Fire & Marine Insur. Co.*, 2005 U.S. Dist. Lexis 37197 (D.C. Utah 2005).

9. Example based on *Ladner Testing Laboratories, Inc. v. United States Fidelity & Guaranty Company*, 2000 U.S. App. Lexis 29208 (6th Cir. 2000).

10. Example based on *Appeals of—B. J. Larvin, General Contractor, Inc.*, 77-2 B.C.A. (CCH) P12,717; 1977 ASBCA LEXIS 109 (1977).

11. *Tecom, Inc. v. U.S.*, 66 Fed. Cl. 736 (Ct. Cl. 2005) citing *In re: Binghamton Bridge*, 70 U.S. (3 Wall.) 51, 74, 18 L. Ed. 137, 30 How. Pr. 346 (1865); see also *Intergraph Corp. v. Intel Corp.*, 241 F.3d 1353, 1354 (Fed. Cir. 2001).

12. Corbin on Contracts § 578. *Effect of Express Written Statement that There Have Been No Extrinsic Representations, Warranties, or Other Provisions.* Matthew Bender & Co., Inc. (2006).

13. See Harry G Prince. *Contract Interpretation in California: Plain Meaning, Parol Evidence and Use of the "Just Result" Principle.* 31 Loy. L.A. L. Rev. 557 (1998).

14. Example based on *Nucla Sanitation District v. Rippy*, 344 P.2d 976, 140 Colo. 444 (Co. 1959).

15. Example based on *Sprucewood Investment Corp. v. Alaska Housing Finance Corp.*, 33 P.3d 1156 (Alaska 2001).

16. Restat 2d of Contracts, § 209.

17. *Mies Equipment, Inc. v. NCI Building Systems, LP*, 167 F. Supp. 2d 1077 (Minn. 2001).

18. See Corbin on Contracts § 24.30, *Is Interpretation A Question of Fact or a Question of Law?* Matthew Bender & Co., Inc. (2006). See "Our appellate court has, indeed, reached different conclusions as to whether integration is a factual or legal question and whether it is to be answered against only the subject writing. *J&B Steel Contractors, Inc. v. C. Iber & Sons, Inc.*, 162 Ill. 2d 265, 642 N.E.2d 1215 (Ill. 1994).

[19] The parol evidence rule is not a rule of evidence but a rule of substantive contract law. Today the term "evidence law" deals specifically with rules evidence admitted at trial. See *The Rule Is a Rule of Substantive Contract Law, Not a Rule of Evidence.* Corbin on Contracts § 573 *The Rule Is a Rule of Substantive Contract Law, Not a Rule of Evidence.* Matthew Bender (2006).

[20] *P.A. Bergner & Co. of Illinois v. Lloyds Jewelers, Inc.,* 112 Ill. 2d 196, 203, 492 N.E.2d 1288, 1291, 97 Ill. Dec. 415 (1986).

[21] Example based on *Alaska Diversified Contractors, Inc. v. Lower Kuskokwim School District,* 778 P.2d 581 (Alaska, 1989) cert. denied, 493 U.S. 1022, 110 S. Ct. 725, 107 L. Ed. 2d 744 (1990).

[22] *Still v. Cunningham,* 94 P.3d 1104 (Alaska 2004).

[23] Example based on *Teg-Paradigm Environmental, Inc. v. U.S.,* 465 F.3d 1329 (Fed. Cir. 2006).

[24] Example based on *Batzer Construction, Inc. v. Boyer,* 204 Ore. App. 309, 129 P.3d 773 (Ore. App. 2006).

[25] Example based on *City of Houston v. Howe & Wise,* 323 S.W.2d 134 (Tex. Civ. App.1959).

[26] Example based on *J&B Steel Contractors, Inc. v. C. Iber & Sons, Inc.,* 162 Ill. 2d 265, 642 N.E.2d 1215 (Ill. 1994).

[27] Witkin on California Evidence.

[28] Example based on *Caputo v. Continental Constr. Corp.,* 162 N.E.2d 813, 340 Mass. 15 (1959).

[29] *Levy v. Leaseway System, Inc.,* 154 A.2d 314, 190 Pa.Super. 482 (1959).

[30] See *Lowell O. West Lumber Sales, Inc. v. U.S.,* 270 F.2d 12 (C.A. 9th, 1959).

[31] *Fisher v. J.A. Jones Constr. Co. Inc.,* 87 Ga. App. 317, 73 S.E.2d 587 (Ga. App. 1952). 31 *City of Houston v. Howe & Wise,* 323 S.W.2d 134 (Tex. Civ. App.1959). Corbin on Contracts § 583. *Oral Testimony of Additional Terms Not Expressed in Writing.* Matthew Bender & Co., Inc. (2006). **[There appear to be two footnote 31s.]**

[32] Example based on *All Angles Construction & Demolition, Inc. v. Metropolitan Atlanta Rapid Transit Authority,* 246 Ga. App. 114, 539 S.E.2d 831 (Ga. App., 2000).

[33] "Evidence to prove that the instrument is void or voidable for mistake, fraud, duress, undue influence, illegality, alteration, lack of consideration, or other invalidating cause is admissible. This evidence does not contradict the terms of an effective integration, because it shows that the purported instrument has no legal effect." 2 **[Should "2" appear here?]** Witkin Cal. Evid. Doc Evid § 95 citing Cal. Civ. Pro. 1856(e), (f), and (g), Corbin § 580; and Rest.2d, Contracts §214(d).

[34] Example based on *Lewis and Queen v. N. M. Ball Sons,* 48 Cal. 2d 141, 308 P.2d 713 (Cal. 1957). See also Corbin on Contracts § 580 *Oral Proof of Fraud, Illegality, Accident, or Mistake.* Matthew Bender & Co., Inc. (2006).

[35] *LBL Systems (USA) Inc. v. APG-America, Inc.,* 2005 U.S. Dist. LEXIS 19065 (E.D. Pa. 2005) quoting *Atlantic Richfield Co. v. Razumic,* 480 Pa. 366, 390 A.2d 736, 741 (Pa. 1978), citing Restatement (Second) of Contracts § 228.

[36] Example based on *LBL Systems (USA) Inc. v. APG-America, Inc.*, 2005 U.S. Dist. LEXIS 19065 E.D. Pa. 2005).

[37] Example based on *Blinderman Construction Co. v. U.S.*, 695 F.2d 552 (Fed. Cir. 1982). The court used several rules to come to this conclusion.

[38] Example based on *Batzer Construction, Inc. v. Boyer*, 204 Ore. App. 309, 129 P.3d 773 (Ore. App. 2006).

[39] *Newsom v. United States*, 676 F.2d 647, 650-651 (Ct. Cl. 1982).

[40] Example based on *Triax Pacific, Inc. v. West*, 130 F.3d 1469 (Fed. Cir. 1997).

[41] Example based on *J.H. Electric of New York, Inc. v. NYCHA*, 2004 WL 439886.

[42] *R.S. Neal v. Camden Ferry Co.*, 166 N.C. 563, 82 S.E. 878 (N.C. 1914).

[43] Example based on *Arb Construction, LLC v. Pinney Constr. Corp.*, 75 Conn. App. 151, 815 A.2d 705 (Conn. App. 2003).

[44] Example based on *Glover Constr. Co., Inc. v. Hampton Roads Sanitation District*, 1998 U.S. App. Lexis 31622 (4th Cir. 1998) (unpublished opinion).

[45] Example based on *General Plumbing & Heating, Inc. v. American Air Filter Co.*, 696 F.2d 375 (5th Cir. 1983).

[46] Example from *Jowett, Inc. v. US*, 234 F.3d 1365 (Fed. Cir. 2000).

[47] Example loosely based on *C.P. Apparel Manufacturing Corp. v. Microfibres, Inc.*, 210 F. Supp. 2d 272 (D.C. S.N.Y.).

[48] *Minor v. Minor*, 863 S.W.2d 51, 54 (Tenn. Ct. App. 1993); see also *Hathaway v. Hathaway*, 98 S.W.3d 675, 679 (Tenn. Ct. App. 2002).

[49] *Trompeter Construction Co. v. Higby*, 161 N.E.2d 159, 22 Ill. App. 420 (1959).

[50] Example based on dicta in *Johnson v. Welch*, 2004 Tenn. App. LEXIS 86 (2004).

[51] *Centex Corp. v. U.S.*, 395 F.3d 1283, 1304 (Fed. Cir. 2005); see also *Restatement (Second) of Contracts § 205* (1981).

[52] Example based on *Lewis-Nicholson, Inc. v. U.S.*, 550 F.2d 26, 32, 213 Ct. Cl. 192 (Ct. Cl. 1977).

[53] Example based on *Blake Construction Co., Inc. v. C.J. Coakley Co., Inc.*, 431 A.2d 569 (D.C. App. 1981).

[54] *Telex Corp. v. Data Prods. Corp.*, 271 Minn. 288, 135 N.W.2d 681, 685 (Minn. 1965), *Tecom, Inc. v. US*, 66 Fed. Cl. 736 (2005).

[55] Example from *Seabury Constr. Corp. v. Jeffrey Chain Corp.*, 289 F.3d 63 (2d Cir. 2002).

[56] Example based on *Vermont Marble Company, v. Baltimore Contractors, Inc.*, 520 F. Supp. 922, 29 Cont. Cas. Fed. (CCH) P81,786 (D.C. D.C. 1981).**[Is "D.C. D.C." intentional?]**

[57] Example based on *Otto Interiors, Inc. v. Nestor*, 196 Misc. 2d 48, 763 N.Y.S.2d 439 (Civ. Ct. 2003).

[58] *Semmes Motors v. Ford Motor Co.,* 429 F.2d 1197 (2d Cir. 1970).

[59] If not found to be patent, then an ambiguity is latent. The Court will adopt a contractor's reasonable interpretation of a latent ambiguity under the *contra proferentem* rule, that is, construing an ambiguity against the drafter. See *Peter Kiewit Sons' Co. v. United States,* 109 Ct. Cl. 390, 418 (19470, *U.S. v. Seckinger,* 397 U.S. 203, 216, 25 L. Ed. 2d 224, 90 S. Ct. 880 (1970). If the contractor's interpretation of such a contract provision is determined to be reasonable the contractor will prevail against the author of the contract. See e.g., *Underground Const. Co. v. U.S.,* 16 Cl. Ct. 60, 69 (1988), *Reliable Bldg. Maint. Co. v. U.S.,* 31 Fed. Cl. 641, 644 (1994).

[60] *U.S. v. Haas and Haynie Corp.,* 577 F.2d 568, 27 U.C.C. Rep. Serv. (Callaghan) 32 (9th Cir. 1978).

[61] Example based on *Washtenaw Asphalt Co. v. Michigan,* 42 Mich. App. 132, 201 N.W.2d 277 (Mi. App. 1972). In that case, the court did not discuss patent and latent ambiguity.

[62] Example based on *R. J. Crowley, Inc. v. U.S.,* 37 Cont. Cas. Fed. (CCH) P76,063 (1990).

[63] Example based on *WDC West Carthage Associates v. U.S.,* 324 F.3d 1359 (Fed. Cir. 2003).

[64] Example based on *Mellon Stuart Const., Inc. v. Metropolitan Water Reclamation Dist.,* 1995 U.S. Dist. LEXIS 5376 (ND Il. 1995).

242

CHAPTER 10

Environmental, Real Property, and Intellectual Property Law

■ Environmental Law

In 1990, the U.S. Congress passed the Pollution Prevention Act,[1] which formalized society's desire for a more environmentally conscious business community. In the preamble to the Act, Congress found that "(1) The United States of America annually produces millions of tons of pollution and spends tens of billions of dollars per year controlling this pollution. (2) There are significant opportunities for industry to reduce or prevent pollution...The Congress hereby declares it to be the national policy of the United States that pollution should be prevented or reduced at the source whenever feasible; pollution that cannot be prevented should be recycled in an environmentally safe manner, whenever feasible; pollution that cannot be prevented or recycled should be treated in an environmentally safe manner whenever feasible; and disposal or other release into the environment should be employed only as a last resort and should be conducted in an environmentally safe manner."

In accordance with the above and other government laws, government agencies have broad powers to halt construction or impose restrictions on construction to protect the environment. Government agencies have great control over environmental decisions, and courts are required to uphold administrative agency decisions unless agency action is "arbitrary, capricious, an abuse of discretion, or otherwise not in accordance with law."[2]

EXAMPLE

The government agency ordered a moratorium on land development while the agency developed a comprehensive land-use plan for Lake Tahoe. The developers sued, claiming that this was a taking by the government requiring compensation; however, the court found this did not constitute a taking by the government requiring compensation to the owners.[3]

However, environmental law is one area where the amount of protection of the environment is proportional to the liberal-mindedness of the government agencies. It is more likely that an agency in California will issue orders and decrees protecting the environment and less likely that an agency in Florida will do so.

EXAMPLE

The U.S. Army Corp of Engineers issued a permit for construction of a municipal landfill in Sarasota County, Florida. The site was home to the highly endangered Florida panther and also home to the threatened eastern indigo snake. Several environmental groups brought suit to enjoin construction of the landfill, but the court upheld the agency's decision and did not find that it was arbitrary, capricious, an abuse of discretion, or otherwise not in accordance with law.[4]

The construction industry is a source of many potential environmental concerns, and attention must be paid to the following:

- Storm water runoff
- Solid and hazardous materials including asbestos, lead paint, fuels and other carbon-based products, pesticides, polychlorinated biphenol (PCB) wastes, and construction debris
- Dust and motor vehicle emissions
- Existence of threatened or endangered species on the site
- Brownfields, or land that is or may be contaminated

The Environmental Protection Agency (EPA) has the authority to implement environmental regulations and standards. In addition, it has broad powers to prohibit or stop construction projects that impact water quality or have other unacceptable environmental consequences. As a practical matter, the EPA has delegated this authority to state environmental agencies which, in addition to having delegated federal power, also exercise additional state power to regulate the environment.

What is a Pollutant?

The terms "pollutant" and "hazardous waste" can cover a variety of substances commonly encountered on construction sites, and there is no single definition of either. For example, under the Resource Conservation and Recovery Act (RCRA), "hazardous wastes" are materials that are ignitable, corrosive, reactive, or toxic. The EPA has authority to create regulations "designating as hazardous substances...such elements, compounds, mixtures, solutions, and substances which, when released into the environment may present substantial danger to the public health or welfare or the environment."[5]

The term "pollution" has been broadly defined in insurance litigation. In litigation involving pollution exclusion clauses in insurance policies, several substances commonly found on construction sites have been determined to be "pollutants." Xylene, a solvent contained in sealants used on construction sites, is a pollutant.[6] Asbestos is a pollutant.[7] MC-30 prime oil sprayed on a parking lot by a paving subcontractor is pollution.[8]

Various types of toxic fumes have been considered "pollutants."[9] Where vapor degreasers TCE and TCA, used to clean equipment, were released into storm drain and sewer lines, the company responsible for the release was required by the EPA to perform remedial action.[10] Odors coming from a roofing materials plant were considered pollutants.[11] A court decided that the paint and glue were "pollutants."[12]

While not specifically related to the construction industry, a case involving restaurant kitchen wastes is interesting, because grease and scouring pads were deemed to be "pollutants." A construction site produces a prodigious amount of waste, and so it is possible that this waste could be deemed a pollutant. Pollution was defined in that policy as "any solid, liquid, gaseous, or thermal irritant or contaminant including smoke, vapor, soot, fumes, acid, alkaline, chemicals, and waste. Waste includes material to be recycled, reconditioned, or reclaimed."[13]

CERCLA and RCRA

The Comprehensive Environmental Response, Compensation, and Liability Act of 1980 (CERCLA) requires an entity causing environmental damage to bear the cost of cleanup or other remediation of the damage. Responsible parties can include contractors, owners, and any others who exercise control over the construction site, an operation on the site, or the contamination. A related statute is the Resource Conservation and Recovery Act of 1976 (RCRA),[14] which deals with the disposal, recycling, and reuse of wastes.

Contractors are liable for exacerbation of contamination and must exercise caution at the site. In the event that the contractor discovers environmental damage or products that could cause environmental damage at the site, the contractor should stop work and remedy the damage. Many contracts contain provisions outlining what the contractor is to do in this situation.

EXAMPLE

An excavating contractor was hired by the contractor to excavate and grade a portion of the land for a proposed housing project. The excavator mixed contaminants with soil and other fill materials and then dispersed the resulting contaminated mixture over the land in the project. The excavating contractor was liable for damages.[15]

CERCLA is a form of strict liability;[16] parties bear the responsibility without reference to intent, fault, or reasonableness of actions.

A fire destroyed a building containing hazardous substances. The owner is liable for the cleanup despite the fact the fire was accidental.[17]

However, several defenses to liability exist. Defenses include the bona fide prospective purchaser a.k.a. the innocent purchaser defense,[18] the contiguous landowner defense, the security interest holder's defense,[19] the common carrier exclusion,[20] the application of pesticides pursuant to the Federal Insecticide, Fungicide, and Rodenticide Act,[21] and federally permitted releases. Most recently, an exemption has been created for arrangers and recyclers of certain materials under the Superfund Recycling Equity Act of 1999. In addition, because private businesses are leery of the potential liability related to developing land that may be contaminated, the federal government has passed the Small Business Liability Relief and Brownfields Revitalization Act, which shields developers from liability.[22]

Defenses to CERCLA and RCRA actions include the following:

- Useful product and recycling defense. The contractor is not liable if the waste, hazardous or not, was properly recycled.[23]
- Innocent landowner defense. A private owner can escape liability if it established the hazardous substances were placed at the site before its acquisition, and it exercised due diligence by having comprehensive environmental testing done on the site.
- Releases caused by third parties, acts of God, or acts of war.

Waterways and Wetlands

The Clean Water Act regulates the discharge of foreign materials into water or waterways either by direct discharge or merely through storm water run-off rain. **Water** is broadly defined and includes lakes, rivers, streams, including intermittent streams, mudflats, sand flats, wetlands, sloughs, prairie potholes, wet meadows, playa lakes, and natural ponds. **Navigable waterways** are also broadly defined and include all waters of the United States and, territorial seas. They include all navigable waters of the United States, intrastate lakes, rivers, and streams which are utilized by interstate travelers for recreational or other purposes such as harvesting fish or shellfish sold in interstate commerce, tributaries of such waters, and dry drainages (ephemeral streams) with an ordinary high water mark that are eventually tributary to any interstate water. **Foreign materials** include hazardous materials, dredged soil, solid waste, incinerator residue, sewage, garbage, sewage sludge, munitions, chemical wastes, biological materials, radioactive materials, heated, wrecked or discarded equipment, rock, sand, cellar dirt and industrial, municipal, and agricultural waste discharged into water.[24]

The Rivers and Harbors Act of 1899[25] "prohibits the creation of any obstruction to the navigable capacity of any of the waters of the United States without specific approval of the Chief Engineer of the U.S. Army Corps of Engineers..." Section 404 of the Clean Water Act extends the requirement U.S. Army Corps of Engineers approval to building impacting wetlands. This means the construction of any bridge, dock, or other project on navigable waterways or wetlands must be approved by the U.S. Army Corps of Engineers.

Regulations require the U.S. Army Corps of Engineers to consult with the EPA when permits are requested. While it is unusual for the court to interfere with the decisions of the U.S. Army Corps of Engineers or the EPA, it is not unheard of.

EXAMPLE

The Port Authority sought a section 404 Clean Water Act approvals from the U.S. Army Corp of Engineers for dredging of docks 14–20 at a site containing an abandoned Navy installation transferred to the Port Authority. The dock areas contained debris and contaminated sediment that made the docks useless for modern shipping. The Port Authority had a comprehensive design to turn the site into a commercial shipping center. At the end of the project, it was expected that the port's current size would be tripled and that the level of ship traffic would double.

Following required procedure, the U.S. Army Corp of Engineers sought EPA consultation for each request because the area is home to five federally listed endangered or threatened species: the Sacramento River winter-run Chinook salmon, the delta smelt, the green sturgeon, the Central Valley steelhead trout, and the Central Valley spring-run Chinook salmon.

After three years of negotiations, public hearings, and consultations with the EPA, the required permit had not been issued. The EPA said the project had to be reviewed in total for its environmental impacts and not piecemeal.

The Port Authority then amended its section 404 Clean Water Act permit request to dredge docks 14 and 15 only. The U.S. Army Corp of Engineers approved the request as an environmentally benign demonstration project that was independent of the development of the port and, as such, did not require EPA approval or an Environmental Impact Statement.

The Port Authority began dredging on the day the permit was issued. On the next day, the plaintiffs sought a temporary injunction to halt the dredging. The court granted the injunction preventing the continued dredging pending the final resolution of the case.[26]

Endangered Species

The Endangered Species Act of 1973 provides for the protection of endangered species or species threatened by extinction. Activities that significantly modify the habitat of endangered species are subject to the Act.[27] The Act attempts to promote the conservation of ecological systems in which such species exist. Specific species are listed by federal agencies for protection status. Construction activities that can impact ecological systems in which endangered species or threatened species exist are subject to regulation.

EXAMPLE

A real estate developer prepared a plan for a 202-acre residential community in San Diego County, California. The plan included removal of six feet or more of top soil from the Keys Creek stream bed, the home of the Arroyo toad, an endangered species protected by the Endangered Species Act. The developer applied for a section 404 permit from the Army Corps of Engineers. The Corps and the Federal Wildlife Service reviewed the application and suggested that fill dirt be brought in from off site and not from the stream bed in order to protect the toad's habitat. The developer filed suit, challenging the application of the Endangered Species Act, but lost.[28]

It is common for the Department of Interior to negotiate habitat conservation plans (HCPs) with developers and/or property owners who wish to engage in property development.[29] In addition, agencies can impose limitations on development to protect endangered species.

EXAMPLE

Marina Point Development wanted to develop a condominium project on the north shore of Big Bear Lake, California. The site is a key foraging and perching habitat for the bald eagle, at the time a threatened species protected by the Endangered Species Act. The eagles like to perch on the 200-year-old Jeffrey pine trees on and around the project site. In addition, the site contains wetlands.

The permit forbade developers from (1) placing rip-rap below the then-present contours of the lake bottom; (2) depositing sand below the high water line; (3) transferring fill or structures to the neighboring wetlands; and (4) working at all from December 1 to April 1 of any year, out of deference to the seasonal habits of the bald eagle population. Other state and local permits had to be obtained.[30]

Clean Air

The Clean Air Act and similar state laws protect the air from pollution-causing activities. Dust, fumes, and open burning can occur on construction sites and are covered by this act. Contractors are required to implement management practices such as watering, wheel washing, reduced vehicle speed, removal of dirt on roadways, and other measures to reduce air pollution.

Diesel engines emit particulate matter controlled by clean air laws. At the present time, existing equipment need not be retrofitted, but regulations are increasing for diesel engines. For example, some states have anti-idling laws.

OSHA Requirements

OSHA regulates hazardous materials in the workplace. See Chapter 3.

Reporting Requirements and Penalties

Contractors are required to prepare and submit many environment-related reports. The following is a list of the common reports needing to be compiled:

- Oil and chemical spills must be reported to the National Response Center (NRC). The NRC is open 24 hours a day with information available on the Web or in local telephone books.
- An Environmental Impact Statement for federal construction projects. Some states may have similar requirements for public or private projects.
- A Spill Prevention Control Countermeasures Plan (SPCC). This is a plan outlining procedures to prevent spills, including proper training and containment procedures.
- A Wetlands or §404 permit. Application for this permit will trigger the government to review the project for endangered species and may require a Habitat Conservation Plan.
- A Storm Water Construction Permit. Application for this permit will trigger the government to review the project for endangered species and may require a Habitat Conservation Plan.
- A Habitat Conservation Plan for the protection of endangered species.
- A Clean Air Act Risk Management Plan if the contractor releases more than the regulated threshold of regulated substances.
- Hazardous materials on the workplace. OSHA requires employers to compile information involving chemicals encountered on the worksite and make this information available to employees.

  OSHA requires employer monitoring, reporting, and informational packets related to environmental or potential environmental hazards.
- An Emergency Action Plan or Emergency Response Plan. Employers must have such a plan and make it accessible to employees.
- A Hazardous Waste Contingency Plan for owners or operators of hazardous waste facilities.

Violators of environmental laws are subject to administrative, civil, and/or criminal prosecutions. Penalties and sanctions include (1) prison sentences, (2) fines, (3) cleanup costs and natural resource damages, and (4) property damage and bodily injury caused by the release.

Ab-Remove, Inc. was an asbestos-abatement consulting firm owned by Maz. The company was hired to oversee the removal of asbestos-containing materials from the Landmark Hotel and Casino in Las Vegas prior to its demolition. Hagen was the industrial hygienist employed by Ab-Remove, Inc. as the on-site inspector for the job.

Prior to its removal, asbestos must be "adequately wet," and it must remain wet until collected and contained in leak-tight containers for proper disposal.[31]

Maz and Hagen were indicted by the grand jury and criminally charged with knowingly conspiring to violate the Clean Air Act by removing asbestos-containing materials without wetting them, leaving scraped asbestos-containing debris on floors and other surfaces, and causing asbestos-covered facility components to fall from the ceiling to the floor rather than carefully lowering such components so as not to dislodge asbestos. One government inspector described the removal project as "the worst abatement job I've seen."

At trial, the jury was presented with the facts and determined that Maz and Hagen were guilty of the crime.[32]

Various governmental agencies such as the EPA and the U.S. Army Corps of Engineers, have specialized processes and procedures for enforcing environmental laws and regulations. Government agencies may impose criminal sanctions and/or fines through internal hearings and processes.

EXAMPLE

Pozsgai purchased a lot next to his truck repair business for the purposes of expansion. The tract was legally classified as wetlands, and he was notified by local authorities that he needed a permit to develop the lot. He continued development despite repeated warnings that the construction activity was illegal. Eventually, he was indicted, convicted to a three-year prison term, and fined $200,000.[33]

EXAMPLE

In addition to criminal prosecution, the appropriate agency can file a civil action in a federal district court against owners, operators, or persons in charge for the prohibited discharge. An owner, operator, or person in charge may be fined up to $31,500 per day for a prohibited discharge or up to $1,300 per barrel of oil discharged.

In the event that a discharge is the result of gross negligence or is a willful discharge, penalties range from a minimum of $125,000 to a maximum of $3,700 per barrel of oil. In addition, violators can also be imprisoned for up to one year.

EXAMPLE

Hanousek was employed by the Pacific & Arctic. One of the projects under his control was the Six-Mile Project, designed to realign a sharp curve in the railroad. The project involved blasting rock outcroppings alongside the railroad. A high-pressure petroleum products pipeline owned by a sister company ran adjacent to the project. Prior to Hanousek taking over the supervision of the project, the pipeline was protected from damage. However, Hanousek ordered a different procedure, which left the pipeline unprotected. A few months thereafter, a backhoe operator struck the pipeline, causing a rupture. An estimated 1,000 to 5,000 gallons of oil were discharged over the course of many days into the adjacent Skagway River, a navigable water of the United States.

Following an investigation, Hanousek was charged with one count of negligently discharging a harmful quantity of oil into a navigable water of the United States, in violation of the Clean Water Act. Hanousek was also charged with one count of conspiring to provide false information to U.S. Coast Guard officials who investigated the accident, in violation of *18 U.S.C. § § 371*. Hanousek was convicted by the jury only on the count of negligently discharging a harmful quantity of oil into a navigable water of the United States. The court imposed a sentence of six months of imprisonment, six months in a halfway house, and six months of supervised release, as well as a fine of $5,000. The conviction was upheld on appeal.[34]

Violators are subject to criminal prosecution at both the state and federal levels for the same act. This is not a violation of double jeopardy. **Double jeopardy** is the legal principal that a person can only be subject to criminal prosecution once by any sovereign. Since the state and federal governments are independent sovereigns, each can prosecute.

EXAMPLE

Luthar Construction intentionally disposes of used oil in a nearby stream rather than properly disposing of it. The Clean Water Act provides for criminal penalties for willful violations of the act. Assuming the state has a similar clean water statute, Luthar Construction can be criminally prosecuted by both the federal and the state government.

Many federal and state environmental acts allow individuals to enforce environmental laws; such suits are called **citizen suits.**

EXAMPLE

"Each person has the right to a clean and healthful environment as defined by laws relating to environmental quality including conservation, protection, and enhancement of natural resources. Any person may enforce this right against any party, public or private, through appropriate legal proceedings." Hawaii Constitution, Art. XI, § 9.

At the federal level, citizen suits in the environmental arena are allowed only if the government is not prosecuting the violation and if the violation is continuing. The person may recover attorney fees, but statutory penalties are paid to the appropriate government entity.

In addition, if an entity has been damaged by hazardous waste or other acts, that entity may sue under tort law for damages. Citizens may be able to collect under nuisance, trespass, negligence, and/or and inverse condemnation theories. A trespass can be any physical invasion of the property of another—for example by dumping waste on another's land. A party can be liable for negligence by acting unreasonably. If a government agency is involved in illegal dumping or another violation of environmental law, a citizen could sue for inverse condemnation to recover the value of the property taken by the government. See Chapter 12.

■ Real Property

Ownership Interests and Estates

Land can be owned in several different ways. The most common ownership type is **fee simple** or **fee simple absolute,** which means that the owner or owners have the most extensive interest in the property currently recognized by the law. The owner's interests under this form of ownership include the ability to transfer and encumber the property as desired.

A **life estate** is an estate that terminates upon the death of the holder of the life estate. During their life, the owner of a life estate may occupy or rent the property to others but may not sell the property. The life estate holder must maintain the property during their life. Every life estate is matched with an owner of the future interest or remainder interest in the land. A **future interest or remainder interest** is an interest that becomes possessory at some point in the future—for example, upon the death of the life tenant.

EXAMPLE

Betty and Franklin have three children: Caitlyn, Todd, and Mark. Caitlyn is unmarried and not likely to marry, Todd is married but has no children, and Mark is married with three children. Both Mark and Todd own their own homes, and Caitlyn rents a small apartment. Betty and Franklin own a house, and in their will, they instruct the executor to transfer the property to Caitlyn for life with the remainder to Mark's children alive on Caitlyn's death. Caitlyn is the owner of a life estate in the house, and Mark's children are the future or remainder interest holders. During her life, Caitlyn can live in the house, but upon her death, it is transferred to the living children of Mark. During her life, Caitlyn must maintain the property in a livable condition.

Two or more people or entities may have co-ownership interests in land. **Joint tenancy or joint tenancy with right of survivorship** means that the joint tenants are both co-owners of the property with equal rights to the use and enjoyment of land. In addition, this form of ownership transfers ownership to the surviving joint tenant upon the death of the other joint tenant. The surviving joint tenant may have to record a copy of the death certificate or other paperwork.

EXAMPLE

Joe and Helen Millar own their home as joint tenants. They have two children. Joe dies, and several years later, Helen remarries. The house belongs solely to Helen, and the children have no interest in the property.

An alternative form of co-ownership is tenancy in common. **Tenancy in common** means that the co-owners have equal rights to the use and enjoyment of the land but also the power to transfer their share by will. If no will is left, the property goes by intestate succession. **Intestate succession** is the state law that outlines which relatives receive a person's estate upon death.

EXAMPLE

Ling Lee and Han Lee own their home as tenants in common. They have two children. Ling has a will that gives all of her property in equal shares to her children upon her death as tenants in common. Upon the death of the Ling, Han is still a half owner of the home, but now the other half is co-owned by his children.

EXAMPLE

Taylor and Sarah Ungang own their home as tenants in common. They have two children, and neither has a will. The law of intestate succession in their state divides a married person with two children's property as follows: one-half of the deceased spouse's property goes to the surviving spouse, and the remaining half is divided among the surviving children. Upon the death of Taylor, Sarah retains her half interest in the property, inherits half of Taylor's interest (one-fourth of the total), and the children split Taylor's other half. Sarah now owns three-fourths of the property, and children each own one-eighth.

Tenancy by the entirety is a form of ownership recognized in some states. **Tenancy by the entirety** is a type of co-ownership similar to joint tenancy except that the co-owners must be married. Neither party may alienate their interest in the property without the permission of the other. Both spouses must act together to sell the property or subject it to a lien. Property held by the entireties is not subject to an individual debt of one of the spouses.

When holding property as tenants by the entirety, upon the death of either, the survivor becomes the sole owner.

254

Nonpossessory Rights in Land

Nonpossessory rights are rights to use the land belonging to another. The most common nonpossessory right is an easement. An **easement** allows an entity to use the land of another. Other nonpossessory rights are profit à prendre and license. A **profit à prendre** allows an entity to take something, such as gravel, from the land of another. A **license** allows another to use the land for a limited purpose such as temporary parking.

Easements can be held by other property owners, government entities, and utility companies for a variety of purposes. Other property owners may have rights to cross the property of others to get to their own property or to lakes or oceans. Government entities may have rights to install sidewalks or driveways. Utility companies may have rights to install pipes or electric poles and wires.

The rights of easement holders are protected by the law, and the owner of the land must respect them. This does not mean that the owner entirely gives up its rights to the land.

In recent years, landowners who wish to protect the environment and prevent development of their property by future generations have employed the conservation easement. A **conservation easement** is an easement that prevents the owner of the land from certain uses such as industrial development. The owner of the land gives an easement to a nonprofit organization such as the Nature Conservancy or other entity. When the owner sells or gifts the property to another, the new owner is limited in what can be done with the property.

EXAMPLE

Georgia. Code Ann. *§ 44-10-2* (Supp. 2001). "Conservation easement" means a nonpossessory interest of a holder in real property imposing limitations or affirmative obligations, the purposes of which include retaining or protecting natural, scenic, or open-space values of real property; assuring its availability for agricultural, forest, recreational, or open-space use; protecting natural resources; maintaining or enhancing air or water quality; or preserving the historical, architectural, archaeological, or cultural aspects of real property."

Land Use and Regulation

The use of land can be limited by public and private entities. Limits placed by governments are called **zoning regulations.** Government entities need not reimburse owners for any diminution in value of the land resulting from a zoning regulation.

EXAMPLE

The Village of Euclid adjoins and is practically a suburb of the city of Cleveland. Over the years, industry has expanded into the Village, and in an attempt to stop industrialization of the Village, the Village adopted a comprehensive zoning plan for regulating and restricting the location of trades, industries, apartment houses, two-family houses, single-family houses, the lot area to be built upon, the size and height of buildings, and so on. Amber Realty Company owned a certain lot that it intended to develop for industrial purposes but would be unable to because of the new zoning regulations. Amber offered proof that for industrial purposes, the lot was worth about $10,000 per acre, but for residential purposes, it was worth only about $2,500 per acre.

Amber filed a suit to overturn the ordinance, claiming that it was a violation of the Fourteenth Amendment to the Federal Constitution, because it deprived it of property without due process of law and denied it the equal protection of the law. In addition, Amber claimed violations of the Ohio Constitution.

The Village's right to enact the zoning ordinance was upheld by the U.S. Supreme Court. In addition, the Village did not need to reimburse Amber for the diminution in value of its property.[36]

A **nonconforming use** is a particular landowner's use that continues after zoning regulations require a different use.

> ### EXAMPLE
>
> The county limits use of land in a certain quadrant to single-family residences. However, an existing dog kennel uses one of the lots and can continue to use the lot for its business.

A **variance** is an exemption or change to a zoning regulation given to a particular landowner. It usually involves minor changes in the zoning regulation.

> ### EXAMPLE
>
> The county requires 15 feet of open space between the property line and buildings. Mr. Gough wants to build a garage within 10 feet of his property line. The adjoining landowner, Ms. Dillard, has no objection. After notice and a hearing, the county grants a variance and allows Mr. Gough to build the garage within 10 feet of the property line.

Many land-use plans prevent certain types of buildings, such as hospitals, schools, and churches, in certain area unless by special use permit. A **special use permit** is granted to an entity after an application is made. This process gives the county or other government agency more control over the building.

Limits placed by private entities are called **covenants** or **restrictive covenants.**

> ### EXAMPLE
>
> A developer subdivides an area into one-acre lots and puts a restriction on the deed that all lots must be a minimum of one acre and cannot be further subdivided. This is an example of a covenant.

Many states have enacted historic building codes and statutes to protect these buildings from damage and/or demolition. Owners can be prevented from demolishing or making changes to buildings that destroy the historic character of the buildings. Power may exist in a state agency to purchase a building or site or to require conformance to specifications to maintain the historic character of the building. As with zoning regulations, the government entity need not compensate the owner for any diminution in value of a property designated as a historic building or site.

EXAMPLE

29 § 551 (Delaware). "Powers with respect to historical buildings, sites, objects, and archaeological resources.

"To prevent the further loss of part of our national heritage and culture through the deterioration or neglect of historic buildings, sites, objects, or archaeological resources within this State, including archaeological resources in or on subaqueous lands pursuant to Chapter 53 of Title 7, the Department of State may survey, examine, select for preservation, acquire, repair, restore, operate, and make available for public visitation and use such historic buildings, sites, objects, or archaeological resources as it may deem worthy of preservation in the best public interest for the fulfillment of the purposes of this subchapter."

EXAMPLE

The Penn Central Transportation Company wanted to alter Grand Central Terminal, an historical building in New York, by constructing a tall office building over the top of the existing terminal building. The owner was not allowed to alter the building due to the building's protection under New York City's Landmarks Preservation Law.[37]

EXAMPLE

A developer purchased land with an accompanying apartment complex and held onto it for several years with the expectation of eventually developing the property. In the meantime, the land was declared a historic landmark. When the developer sought a permit to develop the site, it was denied due to the historic preservation status of the land. The owner fought the designation as an unconstitutional taking, because the historic designation severely limited the future economic benefit to the owner. The court upheld the historic designation and thus the denial of the developer's ability to develop the site. The court stated that the denial of a building permit did not deprive the owners of the land of all possible economic benefit of the property and therefore was permissible.[38]

In an attempt to encourage historic preservation, the federal government and many states offer tax credits to renovators of historic buildings. At the present time, the federal government offers a 20% investment tax credit for qualified rehabilitation of certain historic buildings when the buildings are rehabilitated and then used to produce income.

■ Intellectual Property

The types of intellectual property of most concern to people operating in the construction industry are trade secrets, copyrights, and trademarks. Patents are also important to engineers and others who invent new processes.

Trade Secrets

Image courtesy © istockphoto.com/jskiba

A **trade secret** is a piece of confidential information owned by a business. This information gives the business an economic advantage over other competing businesses that do not have the information. The information can be anything from a customer list to a secret formula or process. The Florida statute defines a trade secret as "information, including a formula, pattern, compilation, program, device, method, technique, or process that derives independent economic value, actual or potential, from not being generally known to, and not being readily ascertainable by proper means by, other persons who can obtain economic value from its disclosure or use; and is subject of efforts that are reasonable under the circumstances to maintain its secrecy."[39]

A trade secret is protected for as long as the entity owning the trade secret takes reasonable steps to keep it a secret. A patent, on the other hand, is protected for only 20 years in the United States. The Economic Espionage Act of 1996 (18 USC 1831–1839) makes it a federal crime to take another's trade secret. Fines of up to $5 million and prison terms of up to 15 years per violation are allowable under the act. In addition, most states have adopted a version of the Uniform Trade Secrets Act.

The Rogers developed a concept for EarthDweller, a retail business that would market environmentally sound products to the public. They developed a unique marketing strategy for EarthDweller. During the development phase, they met Rothnagel, a businessman with 40 years of experience in developing and running various retail businesses. The Rogers retained Rothnagel to assist in building the store. Rothnagel stated that he needed the business plan, architectural drawings, and blueprints to complete the project. The Rogers gave him the information with the provision that the information was a trade secret.

After the store was opened and successful, Rothnagel suggested that the parties enter into a licensing agreement whereby Rothnagel would be allowed to use the business plan, architectural drawings, and blueprints to open another store. The parties entered into negotiation of the licensing agreement and, during that time, the Rogers provided more detailed and confidential information to Rothnagel, always with the proviso that the information was a trade secret. Negotiations broke down, and Rothnagel opened another store based on the Rogers' information anyway. Rothnagel was liable for misappropriating the Rogers' trade secrets.[40]

U.S. Surveyor provides surveying services and survey coordination services nationwide and has been in operation for more than 10 years. It essentially acts as an intermediary between its customers and surveyors. A potential customer contacts U.S. Surveyor. U.S. Surveyor then identifies a potential surveyor from its database, contacts the potential surveyor, prepares the bid, coordinates the paperwork, and confirms that the job has been performed to the customer's satisfaction.

U.S. Surveyor has, over the years, compiled a detailed database of surveyors. Its database includes an internal rating system and other information about the surveyors it recommends.

Harding was employed by U.S. Surveyor as a national sales manager and then as vice president of its national sales but was terminated. Shortly after his termination, he went to work for Land Services, a competitor of U.S. Surveyor. He used the information from U.S. Surveyor's database, and Land Services made approximately $500,000 based on U.S. Surveyor's data. U.S. Surveyor sued Harding for infringement of trade secrets.

Harding claimed that the information in the database was available on the Web and therefore was not a trade secret. However, the court said that because the database contained U.S. Surveyor's internal ratings and other data compiled by U.S. Surveyor, the information was a trade secret.[41]

Copyright

Copyright protects writings, drawings, maps, technical drawings, contracts, artwork, musical scores, and other types of materials. Whether or not a copyrighted work has been illegally copied is a question of fact for a jury.[42] The Architectural Works Copyright Protection Act[43] specifically recognizes "architectural works" as copyrightable. Copyright protection for works created after January 1, 1978 is the life of the author plus 70 years. A copyrightable work owed by a corporation is 95 years from the date of first publication or 120 years from date of creation, whichever is shortest. Damages are typically lost profits due to the infringement, although other damages such as attorney fees and punitive damages may also lie.

EXAMPLE

T-Peg sells packages for the construction of timber frame homes. The package contains architectural designs and materials needed for the construction. It creates a *partial* plan for Isbitski. T-Peg copyrights the design. Isbitski does not complete the purchase from T-Peg but instead hires Vermont Timber Works to erect the home using T-Peg's partial design together with additional design work provided by Vermont Timber Works. Both Isbitski and Vermont Timber Works have violated T-Peg's copyright.[44]

EXAMPLE

Cherveny designed a home for a homeowner, and the home was built by Winmar Homes. Cherveny copyrighted the design. Later, Winmar Homes used the same plans without Cherveny's permission. Winmar Homes was in violation of Cherveny's copyright.[45]

The employer, not the employee, owns the copyright of a work created during the course of employment.[46] This is called the "works for hire" doctrine. The parties can agree otherwise.

A bid proposal including estimates of scope of work, price, and quantities is not copyrightable.

Olin, the contractor, solicited a bid proposal from PDG for asbestos removal. After receiving PDG's bid and several others, Olin decided to self-perform the work, copied some of the information from PDG's bid proposal, and submitted it to the owner. Olin then self-performed the asbestos removal.

The bid proposal is not copyrightable because "no author may copyright his ideas or the facts that he narrates."[47] "The source of Congress's power to enact copyright laws is Article I, § 8, cl. 8, of the Constitution, which authorizes Congress to 'secure for limited times to authors...the exclusive right to their respective writings.'[48] Originality is therefore a constitutional requirement. Because facts do not owe their origin to an act of authorship, no one may claim originality as to facts."[49]

A site plan is not copyrightable. "To the extent that the site plan sets forth the existing physical characteristics of the site, including its shape and dimensions, the grade contours, and the location of existing elements, it sets forth facts; copyright does not bar the copying of such facts."[50] A compilation of facts, no matter how much work has gone into the compilation, is not copyrightable. The U.S. Supreme Court has attempted to clarify this difference by explaining that copyright protection can extend only to original authorship. The facts are not created by the author's act and therefore are not copyrightable.[51] Other aspects of the map, for example its color or original presentation, can be copyrighted.[52]

The law is not copyrightable. Statutes and cases can be posted in writings and Web pages.[53]

Trademark

A **trademark** is a name, picture, or drawing that distinguishes a product or service from other products and services. A trademark is protected as long as it continues to be used. A party violates another's trademark when there is a likelihood of confusion in the minds of consumers when confronted by the two marks. The second comer to a market is required to "so name and dress [their] product as to avoid all likelihood of consumers confusing it with the product of the firstcomer."[54] "The gist of a claim for trademark infringement...is a sanction against one who trades by confusion on the goodwill or reputation of another, whether by intention or not."[55]

Patent

Patents are protection for machines and processes. Recently, patent protection has been expanded to intangible processes such as computer programs.

EXAMPLE

Signature Financial Group received a patent for a computerized accounting system that determines share prices. It is used in managing mutual fund accounts; see U.S. Patent No. 5,193,056. State Street sued to have the patent declared invalid but lost because the court held that the computer program was a process that produced a tangible result.[56]

Business models and other forms of computer business operation may be patentable. However, models cannot be obvious. Software is usually copyrighted. This area of the law is not settled.

EXAMPLE

Most online buyers abandon their purchase before completing the transaction. To increase sales, Amazon.com developed a "one-click" process. Customers who had registered to use this process could make a purchase with a single click. Amazon patented the process, but the patent was not valid because the process was obvious and had been used and suggested by others prior to Amazon filing its patent.[57]

It is extremely expensive to protect a patent.

EXAMPLE

Donald BonAsia invented "fork-chops": two eating utensils with chopsticks on one end and a knife and fork on the other. He spent two years and $7,500 to receive a patent. Others began to manufacture his product without his permission. He went to a patent lawyer and found out that it normally costs the client about $1.5 million to defend a patent suit through trial..[58]

Once a patent expires, competitors can copy the product.

EXAMPLE

Marketing Display held two patents for a dual-spring sign mechanism that kept temporary road and other outdoor signs upright in strong winds. After the patents ran out, a competitor, Traffix, began producing signs with the same design. Marketing Display sued, claiming its dual-spring design was protected as trade dress because it was visible near the base of the stand and recognizable to customers. The dual-spring design was not a trade dress or mark because it was functional. It was no longer protected by patent, and competitors could copy it.[59]

Endnotes

1 42 USC § 13101 *et seq.*

2 5 U.S.C. § 706(2)(A).

3 Example based on *Tahoe-Sierra Preservation Council v. Tahoe Regional Planning Agency*, 535 U.S. 302 (2002).

4 Example based on *The Fund for Animals, Inc. v. Rice*, 85 F.3d 535 (11th Cir. 1996).

5 42 U.S.C. § 9603(a) (2000). Reportable quantities are regulated and defined under 42 U. S.C. §9602(a) and (b).

6 *Cincinnati Ins. Co. v. Becker Warehouse, Inc.*, 262 Neb. 746, 635 N.W.2d 112 (2001).

7 *Selm v. American States Ins. Co.*, 2001 WL 1103509 (Ohio Ct. App. 1st Dist. Hamilton County 2001) (unpublished opinion).

8 *Tri County Service Co., Inc. v. Nationwide Mut. Ins. Co.*, 873 S.W.2d 719 (Tex. App. San Antonio 1993).

9 *White v. Freedman*, 227 A.D.2d 470, 643 N.Y.S.2d 160 (NY Sup. Ct. 1996).

10 *U.S. v. Dravo Corp. et al.*, 202 U.S. Dist LEXIS 22521, 54 ERC (BNA) 1539 (D. Neb. 2002).

11 *Iko Monroe, Inc. v. Royal & Sun Alliance Ins. Co. of Canada, Inc.*, 2001 WL 1568674 (D. Del. 2001) and *Zell v. Aetna Cas. & Sur. Ins. Co.*, 114 Ohio App. 3d 677 (Ohio Ct. App. 1996).

12 *American States Ins. Co. v. Nethery*, 79 F.3d 473 (5th Cir. 1996).

13 *Boulevard Investment Co. v. Capitol Indemnity Corp.*, 27 S.W.3d 856 (Mo. App. 2000).

14 42 U.S.C. 6901–6992k (1994).

15 Example based on *Kaiser Aluminum & Chemical Corp. v. Catellus Dev. Corp.* 976 F.2d 1338, 92 CDOS 8405, 92 Daily Journal DAR 13871, 35 Envt Rep Cas 1689, 23 ELR 20020 (9th 1992).

16 See Ian Erickson, *Reconciling the CERCLA Useful Product and Recycling Defenses*, 80 N.C.L. Rev. 605 (2002) for discussion of intent.

17 Example based on *U.S. v. Wedzeb Enters., Inc.*, 844 F. Supp. 1328, 1337 (S.D. Ind. 1994).

18 42 U.S.C. § 9601(35), CERCLA § 101(35).

19 42 U.S.C. § 9601(20)(A), CERCLA § 101(20)(A).

20 42 U.S.C. § 9601(20)(B), CERCLA § 101(20)(B).

21 42 U.S.C. § 9601(10), CERCLA § 101(10).

22 Pub. L. No. 107-118, 115 Stat. 2356 (2002), 42 U.S.C. § § 9601(40) and 9607(r).

23 See *Catellus Dev. Corp. v. U.S.*, 34 F.3d 748, 751 (9th Cir. 1994) (spent batteries not waste because they were being recycled), *Morton Int'l, Inc. v. A.E. Stalely Mfg. Co.*, 106 F. Supp. 2d 737, 747 (D.N.J. 2000) (mercury recycling), *U.S. v. Atlas Lederer Co.*, 97 F. Supp. 2d 830 (S.D. Ohio 2000) (recycling arrangements in general).

24 33 U.S.C. 1362(6) (2000)

25 33 USC 403.

[26] Example based on *Baykeeper v. U.S. Army Corps of Engineers*, 2006 U.S. Dist. Lexis 67483 (U.S. Dist. Ct. Eastern Dist. of Ca., 2006); this is a pending matter.

[27] *Babbitt v. Sweet Home*, 515 U.S. 687 (1995).

[28] Example based on *Rancho Viejo, LLC v. Norton,* 334 F. 3d 1158, 1160 (D.C. Cir. 2003), cert. denied, 124 S. Ct. 1506 (2004).

[29] See John F. Turner & Jason C. Rylander, *Conserving Endangered Species on Private Lands,* 32 Land & Water L. Rev. 571, 577 (1997). Lin, Albert C., *Participants' Experiences with Habitat Conservation Plans and Suggestions for Streamlining the Process*, 23 Ecology L.Q. 369, 372 (1996). See also Jean O. Melious & Robert D. Thornton, *Contractual Ecosystem Management Under the Endangered Species Act: Can Federal Agencies Make Enforceable Commitments?*, 26 Ecology L.Q. 489, 495 (1999).

[30] Example based on *Center for Biological Diversity v. Marina Point Development Asso.*, 434 F. Supp. 2d 789 (D.C. C. Ca. 2006).

[31] 40 C.F.R. § 61.145(c)(6)(i).

[32] Example based on *U.S. v. Price*, 314 F.3d 417, 2002 U.S. App. LEXIS 26745, 2002 Cal. Daily Op. Service 12404, 2002 D.A.R. 14622, 56 Env't Rep. Cas. (BNA) 1028, 33 Envtl. L. Rep. 20146 (9th Cir. Nev. 2002). For procedural reasons, the defendants' convictions were overturned and the matter remanded to trial court for a new trial. See also *U.S. v. Buckley*, 934 F.2d 84 (6th Cir. 1991).

[33] Example based on *U.S. v. Pozsgai,* 757 F. Supp. 21 (E.D. Pa. 1991).

[34] Example based on *U.S. v. Hanousek*, 176 F.3d 1116 (9th Cir. 1999), cert. denied, 528 U.S. 1102, 120 S.Ct. 860, 145 L.Ed.2d 710 (2000).

[35] Example based on *Lewis v. Young*, 705 N.E.2d 649 (N.Y. App. 1998).

[36] Example based on *Village of Euclid v. Ambler Realty Co.*, 272 U.S. 365, 47 S. Ct. 114, 71 L. Ed. 303, 54 A.L.R. 1016 (1926).

[37] Example based on *Penn Cent. Transp. Co. v. City of New York*, 438 U.S. 104 (1978).

[38] Example based on *Intown Properties, LP v. District of Columbia*, 198 F.3d 874 (D.C. Cir. 1999).

[39] Fla. Stat. § 688.002.

[40] Example based on *Earthdweller, Ltd. v. Rothnagel*, 1993 U.S. Dist. Lexis 16531 (ND Ill. 1993).

[41] Example based on *U.S. Land Services, Inc. v. U.S. Surveyor, Inc.*, 826 N.E.2d 49, 2005 Ind. App. Lexis 666, 22 I.E.R. Cas. (BNA) 1405 (Ind. App. 2006).

[42] *Leigh v. Warner Bros. Inc.,* 212 F.3d 1210, 1213 (11th Cir. 2000).

[43] 17 U.S.C. § 102(a)(8).

[44] Example based on *T-Peg, Inc. v. Vermont Timber Works, Inc.*, 459 F.3d 97, 79 U.S.P.Q.2d (BNA) 1919, Copy. L. Rep. (CCH) P29, 238 (1st Cir.).

[45] Example based on *Axelrod & Cherveny Architects, P.C. v. Winmar Homes*, 2007 U.S. Dist. Lexis 15788 (ED NY 2007).

[46] 17 U.S.C. § 201 (b).

[47] *Harper & Row Publishers, Inc. v. Nation Enterprises*, 471 U.S. 539, 556, 85 L. Ed. 2d 588, 105 S. Ct. 2218 (1985).

48 United States Constitution, Article I, § 8, cl. 8.

49 Example based on *Project Development Group, Inc. v. O.H. Materials Corp.*, 766 F. Supp. 1348, Copy. L. Rep. (CCH) P26,799 (WD Penn. 1991).

50 *Sparaco v. Lawler, Matusky, Skelly, Engineers LLP*, 303 F.3d 460, 64 U.S.P.Q.2d (BNA) 1363, Copy. L. Rep. (CCH) P28,476 (2d Cir. 2002).

51 *Feist Publications, Inc. v. Rural Telephone Service Co.*, 499 U.S. 340, 347-48, 113 L. Ed. 2d 358, 111 S. Ct. 1282 (1991).

52 *County of Suffolk, New York v. First American Real Estate Solutions*, 261 F.3d 179, 188 (2d Cir. 2001).

53 *Veeck v. Southern Building Code Congress International, Inc.*, 293 F.3d 791, 63 U.S.P.Q. 2d (BNA) 1225, Copy. L. Rep. (CCH) P28, 448 (5th Cir. 2002).

54 *Osem Food Indus. Ltd. v. Sherwood Foods, Inc.*, 917 F.2d 161, 165 (4th Cir. 1990).

55 *Quality Inns Int'l, Inc. v. McDonald's Corp.*, 695 F. Supp. 198, 209 (D. Md. 1988).

56 Example based on *State Street Bank & Trust Co. v. Signature Financial Group, Inc.*, 149 F.3d 1368 (Fed. Cir. 1998).

57 Example based on *Amazon.com, Inc. v. Barnesandnoble.com, Inc.*, 239 F. 3d 1343 (Fed. Cir. 2001).

58 See Forkchops, Inc., http://www.theforkchopstore.com (accessed May 30, 2007).

59 Example based on *Traffix Devices, Inc. v. Marketing Displays, Inc.*, 532 U.S. 23, 121 S. Ct. 1255, 149 L. Ed. 2d 164 (2001).

Insurance

■ Introduction

The costs associated with construction insurance have soared in recent years.[2] The numbers of carriers have declined as premiums have not kept up with claims paid. Insurance companies agree to pay to policyholders certain sums in the event that a covered event occurs, but the tremendous increase in construction defect litigation has led to much litigation, often unfavorable to insurance companies.

One author writes, "[C]onstruction defect litigation is rapidly growing, with signs of even more growth on the horizon. Attorneys and clients new to construction defect litigation will find more questions than answers as they evaluate new cases."[3] The basic problem is that insurance companies do not want to accept the risk for defective construction and attempt to clearly state so. While professionals such as architects and engineers can obtain insurance for their negligent acts, the insurance industry has not been willing to cover contractors for negligent construction. The pressure to find coverage for negligent construction is great.

■ Common Insurance Issues

This section discusses insurance issues common to all policies. Later, issues common to specific types of insurance are discussed. Issues common to all polices include the following:

- Who is covered, including insured, additional insured, and indemnity agreement
- Duty to defend and reservation of rights
- Whether or not an event is an "occurrence" under the policy and potentially covered
- Exclusions
- Old and or multiple policies
- Insurance company bad faith

Indemnity Agreement Issues

An entity purchases an insurance policy; if a claim is made against that entity, potential coverage exists. If a company changes its name or incorporates, it must inform its insurance company.

EXAMPLE

Bill and Bob are doing business as Advanced Architects. They decide to incorporate, and the name of the new corporation is Advanced Architects, Inc. Advanced Architects, Inc. is a new entity, and either additional policies must be purchased or the corporation can be named as an additional insured on Advanced Architects' policies.

Advanced Architects, Inc. decides to work with Contractor, Inc. as a design build firm. The principals in the two corporations form another corporation, Design-Build, Inc. The principals must be careful and get insurance for the new corporation.

MANAGEMENT TIP

Always contact your insurance company if the name of your business changes or you form another business.

In the construction industry, it is common for multiple businesses to work together on one project. Because of the possibility that one entity may be legally responsible to pay claims for which another is at least partially liable, it is common for the parties to enter into indemnity agreements. However, because of the problems inherent with indemnity agreements that the preferred method of allocating the risk has been for entities to be listed as additional insureds on policies covering the work.

Three types of indemnity agreements exist: limited, intermediate, and broad form. Because limited form indemnity agreements do not obligate the indemnitor to any liability other than what is already legally required, the signer's existing insurance will defend and cover claims. The **indemnitor** is an entity who agrees to pay damages and usually defense costs that another entity is legally obligated to pay.

However, signing an intermediate or broad form indemnity agreement increases the liability of the signer; that is, the signer is agreeing to take on liability for which the signer is not at fault. The indemnitor's insurance coverage will only partially apply, and the signer is personally obligated for the additional defense and coverage costs. See further discussion below.

The limited indemnity agreement requires Business B, often a subcontractor, to pay Business A, often a general contractor, any damages and defense costs Business A must pay but which are caused to Business B's acts or omissions. See the form below. The law actually requires this, and the existence or lack thereof of such an agreement between the parties is irrelevant. If not based upon an indemnity agreement but upon the law, this type of claim is called "a **contribution** claim."

FORM

Limited Form of Indemnity: Subcontractor shall at all times indemnify, protect, defend, and hold harmless Contractor and Owner from all loss and damage, and against all lawsuits, arbitrations, mechanic's liens, legal actions, legal or administrative proceedings, claims, debts, demands, awards, fines, judgments, damages, interest, attorney's fees, and any costs and expenses which are directly or indirectly caused or contributed to, or claimed to be caused or contributed to by, any act or omission, fault or negligence, whether passive or active, of Subcontractor, its agents, or employees, in connection with or incidental to the work performed under this Agreement.

EXAMPLE

A general contractor hires a subcontractor to do electrical work. The contract contains the above indemnity agreement. No additional insured coverage exists, only the indemnity agreement. The subcontractor's work is defective and injures the homeowner. The homeowner sues the general contractor. The general contractor sues by filing a counterclaim against the subcontractor on the indemnity agreement.

The general contractor is legally liable to the homeowner for the subcontractor's work. The subcontractor is legally liable to general contractor for any sums the general contractor must pay resulting from the subcontractor's defective work.

Under the indemnity agreement, the subcontractor is required to pay for defense costs and damages. Of course, when the lawsuit is started, it may not be clear that the subcontractor's work was defective. The subcontractor, or its insurance company, will

continued

refuse to defend the general contractor. The general contractor must turn over the claim to its own insurance company. Once liability is determined, it will be determined who pays defense costs and damages. As most cases are settled, it is likely that the insurance companies will reach a settlement rather than litigate actual liability.

EXAMPLE

The general contractor hires the subcontractor to do electrical work. The contract does *not* contain an indemnity agreement. No additional insured coverage exists. The subcontractor's work is defective and injures the homeowner. The homeowner sues the general contractor. The general contractor sues by filing a counterclaim against the subcontractor for contribution. The result is the same as in the prior example, because the law recognizes the responsibility of the subcontractor for damages it causes. This responsibility includes repaying another party who pays or is responsible for the subcontractor's damages.

EXAMPLE

An employee works for Company A. The employee's negligent acts cause damage to the customer. The customer sues Company A, and Company A is legally obligated to pay damages to the customer. Company A can sue the employee for contribution, as employees are legally liable for the damages they cause. Of course, in the real world rather than the legal world, the employee may have no money or may have left the area, and it is not worth it for Company A to sue the employee.

MANAGEMENT TIP

Sign a limited form indemnity agreement. The law imposes this liability on all entities whether or not such an agreement has been signed. Limited form indemnity agreements do not obligate the signer to any more liability than the signer already possesses.

Even if you do *not* have an indemnity agreement with another party but you are legally obligated to pay damages caused by that other party, you can sue the party for contribution.

If the other entity no longer exists or has no money, it may still be advisable to pursue the contribution lawsuit. This can be done as part of the original lawsuit, and it is often relatively inexpensive to get a judgment against the liable party. If the judgment is uncollectible, seek tax advice on whether or not it may be written off as a bad debt.

EXAMPLE

A general contractor has an agreement with a roofing subcontractor. No indemnity agreement exists, and the general contractor is not an additional insured on the subcontractor's insurance. The homeowner sues the general contractor, claiming that the roofing subcontractor's work was negligent and caused injury to the homeowner. Assume this to be true. Even if the homeowner does not sue the roofing subcontractor, the general contractor can file a counterclaim against the roofing subcontractor in the lawsuit brought by the homeowner. This will not relieve general contractor from liability to the homeowner, but the general contractor can get a judgment against the roofing subcontractor and either try to collect it or write it off as a bad debt.

The intermediate form of indemnity is one requiring the indemnitor to pay *all* of the damages and defense costs should the indemnitor be even *partially* at fault for the damages. This clause dramatically increases the risk of the entity agreeing, and it is unlikely that an insurance policy will automatically cover this additional risk. Always contact your insurance company if your contract contains an intermediate or broad form (discussed below) indemnity agreement.

EXAMPLE

A general contractor signs an intermediate form indemnity agreement with the owner (see form below) by which the general contractor, now the indemnitor, agrees to indemnify the owner for all damages and defense costs should the general contractor be partially at fault for the damage. A roof collapses and injures several people. It is determined that the collapse was partially caused by the contractor's defective work and partially caused by the owner's designer's defective work. The general contractor is responsible for *all* of the damages to the people and the defense costs.

At the time the lawsuit is filed, the cause of the the collapse is unlikely to be clear, and the insurance companies of all of the involved parties will become involved in the litigation.

To the fullest extent permitted by law, Contractor shall indemnify and hold harmless Owner, its consultants, agents, and/or employees from and against claims, damages, losses, and expenses, including but not limited to attorney's fees and court costs, arising out of or resulting from performance of the Work, provided that such claim, damage, loss, or expense is attributable to bodily injury, sickness, disease, or death, loss of use of property, loss of use of the Work, or to injury to or destruction of tangible property other than to the Work itself, *caused in whole or in part by* the Contractor, its employees, subcontractors, sub-subcontractors, and/or anyone directly or indirectly paid by Contractor or anyone for whose acts Contractor may be liable. (Italics added to emphasize important wording.)

272

MANAGEMENT TIP

The risks in signing such an agreement are, in most cases, too great to warrant signing. If the limited form indemnity agreement cannot be negotiated as a substitute for the above, it is probably better to name the other party as an additional insured on your insurance policy.

If you sign such an agreement, contact your insurance carrier immediately. Failure to do so will mean that your insurance company will pay for damages and defense costs attributable to your negligence, and you will be liable for damages and defense costs attributable to the indemnitee's negligence or defective work. The indemnitee is the entity who has the protection of an indemnity agreement. An indemnitor agrees to pay at least some of the damages and usually defense costs that the indemnitee is legally obligated to pay.

EXAMPLE

A general contractor signs an intermediate form indemnity agreement with the owner by which the general contractor, now the indemnitor, agrees to indemnify the owner and its agents for all damages and defense costs should the general contractor be partially at fault for the damage. The general contractor does not obtain any additional insurance coverage. A roof collapses and injures several people. It is determined that the collapse was partially caused by the contractor's defective work and partially caused by the owner's designer's defective work. The general contractor's insurance is obligated to pay defense costs and damages related to the general contractor's act only; the general contractor must pay defense costs and damages related to the owner's acts because she did not notify the insurance carrier.

The final form of indemnity agreement is called a "broad form" indemnity agreement. It requires the indemnitor to pay for *all* damages and defense costs even if the indemnitor is *not responsible for any portion of the damages*. In effect, this clause makes the indemnitor the insurance company of the indemnitee. These clauses are rare, and many states refuse to uphold them. It is most common to see these forms in government contracts.

To the fullest extent permitted by law, Contractor shall indemnify and hold harmless Owner, its consultants and agents and employees of any of them from and against claims, damages, losses, and expenses including but not limited to attorney's fees, arising out of or resulting from performance of the Work, *whether or not* such claim, damage, loss, or expense *is caused in whole or in part by a party identified hereunder* and (Italics added to emphasize important wording.)

EXAMPLE

A general contractor subcontracts the manufacture of cabinets to a cabinetry subcontractor. The subcontract contains the above clause. The general contractor's employees install the cabinets. Later, one of the cabinets falls off of the wall and injures the owner. The cabinetry subcontractor is, at least under the subcontract, liable to the owner for all of the owner's damages and defense costs.

The above form is actually simplified to show important wording, and you should realize that these types of clauses are often some of the most complicated in a contract. The important words to look for in an indemnity clause are "is caused in *whole or in part* by a party indemnified hereunder." If these or similar words appear, it is a broad form indemnity clause, and care must be taken to obtain additional insurance coverage or otherwise manage the risk.

The law does not favor broad form indemnity clauses,[4] and it likely that such a clause will not be enforced in a court.[5] Most states have enacted legislation banning these types of clauses in construction contracts.[6]

MANAGEMENT TIP

It is inadvisable to sign a broad form indemnity agreement. If you do, purchase additional insurance to cover the indemnitee's liability on the job. It is inadvisable to rely upon a broad form indemnity agreement for protection before researching whether or not your particular state will uphold the provision.

Additional Insured Issues

Because of the problems inherent with indemnity agreements, it has become common for parties higher in the chain to require parties lower in the chain to add them as additional insureds on policies. This has raised different issues as insurance companies do battle over who must pay for defense costs and claims.

Even if an entity is an additional insured, the insurance company is only liable if the claim arises out of the work of the named insured.

EXAMPLE

A Construction, Inc. is a solely owned corporation owed by A. A has a friend, B, who is starting up an electrical subcontracting business, and B has the chance to get a big job for C but needs insurance. A is not on the C job but puts his friend B on A Construction, Inc.'s insurance as an additional insured. Later, a claim is made against B. A Construction, Inc.'s insurance company is *not* required to defend B nor pay the claim, as the claim does not arise out of A Construction, Inc.'s operations.

MANAGEMENT TIP

Do not put a friend's start-up company or a different corporation on your insurance, thinking that it will help them or save money. Your policy will not cover your friend's losses, because those losses do not arise out of your work; policies cover only the work of the named insured.

It is now common in the industry for general contractors to require that they be listed as additional insureds on subcontractors' general liability policies. Subcontractors should realize that they are absorbing a significant risk of increased policy premiums in the future should a claim be made against the general contractor. In the event of litigation against the general contractor, the *subcontractor's* insurance will be asked to defend and pay any claims. The general contractor may not even inform its insurance of the claim, thus shielding the general contractor from increases in its own policy premiums. In addition, subcontractors may have to pay deductibles.

Putting an entity on a policy as an additional insured is, from a legal standpoint, very different than agreeing to indemnify that entity. Even though the dollar amount of liability under a broad form indemnity agreement and an additional insured agreement may be the same, the law will uphold the duty to pay under and additional insured endorsement when most states will not uphold the duty to pay under a broad form indemnity agreement. This is likely because insurance exists for additional insured liability but not for broad form indemnity agreement liability.

Because the duty to defend, discussed in more detail below, is separate from the duty to pay claims, the insurance company must defend an additional insured should a claim arise that could even *potentially* lead to the additional insured's legal obligation to pay damages.[7]

Because the practice of naming other entities involved in the construction is relatively new, many issues have arisen. Since cases tend to find insurance coverage whenever possible, it is likely that changes in insurance language and additional insured endorsements will arise over the next few years. The example below shows a case in which the coverage for the additional insured was actually greater than the coverage of the named insured, something the insurance company probably did not predict.

EXAMPLE

A general contractor is a named insured on a subcontractor's general liability policy. As is common, the subcontractor's policy contains an exclusion of coverage for the subcontractor's work. A general liability policy does not insure or warrant the quality of the work and will not pay to replace faulty or defective work. The endorsement by the subcontractor's carrier provides the general contractor with coverage for "liability arising out of the subcontractor's work for which the general contractor is held liable." Later, when both the subcontractor and the general contractor are sued for construction defects related to the subcontractor's work, the subcontractor's carrier refuses coverage to the general contractor because the suit seeks damages for repair or replacement of defective work, something clearly not covered by the policy. The court holds that the subcontractor's endorsement covers the general contractor for *all* liability the general contractor might incur arising out of the subcontractor's work; since the general contractor is liable to the homeowner for the subcontractor's defective work, the carrier is obligated to provide defense and coverage.[8]

EXAMPLE

The general contractor was named as an additional insured on the subcontractor's comprehensive general liability insurance policy provided by Northwestern. Northwestern limited its coverage of the contractor to liability imputed to the general contractor as a result of the subcontractor's acts. The general contractor was sued by an employee of the subcontractor who claimed that the general contractor or its agents had failed to keep the work site reasonably safe. The complaint also sued the subcontractor and claimed that it or its agents had failed to keep the work site reasonably safe. However, the complaint failed to make any allegation that the subcontractor was an agent of the general contractor. The subcontractor's insurance company refused to defend the contractor. The contractor tendered the defense to its own insurance company, Emcasco, which sued Northwestern. The appeal court held that Northwestern, the subcontractor's insurance, had the duty to defend the contractor because the complaint left open the possibility of coverage.[9]

MANAGEMENT TIP

Expect to see changes in endorsements excluding coverage for the work.

Another advantage of additional insured status is the preclusion of subrogation (discussed below) or contribution counterclaims against any policies the additional insured has as a named insured; that is, the additional insured's policy can never be a source of coverage unless the additional insured files a claim on its own policy.[10] It may do this should it believe sufficient coverage does not exist, but that would be rare.

Many states have adopted the "targeted tender" doctrine, which protects the additional insured's own policy. An additional insured has the right to choose which insurer, its own or the policy on which it is named the addition insured, will defend it in a lawsuit. If an additional insured tenders its defense to the policy on which it is an additional insured, that insurance company cannot seek any type of contribution from the additional insured's own CGI or other policy.[11]

EXAMPLE

A general contractor is an additional insured on a structural steel subcontractor's policy. The general contractor has its own policy also. The structural steel subcontractor's policy has a limit of $10 million dollars. A walkway constructed by the structural steel subcontractor collapses. The general contractor tenders its defense to the structural steel subcontractor's insurance carrier and not its own.

Many policies contain a provision requiring the insured to send the insurance company copies of any claims, complaints, or notice of a claim or complaint "immediately." Failure to do so could eliminate coverage.

EXAMPLE

The general contractor was an additional insured on *both* the subcontractor's insurance and the sub-subcontractor's insurance. The subcontractor's insurance policy contained a provision requiring it to be informed "immediately" of any claims or lawsuits. After the general contractor was sued, it tendered its defense to *only* the sub-subcontractor's insurance. It was not until more than three years after the filing of the original complaint that it sought additional coverage from the subcontractor's insurance. The court held that the contractor was not entitled to coverage under the subcontractor's policy because of the delay in informing the insurance company of the lawsuit.[12]

MANAGEMENT TIP

When a claim arises, all available policies must be ascertained and informed of the claim. Failure to do so within a reasonable time after the claim is made could result in no coverage under the policy.

Duty to Defend and Reservation of Rights Issues

The insurance company's duty to defend is independent of its duty to pay claims and arises before its duty to pay claims. The insurance company must defend the insured if the complaint states even one claim that could *potentially* give rise to coverage.[13] The duty to defend exists even if other insurances may be available. The insurance company must defend, and if it chooses, it may seek contribution from other available insurance. However, the insured or the additional insured is not required to seek contribution from other available sources, either before or after the claim.[14]

EXAMPLE

The general contractor was named as an additional insured on the subcontractor's comprehensive general liability insurance policy provided by Northwestern. Northwestern limited its coverage of the contractor to liability imputed to the general contractor as a result of the subcontractor's acts. The general contractor was sued by an employee of the subcontractor who claimed that the general contractor or its agents had failed to keep the work site reasonably safe. The complaint also sued the subcontractor, and claimed that it or its agents had failed to keep the work site reasonably safe. However, the complaint failed to make any allegation that the subcontractor was an agent of the general contractor. The subcontractor's insurance company refused to defend the contractor. The contractor tendered the defense to its own insurance company, Emcasco, which sued Northwestern. The appeal court held that Northwestern, the subcontractor's insurance, had the duty to defend the contractor because the complaint left open the possibility of coverage.[15]

If, during investigation of the claim, the insurance company or surety believes that the claim is *not* covered by the policy it will usually send a "reservation of rights" letter to the insured. This letter informs the insured that insurance coverage may not exist for the claim and that the insurance company "reserves its rights" not to pay the claim if the claim is later determined in the litigation that it is not covered by the policy.

Many insurance companies send reservation of rights letters as a matter of course to protect themselves. Pennsylvania case law has held that if the insurer does not believe coverage exists, it cannot refuse to defend but must either defend under a reservation of rights letter or file a suit requesting a declaration of rights.[16]

In many states, sending a reservation of rights letter gives the insured the power to hire, at the insurance company's expense, its own attorney to make sure that the insurance company does not try to get out of covering the claim. Once the reservation of rights letter is sent, the insurance company and the insured have opposite interests. The insurance company desires to find evidence that the claim is not covered, and the insured desires to find evidence that the claim is.

MANAGEMENT TIP

Management should always seek independent legal advice when it receives a reservation of rights letter so as to determine the proper approach, which will depend on state law.

Occurrences

If an occurrence exists, it must be determined if the policy covers it, and the language of the policy must be reviewed in order to determine whether or not the claim is covered. Most policies require that the damage have occurred during the policy period, not just that the event that has given rise to the damage occurred during the policy period. However, the law is not settled in this area and not only varies from state to state but is not consistent within some states.

Old Policies and Multiple Policies

Old policies may actually still provide coverage, depending on the wording and type of coverage. In addition, it is common for general contractors to be named as additional insureds on subcontractor's policies. Additional insured is discussed above. The contractor may have its own coverage. All of these policies may result in situations where insurance companies are battling out which insurance company must defend, never mind which covers the occurrence.

MANAGEMENT TIP

Always keep old insurance policies in case a claim arises related to a project finished years ago.

Bad Faith

Virtually all insurance policies contain language preventing the insured from handling any litigation related to a claim or settling a lawsuit.[17] The insurance company maintains total control over claims. However, if the insurance company refuses to defend a claim, pay a claim, or settle a claim, the insured must generally go forward on its own, but the insured may also have a bad faith claim. An insurance bad faith claim arises when an insurance company refuses to pay or defend a covered claim. If the insurance company refuses to pay or defend a valid claim, the insured can proceed with a lawsuit against both the insurance company and the party causing injury.

If no lawsuit has been filed, the insured can settle the case and sue its own insurance company for damages. Because the law only allows recovery of reasonable damages, the insured must take care to settle the claim only for a reasonable amount.

If a lawsuit has been filed against the insured, the insured must defend the lawsuit or risk losing by default. The insured could hire its own attorney and file a cross-claim against its own insurance carrier, forcing its insurance carrier into the lawsuit early before settlement or trial on the underlying claim.

If it is determined that the insurance company's decision not to defend and/or pay a claim violated the insurance contract, the insured is entitled to a reimbursement for any defense costs and/or damages it paid that should have been covered by the policy. In addition, the insurance company may be liable for punitive or bad faith damages if it had no reasonable belief that the claim was not covered. Some states require the insurance company to pay treble damages, attorney fees, punitive damages, and/or mental injury damages. It is usually irrelevant whether or not the refusal to defend is based on an honest but mistaken belief that the claim is not covered.[18] Bad faith law strongly encourages insurance companies to properly investigate and to pay legitimate claims, something they have not always done.

EXAMPLE

An owner purchases a performance bond. Shortly after construction is completed, major defects in the roofing are noted. The owner informs the contractor and requests repairs. The contractor refuses, and the owner contacts the surety. The surety refuses to repair or honor the bond, claiming that the defects are minor. The owner repairs the structure and sues the surety for bad faith. The owner could also sue the contractor. Assuming that the repairs are reasonable and all of the procedural requirements have been met, the owner is entitled to the costs to repair the roof, the costs of the lawsuit, and any statutorily allowed punitive or additional damages.

EXAMPLE

In what has been come to be called the "Ballard mold case," which is in reality a bad faith case, the homeowners tendered a claim to their homeowners' insurance company for damages to their home caused by mold. The court awarded the Ballards approximately $28 million in punitive damages related to their insurance company's bad faith in investigating and handling the claim. The award was reduced to $4 million on appeal for repair costs, replacement costs, and attorney fees. The part of the award relating to punitive damages and emotional distress was stricken.[19]

Interpretation of Insurance Policies

Insurance contracts are a specialized form of contract, and all of the rules applicable to general contract interpretation apply to the interpretation of insurance policies (see Chapter 9). For example, the ordinary meaning rule or law states that words are to be given their ordinary meaning. Ambiguity is resolved against the insurance company. In addition, most courts will interpret the contract as understood by the average person. Disclaimers and exclusions must be clear.

In addition, a subset of common law developed with unique rules applicable only to interpretation of insurance contracts exists. The law tends to read insurance policies in favor of the claimant rather than the insurance company. "While unambiguous provisions of a policy are given their plain and ordinary meaning, where there is ambiguity as to the existence of coverage, doubt must be resolved in favor of the insured and against the insurer."[20]

EXAMPLE

A general contractor entered into a sheetrock and carpentry subcontract. The subcontract contained a provision requiring the subcontractor to obtain general liability insurance with "*minimum* general liability limits of $500,000/1,000,000" (emphasis added) and to name the plaintiff as an additional insured. The subcontractor added the general contractor as an additional insured to its existing policy, which had a limit of $1,000,000 per occurrence and $2,000,000 in the aggregate. The additional insured endorsement stated that "the *Limits* of Insurance applicable to the additional insured are those *specified in the written contract* or agreement or in the Declarations for this policy, *whichever is less*." The court held that the subcontract did not mention any *specify any limits* of insurance but merely contained a *minimum*. The general contractor therefore was entitled to the full benefits of the policy.[21]

EXAMPLE

Adjoining landowners sued the owners of certain land, the general contractor, and the excavating contractor for alleged negligent excavation by removing soil, sand, and other materials beneath the property footings and causing the adjoining landowner's properties to shift and erode. The adjoining landowners claimed that such excavation was performed without bracing or reinforcing the excavation, which caused damage to the foundation, footings, masonry walls, and other structural components of the plaintiff's properties.

The owner's and general contractor's general liability policy contained the following exclusion:

This insurance does not apply to "bodily injury," "property damage," "personal and advertising injury," or "medical payments" caused by, resulting from, contributed to, or aggravated by the "subsidence" of land.

The policy defines "subsidence" to mean "earth movement, including but not limited to landslide, mud flow, earth sinking, rising or shifting."

The court stated that "[s]uch language is ambiguous. We find that the examples given for "subsidence"—landslide, mud flow, earth sinking, rising or shifting—define the type of earth movement, *not the cause. The cause is ambiguous* (emphasis added). The issue was did the policy only exclude coverage for natural earth movement *or* natural and human-caused earth movement. Because the policy was ambiguous it was resolved in favor of the owner who was found to be entitled to coverage for the damage to the adjoining landowner's property." [22]

In some ways, insurance can be thought of as the purchase of a product, and insurance law shares many of the characteristics of product liability law. "Insurance is a product—it is risk protection. It is not an individualized agreement, arrived at by free and open arm's-length negotiations over crucial details such as those which might characterize a 'real' contract."[23] Insurance law may therefore be more closely related to product liability law, which tends to protect the buyer. "This strict construal against the insurer is driven by the fact that the insurer drafts the policy and foists its terms upon the customer. The insurance companies write the policies: we buy their forms, or we do not buy insurance."[24]

Common Issues Relating to Specific Types of Insurance

Builder's Risk Insurance

These policies cover the project during construction and will generally pay for losses due to fire, flood, theft, criminal mischief, or other damage as outlined in the policy. It is a business necessity for the contractor to purchase this insurance, as the law does not relieve the contractor from its obligation to perform the contract even if the project is partially or totally destroyed. The policy may have exclusions, and care must be taken to obtain the desired insurance. Either the contractor or the owner purchases this insurance, and no project should be undertaken without it.

It is now common for general contractors to be named as additional insureds on the subcontractor's policy. If this has been done, the general contractor is likely covered only during the time of the subcontract, not for the entire project.

EXAMPLE

A contractor entered into contract with the U.S. Army to build barges and subcontracted installation of a crane onto one of the barges. The contractor was added as an additional insured, called the "assured," on the subcontractor's builder's risk insurance. The policy between the contractor, and the subcontractor stated that the subcontractor's insurance was applicable until "completion of the *entire project.*" However, this was interpreted to mean the entire project from the subcontractor's point of view, not the contractor's. After completing the installation of the crane, the subcontractor returned the barge to the contractor, and its job was done.

The contractor was required to transport the barge from Louisiana to Virginia and purchased insurance ("transportation insurance") to cover this risk. The crane apparently fell off during transportation due to faulty design by the subcontractor's employees. The transportation insurer sued the subcontractor's insurer, claiming that the subcontractor's insurance applied until the barge was delivered to Virginia. The subcontractor's insurer claimed that the insurance lapsed when the subcontractor completed its portion of the job—the crane installation—and turned the barge over to the contractor. The court agreed. The court stated that the surrounding circumstances supported the interpretation of the insurance contract that the parties intended it to cover the project only so long as the subcontractor was working on it.[25]

In the above example, the contractor's builder's risk insurer, who had paid the claim, tried to get reimbursement from the subcontractor's builder's risk insurer. It is unlikely that the actual parties would be engaged in such litigation, although should your insurance company seek such a reimbursement, called "contribution" under the law, management would be involved in the lawsuit, as insurance policies require the insured to cooperate in such endeavors.

Comprehensive or Commercial or General Liability Insurance

The insurance company obligates itself to pay "all sums" the insured becomes "legally obligated to pay as damages" because of an "occurrence" from which "property damage" results. These policies generally have exclusions for malpractice, defective construction, and pollution; separate policies must be obtained if such coverage is desired. These policies will not usually cover the contractor for faulty work nor the designer for negligent design. Some policies may cover faulty work of a subcontractor. However, it is common for several claims to be made against the contractor, and the insurance company has a duty to defend if even one claim potentially comes under the protection of the policy.

EXAMPLE

The homebuilder was sued for faulty installation of exterior stucco. However, the insurance company was not required to defend nor pay the damages, because faulty work is excluded from coverage in comprehensive general liability policies.

The purpose of these policies is to defend and be a source of funds for tort claims made by third parties. They are not designed to pay any contract-related claims between the parties on the project.

EXAMPLE

A construction worker was allegedly injured when he came into contact with high-voltage wires. He alleged that the engineer breached his duty of reasonable care by failing to report unsafe conditions and failing to warn. The engineer's general liability insurance company was required to defend, even though the policy had a clause excluding coverage for professional services.[26] The construction worker was in a similar position as a passerby, a third party, who was injured.

The explosion in construction defect litigation has led to a search for coverage under any and all policies, including comprehensive general liability policies. Most courts recognize that these polices are not designed to cover defective work. "We begin with the well-settled rule that the issuer of a commercial general liability insurance policy is not a surety for a construction contractor's defective work product."[27]

A general contractor was sued by a school for allegedly supplying defective concrete. The general contractor turned the matter over to its insurer, which refused to defend. While the court recognized this refusal, some states have been successful in manipulating these policies so that they provide a source of funds for construction defect litigation.

EXAMPLE

A contractor entered into contract with the Army to build barges and subcontracted installation of a crane onto one of the barges. Contractor was added as an additional insured (called 'assured') on subcontractor's comprehensive general liability policy. After completing the installation of the crane, the subcontractor returned the barge to the contractor. The contractor was required to transport the barge from Louisiana to Virginia and purchased insurance ("transportation insurer") to cover this risk. The crane fell off during transportation apparently due to faulty design of the subcontractor's employees. The transportation insurer sued the subcontractor's insurer claiming the subcontractor's comprehensive general liability insurance applied. However, the court held that exclusions in the comprehensive general liability policy exempted coverage for faulty workmanship of the subcontractor. The court concluded the exclusions "placed outside the coverage any damages to the crane itself, where they resulted from poor work by either [subcontractor's] skilled laborers or its professional staff. They [the exclusions] function to leave the risk of replacing or repairing defective materials as a commercial risk of the purchaser. In other words, they prevent a commercial liability policy from becoming a product or service warranty." [28]

In addition, the comprehensive general liability policy excluded coverage for professional services. The parties had apparently contemplated purchasing a professional liability policy but had not done so. The comprehensive general liability policy was not a source for recovery.

EXAMPLE

A road building contractor was sued by the developer when roads it constructed deteriorated prematurely. It was determined that the deterioration was due to various negligent acts of the contractor including faulty preparation of the subgrade such as failure to remove tree stumps and to compact the soft clay sufficiently, a road course that was too thin, and an improperly installed drainage system. Excessive traffic was also a minor contributing factor to the deterioration. The commercial general liability company was not liable to the developer for damages, because the policy did not cover contractor negligence.[29]

Coverage under the commercial general liability policy may exist when an insured's defective product is a mere *component* of another product or structure.[30]

Some policies will cover faulty workmanship of a subcontractor. Damage to third parties and property caused by faulty workmanship is usually covered.

Exclusions

Comprehensive general liability policies usually have exclusions for malpractice, defective construction, and pollution; separate policies must be obtained if such coverage is desired. These policies will not usually cover the contractor for faulty work nor the designer for negligent design.

Historically, policies have not provided coverage for injuries to employees working on the site, but in the example below, indirect coverage could possibly exist due to the fact that the parties were listed as additional insureds on the comprehensive general liability policy.

EXAMPLE

A general contractor was an additional insured on the subcontractor's comprehensive general liability policy. This policy covered the subcontractor for negligence. The subcontract contained an intermediate form indemnity clause, which transferred all liability for a negligence claim to the subcontractor if the subcontractor was partially at fault with the general contractor. This indemnification provision had no exclusion for injuries to employees. The subcontractor's comprehensive general liability policy contained an exclusion for bodily injury to an employee.

The subcontractor's employee was injured and sued the general contractor for negligence. The general contractor sued the subcontractor under the indemnification agreement. The subcontractor's insurer refused to defend the subcontractor, and the subcontractor filed a third-party claim against its insurance company.

The court held that the subcontractor's insurance company had to defend the subcontractor, because the indemnification agreement placed additional liability upon the subcontractor for which coverage could potentially exist. The subcontractor's policy provided the subcontractor with coverage for liability under an "insured contract," which, with its intermediate form indemnity agreement, the subcontract was. The subcontract transferred the tort liability of the general contractor onto the subcontractor, and therefore the claim was not for injury to an employee but for the subcontractor's assumption of the general contractor's tort liability under the intermediate form indemnity clause.[31]

MANAGEMENT TIP

Insurance law is very complex. In the event that your insurer is denying coverage, it is best to seek outside legal advice with expertise in the insurance laws of your state.

Other typical exclusions include pollution- and mold-related liability. However, many state courts have limited the applicable of pollution exclusion clauses in comprehensive general liability policies, because the exclusions are too broadly worded.

EXAMPLE

The insurance policy contained a very broadly worded exclusion for pollution coverage designed to be an absolute bar to coverage for any pollutants. The court said that the exclusion was so broadly worded that it would virtually bar coverage of any claim involving any substance. The exclusion read, "Pollutants...any solid, liquid, gaseous or thermal irritant, or contaminant, including smoke, vapor, soot, fumes, acids, alkalis, chemicals, and waste. Waste includes materials to be recycled, reconditioned, or reclaimed."[32]

Insurance companies also have a duty to defend any pollution-related claim if it might come within the language of the policy. The insured submitted a claim for coverage concerning what later became known as the Seymour Superfund Site. The insured had stored many barrels of unidentified/unlabeled liquid wastes at its facility. Barrels were permitted to rust and deteriorate, spilling their contents on the ground. These acts might be construed as defenses to coverage, but the court held that as a matter of law, the insurer had to defend the insured against federal environmental claims.[33]

MANAGEMENT TIP

Do not accept an insurance company's denial of coverage for a claim relating to something that could conceivably be considered pollution. Have the matter reviewed by an attorney familiar with your state laws.

Pollution or Environmental Insurance

Because most general liability insurance products contain pollution exclusion clauses, it is necessary to purchase pollution liability coverage (PLC) or pollution legal liability policies (PLL) if coverage for this type of risk is desired. Policies can be purchased to cover only specific risks, known risks, unknown risks, on-site risks, off-site risks, natural resource damage, transportation-related risks, environmental risks under the Comprehensive Environmental Response, Compensation, and Liability Act (CERCLA),[34] or similar laws. Cleanup cost cap (CCC) or remediation cost cap (RCC) policies cover the "overrun" costs on "known" cleanup projects, where expenses exceed those agreed to in a remediation action plan. They insure against discovery of additional contaminants or unexpected site soil conditions, failure of remedial technology, cost overruns, including unexpected increase in base disposal costs, and changes in regulatory requirements. Specialized insurance options for contractors include contractors pollution occurrence (CPO) and lead abatement contractors liability insurance.

Two types of policies exist: policies covering events during a certain period, even if the claim is made after the effective date of the policy, and polices that cover any claim made while the policy is in force, even if the events giving rise to the claim occurred before the date of the policy.

Because pollution can occur over time, it is possible that any particular entity has had several insurance polices that may cover a claim. It is possible that prior policies may cover damages. In Indiana, if environmental damage occurs over a number of years, all of the insured's policies are available to apply to the loss.[35]

Errors and Omissions or Malpractice Insurance

This insurance covers a person's negligence in performing work. Architects, engineers, CMs, and design-build firms might have this insurance to cover claims of negligent design or negligent supervision.[36]

EXAMPLE

Owners and an arena architect were sued for failure to meet the Americans with Disabilities Act (ADA) requirements for a suite design. The architect had certain duties related to integration of the designs, but the owners maintained control over the suite design and wheelchair spaces. The architect's malpractice insurance covered only the *architect's* negligent actions, if any, and the liability of the owners for the negligent acts of the architect. Since the owners maintained control over this aspect and not the architect, the insurance did not cover the claim.[37]

EXAMPLE

The architect's professional negligence policy contained a cost estimate exclusionary clause. The owner sued the architect for negligence when the architect's cost estimate of $12,500 per unit proved in error. The policy was held to provide coverage despite the exclusion, because the cost estimate was not the result of errors in estimating the cost but from underlying defects in the design of the project.[38]

Umbrella or Excess Coverage Policy

Many companies have an umbrella policy or excess coverage policy that gives them additional liability coverage. This policy does not kick in unless other insurance is exhausted. It is usually fairly inexpensive. These policies often extend both the liability coverage of a CGI policy and/or a vehicular policy.

EXAMPLE

Frank's Casing fabricated a drilling platform for Arco that later collapsed. Arco claimed that Frank's Casing negligently constructed the platform and sued. Frank's Casing had a primary liability policy with limits of $1 million and an excess coverage policy of $10 million from a different company, Lloyd's. Since damages were greater than $1 million, both policies covered the claim.

■ Subrogation

If an insurance company pays a claim to the insured for which a third party could be held liable, the insurance company can attempt to obtain reimbursement from the third party for the sums paid to the insured under the policy.[39]

FORM

Subrogation—We shall be subrogated to any and all rights of recovery which any Insured may have or acquire against any party or the insurer of any party for benefits paid or payable under the Policy. Any Insured who receives benefits from us shall be deemed to have assigned their right of recovery for such benefits to us and agree to do whatever is necessary to secure such recovery, including execution of all appropriate papers to cause repayment to us.

EXAMPLE

An isurance company pays a contractor for fire damage done to the site caused by the subcontractor's employee. The insurance company may sue both the subcontractor and the subcontractor's employee for reimbursement of sums paid to the contractor. This is referred to as a subrogation claim or lawsuit.

Rocker owned an oil well rig that blew over in May 1998. Rocker was insured for wind damage to its rights by Markel. The claim was paid by an affiliate of Markel, Essex. Essex sued Shenandoah and Mason Brothers to recover the amounts it paid Rocker, claiming that Shenandoah and Mason Brothers were negligent in installing the anchors that held the rig in place.[40]

MANAGEMENT TIP

These claims usually take the form of the two insurance companies battling to determine which one pays the claim. The actual construction-related companies may only be peripherally involved in the claim or lawsuit. The construction-related companies and their employees must cooperate in any such claim—for example, by giving testimony.

Construction contracts often have provisions in them designed to prevent subrogation claims, and these are generally upheld.[41] One method is to name each party as an additional insured (discussed above) on the project. This is the most effective method of avoiding subrogation claims. "An insurer is not entitled to subrogation from entities named as insureds in the insurance policy, or entities deemed to be additional insureds under the policy." This insurance principle is called the antisubrogation rule.[42]

Another method is to have all of the parties involved in a project sign waivers of subrogation. A waiver is a release of all or some claims against an entity. See additional discussion of Waivers in Chapter 8. "A waiver of subrogation is useful in [construction] projects because it avoids disruption and disputes among the parties to the project. It thus eliminates the need for lawsuits, and yet protects the contracting parties from loss by bringing all property damage under the all risks builder's property insurance."[43] The purpose of a waiver of subrogation is "in effect [to] simply require one of the parties to the contract to provide [property] insurance for all of the parties."[44] See the form below.

FORM

AIA B141-1997, Article 1.3.7.4 contains the following clause:
"To the extent damages are covered by property insurance during construction, the Owner and the Architect waive all rights against each other and against the contractors, consultants, agents, and employees of the other for damages, except such rights as they may have to the proceeds of such insurance as set forth in the edition of AIA Document A201, General Conditions of the Contract for Construction, current as of the date of this Agreement. The Owner or Architect, as appropriate, shall require of the contractors, consultants, agents, and employees of any of them similar waivers in favor of the other parties enumerated herein."

"Owner and Contractor waive all rights against each other and any of the subcontractors, sub-subcontractors, agents, and employees of each of the other, including the architect, architect's consultants, and employees for damage caused by fire or other perils to the extent covered by insurance obtained pursuant to this contract."

EXAMPLE

The contract between the owners and the general contractor contained a requirement that the owner obtain a builder's risk policy that would insure against risks of physical loss or damage to buildings and structures and would cover the owners, general contractor, and all subcontractors. The contract required the builder's risk policy contain a waiver-of-subrogation clause. The owners obtained this policy. During construction, a 49-story temporary scaffolding collapsed, resulting in approximately $20 million in property losses, including physical damage to real and personal property, emergency and stabilization costs, loss of rental income, and the costs of construction delay, work stoppages, and extra interest. The owner's insurance company paid the owners and sued the general contractor for reimbursement. The owner's insurance claimed gross negligence on the part of the general contractor. It also claimed that this gross negligence precluded the application of the waiver of subrogation clause, as common law does not allow parties to obtain insurance for intentional acts. The case was dismissed, and the waiver of subrogation clause was upheld even in the face of alleged gross negligence of the general contractor. While it is true that the law precludes parties from obtaining insurance coverage for intentional harm done to others, it does not preclude obtaining coverage for grossly negligent acts.[45]

MANAGEMENT TIP

Parties should inform their insurance companies of any waivers, because the insured may not be allowed to waive or destroy the insurance company's right to subrogation. However, case law does exist that is not favorable to subrogation claims on construction projects.[46]

Contracts should always contain clauses waiving subrogation rights, and the insurance company must be so informed.

■ Endnotes

[1] This chapter was published in substantially this form by the author in *The Real Estate Law Journal*, 2007 under the title, *Death, Taxes, and...Insurance: Current Legal Issues Relating to Insurance in the Construction Industry.* All permissions obtained.

[2] Kelly Zito, *Insurance Nightmare—Flood of Lawsuits Alleging Defective Construction Leaves Builders Scrambling to Find Coverage For New Projects*, S.F. Chron., July 11, 2002, Gavin, Robert, *Regional Report: Home Builders Face Insurance Woes: Coverage Scarce, Premiums High after an Increase in Construction Suits*, Wall St. J., Feb. 27, 2002, Allen, Mike, Building Industry Faces Skyrocketing Insurance Premiums, San Diego Bus. J., Feb. 11, 2002.

[3] B. Cox and B. Allen, *Fuzzy Logic: Mold Strives for Immortality in Construction Defects Litigation,* 42 Houston Lawyer 24 (March/April, 2005).

[4] T. Strode, *From the Bottom of the Food Chain Looking Up: Subcontractors Are Finding that Additional Insured Endorsements Are Giving Them Much More Than They Bargained For.* 23 St. Louis U. Pub. L. Rev. 697 (2004).

[5] *U.S. v. M.O. Seckinger,* 397 U.S. 203, 211 (1970), *Bisso v. Inland Waterways Corp.,* 349 U.S. 85, 90 (1955), *Jankele v. Texas Co.,* 54 P.2d 425 (Utah 1936), *Otis Elevator Co. v. Maryland Cas. Co.,* 33 P.2d 974, 977 (Colo. 1934), *Sternaman v. Metro. Life Ins. Co.,* 62 N.E. 763, 766 (N.Y. 1902), *Johnson's Adm'x v. Richmond & D.R. Co.,* 11 S.E. 829, 829-30 (Va. 1890).

[6] American Subcontractors Association, Inc., *Subcontractor's Chart of Anti-Indemnity Statutes,* at http://www.asaonline.com/pdf/AntiIndemnityChart.pdf (last updated May 2, 2003).

[7] *Acceptance Ins. Co. v. Syufy Enters.,* 69 Cal. App. 4th 321 (Cal. Ct. App. 1999).

[8] *Pardee Construction Co. v. Insurance Co. of the West,* 92 Cal. Rptr. 2d 443 (Cal. Ct. App. 2000).

[9] *Illinois Emcasco Insurance Co. v. Northwestern National Casualty Co.,* 337 Ill. App. 3d 356, 785 N.E.2d 905 (1st Dist. 2003).

[10] S. Mehta, *Additional Insured Status in Construction Contracts and Moral Hazard,* 3 Conn. Ins. L.J. 169, 175 (1996).

[11] P. J. Wielinski, et al., *Contractual Risk Transfer: Strategies for Contract Indemnity and Insurance Provisions,* International Risk Management Institute (2003).

[12] *American National Fire Insurance Co. v. National Union Fire Insurance Co. of Pittsburgh,* PA, 72 343 Ill. App. 3d 93, 796 N.E.2d 1133 (1st Dist. 2003).

[13] R. D. Hursh, *Refusal of Liability Insurer to Defend Action against Insured Involving Both Claims within Coverage of Policy and Claims Not Covered,* 41 A.L.R.2d 434 (1955 Updated 2005).

[14] *Presley Homes, Inc. v. American States Ins. Co.,* 108 Cal. Rptr. 2d 686 (Cal Ct. App. 2001). This is a California case, and its applicability has not been litigated in all states. See also D. Ezra, *How Presley Homes Has Changed the Duty to Defend,* 26 L.A. Law. 17 (2003).

[15] *Illinois Emcasco Insur. Co. v. Northwestern National Casualty Co.*, 337 Ill. App. 3d 356, 785 N.E.2d 905 (1st Dist. 2003).

[16] *American Nat. Fire Ins. Co. v. National Union Fire Ins. Co. of Pittsburgh, PA*, 277 Ill. Dec. 767, 796 N.E.2d 1133 (App. Ct. 1st Dist. 2003) (applying Pennsylvania law).

[17] C. T. Drechsler, *Consequences of Liability Insurer's Refusal to Assume Defense of Action Against Insured upon Ground that Claim upon which Action Is Based Is Not within Coverage of Policy*, 49 A.L.R.2d 694 (1956 and updated 2005).

[18] *Pennsylvania Nat. Mut. Cas. Ins. Co. v. Associated Scaffolders and Equipment Co., Inc.*, 579 S.E.2d 404 (N.C. Ct. App. 2003).

[19] *Allison v. Fire Insurance Exchange*, 98 S.W. 3d 227 (Tex. App. 2002). This case is known as the Ballard case.

[20] *Tomco Painting & Contracting, Inc., v. Transcontinental Ins. Co.*, 21 A.D.3d 950, 801 N.Y. S.2d 819 (N.Y. App.2005), referencing *Lavanant v. Gen. Acc. Ins. Co. of Am.*, 79 N.Y.2d 623, 629, 595 N.E.2d 819, 584 N.Y.S.2d 744).

[21] *Tomco Painting & Contracting, Inc. v. Transcontinental Ins. Co.*, 21 A.D.3d 950, 801 N.Y.S. 2d 819 (N.Y. App.2005).

[22] *Nautilus Insurance Co. v. Vuk Builders, Inc.*, 2005 U.S. Dist. Lexis 30263 (N.D. Ill., 2005) (applying Illinois law). Several other courts have found the subsidence clause ambiguous or did not apply to earth movement caused by the construction process: *Murray v. State Farm Fire & Cas. Co.*, 203 W. Va. 477, 509 S.E.2d 1 (W.Va. 1998). *Ins. Co. of State of Pennsylvania v. ALT Affordable Housing Services, Inc.*, 1999 WL 33290622, (W.D. Tex., 1999), *American Motorists Ins. Co. v. R & S Meats, Inc.*, 190 Wis. 2d 196, 526 N.W.2d 791, 796 (Wis.App.Ct. 1994), *Henning Nelson Constr. Co. v. Fireman's Fund American Life Ins. Co.*, 383 N.W.2d 645, 653 (Minn.1986).

[23] G. M. Plews & D. C. Marron, *Environmental Law Developments: Hope and Ambiguity in Achieving the Optimum Environment*, 37 Ind. L. Rev. 1055 (2004).

[24] *Am. States Ins. Co. v. Kiger*, 662 N.E.2d 945, 947 (Ind. 1996) (quoting *Am. Econ. Ins. Co. v. Liggett*, 426 N.E.2d 134, 142 (Ind. Ct. App. 1981).

[25] *Bollinger Shipyards Lockport, LLC v. Certain Underwriters at Lloyd's, London*, 98 Fed. Appx. 311; 2004 U.S. App. Lexis 9001 (5th Cir. 2004) (applying Louisiana law).

[26] *Gregoire v. AFB Construction, Inc.*, 478 So2d 538 (La App 1st Cir, 1985).

[27] *Bonded Concrete, Inc. v. Transcontinental Ins. Co.*, 12 A.D.3d 761; 784 N.Y.S.2d 212 (NY App. 3d, 2004). *Farmington Casualty Co. v. Duggan*, 417 F.3d 1141 (10th Cir. 2005) (applying Colorado law).

[28] *Bollinger Shipyards Lockport, LLC v. Certain Underwriters at Lloyd's, London*, 98 Fed. Appx. 311; 2004 U.S. App. Lexis 9001 (5th Cir. 2004) (applying Louisiana law).

[29] *L-J, Inc. v. Bituminous Fire & Marine Insurance Co.*, 2005 S.C. Lexis 270 (2005).

[30] *Apache Foam Prods. v. Continental Ins. Co.*, 139 A.D.2d 933, 528 N.Y.S.2d 448 (1988), Penn *Aluminum v. Aetna Cas. & Sur. Co.*, 61 A.D.2d 1119, 402 N.Y.S.2d 877 (1978), *Marine Midland Servs. Corp. v. Kosoff & Sons*, 60 A.D.2d 767, 400 N.Y.S.2d 959 (1977).

[31] *West Bend Mutual Ins. Co. v. Subcontractor Masonry Co. Inc.*, 337 Ill. App. 3d 698, 786 N.E.2d 1078 (2d Dist. 2003).

[32] *American States Ins. Co. v. Kiger*, n30 662 N.E.2d 945, 947 (Ind. 1996).

[33] 3 *Seymour Manufacturing Co. v. Commercial Union Ins. Co.*, 665 N.E.2d 891 (Ind. 1996).

[34] 4 42 U.S.C. § 9601 *et seq.*

[35] 5 *Allstate Insurance Co. v. Dana Corp.*, 759 N.E.2d 1049 (Ind. 2001).

[36] 6 J.M. Draper, *Construction and Application of Liability or Indemnity Policy on Civil Engineer, Architect, or the Like*, 83 A.L.R.3d 539 (1978, updated 2004).

[37] 7 *Washington Sports and Entertainment, Inc. v. United Coastal Ins. Co.*, 7 F. Supp. 2d 1 (D.D.C. 1998) relating to Americans with Disabilities Act of 1990, § 2 *et seq.*, 42 U.S.C.A. § 12101 *et seq.*

[38] 8 *Comstock Ins. Co. v Thomas A. Hanson & Associates, Inc.*, 77 Md App 431, 550 A2d 731. (Md. App, 1988).

[39] 9 Am Jur 2d, Insurance, § 1794.

[40] 40 *Essex Ins. Co. v. Mason Brothers Const., Inc.*, 2004 Tex. App. Lexis 5740 (2004) (rehr. den.). In that case, the insurance company did not recover, because it failed to prove Shenandoah and/or Mason Brothers acted unreasonably. However, the lawsuit itself was not precluded.

[41] 41 J.M. Zitter, *Insurance: Subrogation of Insurer Compensating Owner or Contractor for Loss Under "Builder's Risk" Policy against Allegedly Negligent Contractor or Subcontractor*, 22 A.L.R.4th 701 (2005). But see the following cases where the court held there was an issue of fact as to whether or not the insurer would be permitted to bring a subrogation claim. *McGuire v. Wilson*, 72 So 2d 1297 (Ala. 1979), *Atlas Assur. Co. v General Builders, Inc.*, 93 N.M. 398, 600 P2d 850 (N.M. App. 1979), *United Nuclear Corp. v Mission Ins. Co.*, 97 N.M. 647, 642 P.2d 1106. (N.M. App. 1982).

[42] 42 Black's Law Dictionary 104 (8th ed. 2004).

[43] 4 *Tokio Marine & Fire Ins. Co.* 786 F.2d 101, 104–5 (2d Cir. 1986).

[44] 44 *Bd. of Educ. v. Valden Assocs., Inc.*, 46 N.Y.2d 65, 389 N.E.2d 798, 799, 416 N.Y.S.2d 202 (N.Y. 1979). See also *Walker Engineering, Inc. v. Bracebridge Corp.*, 102 S.W.3d 837 (Tex. App. 2003).

[45] 45 *St. Paul Fire and Marine Insur. Co., v. Universal Builders Supply*, 409 F.d 73 (2d Cir, 2005) (applying New York law). *S.C. Nestel, Inc. v. Future Construction, Inc.*, 836 N.E.2d 445 (Ind.Ap, 2005).

[46] 46 See J. Sweet & J. Sweet, **[Is duplication intentional?]** 2 *Sweet on Construction Industry Contracts* (rd ed., Wiley 1996).

292

CHAPTER 12

Torts and Warranties

■ Introduction

The law requires certain behaviors from individuals in their interactions with others. A failure to behave as legally mandated can result in two different types of lawsuits: criminal lawsuit and/or tort. A criminal lawsuit can only be filed by the government, and though the criminal justice system has several goals, the goal of compensating the party injured by the crime is not a major goal of that process. A tort lawsuit is filed in the civil justice system by the victim of the alleged injurious behavior, and the main goal of a tort lawsuit is to compensate parties injured by the tortious acts of others.

The civil justice system also handles contract actions, but the two types of actions, contract and tort, are kept separate. If a contract exists between the parties, the parties are generally limited to a suit on the contract. The law will not allow the transformation of a breach-of-contract claim into a tort claim.[1] Parties attempt to transfer contract actions into tort actions because the recoverable damages for tort are often, though not always, greather than the damages recoverable for breach of contract. Additionally, the determination of whether or not damages exist is different for the two types of actions. The reason for the different treatment of contracts and torts is that contracts represent duties parties voluntarily establish between themselves; the courts will usually, though not always, require performace of contract duties. Torts, on the other hand, are duties the law requires parties to perform and the courts will always require performance of a tort duty. Parties in a contractual relationship can sue each other for torts, but the tort must be outside of the contract relationship.

■ Negligence and Professional Liability (Unintentional Torts)

Very simply, negligence is the failure to act reasonably toward another entity, with resultant injury. More specifically, **negligence** is the failure to act reasonably toward another entity to whom the actor owes a legally recognized duty, with resultant injury to the victim. In order to determine if a party is negligent, the following four issues must be addressed:

1. Does the law require one of the parties to act reasonably toward the other? Does the law recognize a duty to act reasonably between the parties?
2. Did one party, the alleged tortfeasor, breach the above duty to act reasonably? Did one party, the alleged tortfeasor, act unreasonably?
3. Did injury occur?
4. Was the injury complained of by the victim actually caused by the alleged tortfeasor's unreasonable act?

If the injured party, also called the victim in many cases, can prove the above four issues, the tortfeasor is liable for damages.

MANAGEMENT TIP

While the law does not require every entity to act reasonably toward every other entity it encounters, the law is expanding, and to avoid liability for negligence, entities should always take care and act reasonably toward others.

Owner Liability

The law recognizes a duty to act reasonably between the owner of a structure and the users of the structure. This means that the owner must take reasonable care and maintain its structure in a safe condition. If the owner fails in this duty to maintain a safe structure and a user of the structure is injured, the owner is liable. This duty goes by the specific name of **premises liability**.

In limited circumstances, the owner may be liable for injuries to persons working on the construction site. The owner is responsible for the adequacy of the plans and specifications of its agent. The designer and construction workers can expect that if they build according to the plans and specifications, they will not be exposed to unreasonable risk of injury during the construction process. This does not mean that the builders will not be exposed to injury, it just means that they should not be exposed to unreasonable risk of injury.

If a construction site is open to the public because an owner wishes to continue doing business during construction, both the owner and the contractor must take reasonable care to maintain a safe area.

Designer Negligence and Professional Malpractice

The terms **professional liability** or **malpractice** are used differently in different jurisdictions. Some jurisdictions use the term for a lawsuit claiming that a designer failed to perform its contract with the owner. This is the breach-of-contract action rather than a tort action. A design professional commits malpractice by generating a technically unfeasible design or a design that does not meet the client's stated goals.[2] However, many jurisdictions use these terms to include any lawsuit against a professional, whether it is a contract or tort lawsuit, and that is how the term is used herein.

Designers are required to exercise that "degree of skill, care, and learning ordinarily possessed and exercised by members of the profession in good standing in similar circumstances."[3] Failure to do so will result in liability for any injuries caused by that failure. The designer has this negligence liability to both the ultimate users of the project as well as to contractors and subcontractors on the project during the construction process.

EXAMPLE

The court refused to dismiss a wrongful death action against a designer when the designer had actual knowledge of unsafe practices on the job site. He knew of the safety standards requiring shoring in trenching operations, and he failed to warn the owner or the contractor of the dangerous condition. When the trench caved in and killed a worker, the court held that the designer had a duty to warn of the dangerous condition.[4]

EXAMPLE

In this case, the architect was *not* found negligent as a matter of law, meaning that the case was dismissed and the plaintiff could not even present it to a jury. Howard, a veterinary surgeon, had plans to create an animal emergency clinic in Vermont. Usiak, the architect, was a specialized veterinary architect, licensed in New Mexico. Usiak did obtain a reciprocal license in Vermont before entering into a contract with Howard. The contract between the parties contained three phases. Usiak completed phase 1, the preliminary planning, and was paid for that work.

During phase 2, the design phase of the project, the issue arose of whether or not an elevator was needed to access the second floor. The second floor was to be used only by employees and not the public. The architect consulted with a local contractor and a permitting specialist in Vermont and was told that no elevator would be needed. However, this information later proved to be incorrect due to more stringent Vermont codes requiring elevators. The architect informed the owner. At this point, the designs were not complete, and no construction had begun. The architect was willing to work with the owner to remedy the problem, but the owner fired the architect. He sued for return of the sums paid under phase 1 and also lost future profits from delays related to the elevator issue. The client sued the architect for negligence and negligent misrepresentation. The court dismissed all claims, stating that as a matter of law the architect was not negligent nor had he committed negligent misrepresentation given these facts.[5]

It is necessary to have expert testimony from a designer in the same field as the alleged tortfeasor to support any claim of negligence. "It is well settled that in order to prove negligence or malpractice in the design of a structure, the plaintiff must put forth expert testimony that the engineer or architect deviated from accepted industry standards."[6] In fact, in these types of cases, it is common for the expert testimony to conflict, and the jury must decide which expert is more credible.

Two steelworkers were killed when two steel frames collapsed during the erection of a chip storage building at a paper plant. At the end of trial, the jury determined that the engineer had been negligent in the design of the chip storage building. The jury found the legal cause of the deaths to be the designer's negligent failure to design steel to meet the specifications of the structure, thereby creating an unreasonable danger of injury to the plaintiffs engaged in erecting the structure. The court said, "Although the expert evidence presented to the jury on the issue of causation was contradictory, our review of the record discloses that the jury rationally could have found the following facts. After the collapse of the structure, [the engineer] discovered that he had made computational errors while calculating the necessary sizes of certain steel members to be used in the erection of the building. In consequence of [the designer's] arithmetical errors in applying the formula he had chosen to determine the size of rafter beams, his design specifications called for WF36 x 150 rafter beams instead of WF36 x 230 rafter beams. WF36 x 230 rafter beams have greater lateral stiffness and an increased resistance to buckling. The WF36 x 150 rafter beams would not adequately meet the design specifications without sixteen permanent knee braces to provide the necessary strength and stability. Only two knee braces were provided by [the designer]....Based on this evidence, the jury rationally could find that it was more probable than not that the negligence of [the designer caused the deaths of the workers]."[7]

If the injury is not related to negligent design, the architect or engineer is not liable; that is, the injury must always be caused by something the designer did or failed to do.

EXAMPLE

An architect designed a building. The plans required that during construction, the elevator shaft was to have guardrails around the open elevator shaft. These guardrails were removable and, in fact, had been removed to allow the movement of siding on the project. A workman was killed when the siding swung around and pushed him into the elevator shaft. The architect was held not negligent.[8]

EXAMPLE

A roofer died when he fell through a hole in a roof under construction. The court stated that holes are commonly left in the roof of a building under construction in order to permit installation of ducts and air conditioning units. The plans were not defective and therefore the designer was not liable.[9]

If the defect is caused by defective construction and not by defective designs, the designer is not liable.

An architect designed panels on a balcony to be no more than six inches apart in conformance with the building code. However, at one point, the panels were not installed in conformance with the plans and specifications, and the spacing was larger than six inches. Later, an infant fell through and was injured. The architect was not liable for the infant's injuries.[10]

The liability of the designer for injuries for failing to properly supervise depends on the extent to which the designer is responsible for supervision of the project. If the designer's responsibility is limited to making sure that the contractor complies with the plans and specifications, the designer is not liable for injuries caused by unsafe construction sites. However, if the designer has more extensive supervisory duties, the designer will be responsible for injuries caused by the failure to supervise in a reasonable manner.

EXAMPLE

ALSC Architects was hired by a grocery store to provide architectural services related to a remodeling project. ALSC Architects agreed to act as the owner's representative during the construction phase of the project for the purpose of communicating instructions to the contractor and to assure that the project progressed in a manner consistent with the contract plans and other documents.

Part of the remodeling included the removal of a large walk-in freezer, the top of which was accessible through a manager's office and was used to store seasonal displays. The walk-in freezer was removed and replaced with a smaller freezer, and a drop ceiling was installed. The area was no longer suitable for storage, and the architect designed plans and specifications for the removal of the access door and replacement of the drywall in the manager's office. The owner paid the contractor to remove the door, but the door was never removed by the contractor. ALSC Architects noticed the door had not been removed but did not mention it to the owner. In fact, the architect submitted statements to the owner that the work had been completed.

A few weeks after the remodel was finished, a store employee was told to go to the storage area and retrieve some posters. The employee, who had previously accessed the storage area on occasion but not knowing that the remodeling had changed the area, opened the door in the manager's office and stepped through it. He fell through the drop ceiling to the floor below and was injured. The architect was liable for failing to supervise the contractor in a reasonable manner.[11]

MANAGEMENT TIP

It should be clearly stated in the contract what the duties of the general contractor and the designer are in connection with safety and supervision on the job site. Whoever is responsible will be liable.

Contractor, Subcontractor, and Supplier Liability

The general contractor has the duty to maintain a reasonably safe construction site. Failure to do so will result in contractor liability for injuries to most parties coming onto the construction site including subcontractors, designers, owners, and visitors.

In addition to the liability of the general contractor to all parties entering the construction site, the law is expanding to recognize a duty between *all* of the parties on the site, such as contractors, subcontractors, and suppliers, to act reasonably toward all other parties on the site. If a party acts unreasonably and causes injury, the party causing the injury is liable for the damages caused. In other words, liability follows responsibility. The extent of this liability is not established and varies by jurisdiction.

> ### EXAMPLE
>
> Mobil, the owner, entered into a contract with Asia Badger, the general contractor, for the construction of an aromatics plant. Sheek was an employee of Mobil and inspected the project during construction. At one point during the construction, it was determined that certain pumps were overheating due to inadequate cooling water circulation. The pipes had to be replaced with larger-diameter pipes. The replacement of the pipes was performed by Asia Badger. While inspecting the work one evening, Sheek slipped and fell on a piece of pipe left in the aisle by either the contractor or the subcontractor. Sheek tore his rotator cuff and required surgery. The jury determined that Asia Badger was negligent and therefore liable for Sheek's injury.[12]

> ### EXAMPLE
>
> Lee Lewis Construction was the general contractor on a hospital remodeling project and as such retained the right to control fall-protection systems on the project. An employee of a subcontractor fell to his death while working on the project. He was not wearing an independent lifeline. The contractor was liable in negligence for the death of the subcontractor's employee for failing to ensure adequate fall-protection measures. "Ordinarily, a general contractor does not owe a duty to ensure that an independent contractor performs its work in a safe manner," the Court said, but the rule is not applicable when the general contractor retains control over the manner in which the independent contractor performs its tasks.[13]

> ### MANAGEMENT TIP
>
> Because Workers' Compensation recovery is limited, it is not uncommon for injured parties to seek additional compensation from others who may be liable. Workers' Compensation limitations only protect the employer, not any other parties involved in the construction.

The law has recently begun to allow ultimate users of the project to sue the contractor, subcontractor, and/or supplier for injuries they cause. The rule allowing ultimate users to sue is often called the **modern** or **foreseeability** rule, meaning that the ultimate users can sue contractors, subcontractors, or suppliers for negligence. The historic rule, called the **acceptance rule**, relieved the contractor, subcontractor, or supplier of liability to the ultimate users upon acceptance of the work by the owner; that is, only the owner could be sued for injuries once the owner had accepted the project. Many states still adhere to this rule.[14] Under the modern or foreseeability rule, the contractor, subcontractor, or supplier is liable to users of the project if it is reasonably foreseeable that the user will be injured by the work. Contractors, subcontractors, and suppliers are also liable for failing to disclose a known dangerous condition.[15]

One court has explained the modern or foreseeability rule as follows: "An independent contractor may be liable to third parties who may have been foreseeably endangered by the contractor's negligence, even after the owner has accepted the work. The general rule is subject to two limitations:

1. The independent contractor should not be liable if he merely carefully carried out the plans, specifications, and directions given him, at least where the plans are not so obviously dangerous that no reasonable person would follow them, and

2. If the owner discovers the danger, or it is obvious to the owner, the owner's responsibility may supersede that of the contractor."[16]

EXAMPLE

Skinner Pool installed a pool on the Stewarts' property. Cox, a subcontractor, installed the concrete. On the day after the pool was first filled, a crack appeared. The pool was drained, and the crack was repaired. It reappeared almost immediately upon the refilling of the pool. Water loss was two inches per day and later three inches per day. The water undermined the land and damaged the owners' home.

The pool was to have been constructed by completely embedding the reinforcing steel in gunite. However, it was determined that in applying the gunite, Cox permitted the reinforcing steel to lie on the ground. As a result, the gunite did not completely surround the steel and thereby did not properly reinforce the pool. Additionally, the layer of gunite was too thin for its purpose and not of the thickness called for by the plans. Cox was liable for the damage to the owners' home and yard.[17]

Negligence Per Se

Negligence per se is a special category of negligence, and it applies when a victim is injured because of the tortfeasor's failure to follow a statute or regulation. In this special circumstance, the victim is not required to prove that the tortfeasor was unreasonable; the failure to follow the law is deemed unreasonable. In order to recover under this theory, the victim must prove the following:

1. A violation of a statute or regulation
2. The victim belongs to the class of person whom the statute or regulation is designed to protect
3. The injury suffered is of the type the statute was intended to prevent

EXAMPLE

Atkinson, a guest at the hotel, had been drinking alcoholic beverages and was looking for a restroom. She went through a door in a stairwell that took her into a dark construction area. She fell into an excavated pit that was not roped off. She fractured her lumbar spine and incurred medical expenses in excess of $110,000. The construction site was secured by a chain link fence except for the access through the stairwell, which was not locked or secured.

A particular building code, *NRS 455.010*, read as follows:

"Any person or persons, company or corporation, who shall dig, sink or excavate, or cause the same to be done...shall, during the time they may be employed in digging, sinking, or excavating, or after they may have ceased work upon or abandoned the same, erect, or cause to be erected, good and substantial fences or other safeguards, and keep the same in good repair, around such works or shafts, sufficient to guard securely against danger to persons and animals from falling into such shafts or excavations."

The court said, "The plain and unambiguous language of *NRS 455.010* is intended to protect members of the public from falling into excavations. In this case, MGM [the owner] and Marnell Corrao [the general contractor] were required to follow the provisions of *NRS 455.010* and secure the excavation area by erecting a fence or other safeguard. Additionally, Atkinson [the injured party] is within the class of persons protected by the statute, and her injury is the type that the statute was designed to prevent. Further, Atkinson introduced evidence that she was able to access the excavation site through the stairwell, which was not secured by fencing." The owner and the contractor were liable for her injuries.[18]

Res Ipsa Loquitur

Res ipsa loquitur is Latin for "the thing speaks for itself." This maxim permits a jury to infer negligence merely from the fact that an injury occurred. The victim need not show any direct evidence of the alleged tortfeasor's unreasonable conduct; the mere fact that the injury occurred is sufficient to support the negligence. In order to recover under this theory, the victim must show the following: "(1) The situation, condition or apparatus causing the injury must be such that in the ordinary course of events no injury would result unless from a careless construction, inspection, or user. (2) Both inspection and user must have been at the time of the injury in the control of the party charged with neglect. (3) The injurious occurrence or condition must have happened irrespective of any voluntary action at the time by the party injured...Whether the doctrine applies in a given case is a question of law for the court...."[19]

This doctrine is applicable only in rare cases where the injury is of the type that would not occur unless the alleged tortfeasor did something unreasonable but the injured party cannot determine exactly what the tortfeasor did that was unreasonable. The doctrine stems from the case of *Byrne v. Boadle*,[20] decided in England in 1863. In that case, the injured party was walking in front of the defendant's building when a barrel of flour fell from the defendant's window and caused the plaintiff injury. However, the plaintiff could not present any affirmative evidence of what the defendant, or its employees, had done that was unreasonable. The defendant claimed that because the plaintiff could not prove what unreasonable act the defendant or his employees had done, the plaintiff was precluded from recovering. The judge, Chief Baron Pollock, said that when a man passing in front of the premises of a flour dealer is hit by a falling barrel of flour, it is "apparent that the barrel was in the custody of the defendant who occupied the premises, and who is responsible for the acts of his servants who had the control of it; and in my opinion, the fact of its falling is prima facie evidence of negligence, and the plaintiff who was injured by it is not bound to show that it could not fall without negligence, but if there are any facts inconsistent with negligence, it is for the defendant to prove them."

Use of this doctrine does not automatically mean that the injured party will win, but it shifts the burden of proof from the injured party to the alleged tortfeasor. In other words, when the instrumentality causing an injury is in the control of the alleged tortfeasor, and the accident is one that would not occur without negligence, the alleged tortfeasor must prove that it did *not* do anything unreasonable in the situation. The victim is not required to prove that the tortfeasor did something unreasonable.

EXAMPLE

Rosenblum's condominium and personal property were destroyed in a fire. Inspection traced the fire to an electrical ceiling fan in a common area of the project. Rosenblum sued the owner of the ceiling fan, Deerfield Woods Condominium Association, Inc., alleging that it did not exercise due care with respect to the installation, inspection, maintenance, and repair of the electrical wiring of the ceiling fans, that Deerfield was in exclusive control of the wiring system, that the accident was of the sort that would not ordinarily occur in the absence of negligence, and that Rosenblum did not contribute to the accident. The court stated that these facts, if proven, would allow a jury to find negligence based on *res ipsa loquitur,* and the matter must proceed to trial.[21]

EXAMPLE

Goedert was the driver of a truck involved in a one-vehicle accident, and he sued Newcastle for negligence in repair of the truck brakes. Goedert pleaded the doctrine of *res ipsa loquitur*. The evidence supporting *res ipsa loquitur* included that Goedert experienced problems with the brakes on the truck unit of a tractor-trailer rig and took the truck to Newcastle for repairs of the brakes. A Newcastle employee worked on the brakes. Goedert picked up the truck at defendant's repair shop and drove it to a location where the trailer was loaded with wood. The first time that he started to descend a grade with the loaded truck, he attempted to actuate the brakes, but the truck did not slow down. He steered into an embankment to stop the truck from being a runaway and suffered injury as a result. The doctrine of *res ipsa loquitur* applied, and the decision of the jury that Newcastle had been negligent was upheld.[22]

Defenses

Several defenses to a negligence claim exist. A **defense** is a legally recognized excuse or a method through which a party can limit its liability. The two most common defenses are assumption of the risk and comparative negligence.

Assumption of the risk arises when the injured party, knowing the risks inherent in the activity, knowingly assumes the risks of the injury.

EXAMPLE

Andren knew the danger of lighting a cigarette in the presence of propane gas, and he knew that there was a liquid propane gas leak in the basement where he was working on the repair of the propane gas tank. He therefore assumed the risk of the injuries he sustained when he lit a cigarette, the gas exploded, and he was injured.[23]

Contributory negligence is a defense stating that each entity, including the victim, is responsible for the proportion of damages attributable to their own negligent actions. "Contributory negligence is a 'breach of a duty which the law imposes upon persons to protect themselves from injury, and which, concurring and cooperating with actionable negligence for which defendant is responsible, contributes to the injury complained of as a proximate cause.'"[24]

Most states have enacted statutes that outline in detail how contributory negligence operates. For example, a negligent victim cannot usually recover from an entity that was less negligent than the negligent victim.

■ Intentional Torts

Intentional torts are a classification of torts that involve some type of intentional conduct on the part of the tortfeasor. Punitive damages can be awarded in intentional tort cases. The intentional conduct is not necessarily the intent to cause injury but that the tortfeasor's act was deliberate rather than just accidental or careless. If the act was accidental or careless, a cause of action for negligence or strict liability would apply.

Deceptive Trade Practices or Unfair Trade Practices

Because of the difficulty in proving fraud, jurisdictions have adopted the Uniform Deceptive Trade Practices Act or a variation of that act. These acts attempt to prevent the use of unfair methods in doing business and are so broadly worded that almost any unfair or deceptive act in the furtherance of a trade or business is subject to this form of liability. Different jurisdictions may have different names for this type of liability, and "unfair trade practices" is a common variant.

Commonly recognized deceptive trade practices include the following:

- Representing that goods or materials are original or new when they are not
- Representing that goods, materials, or services are of a particular standard, quality, or grade, or of a particular style or model, when they are not

- Representing that goods or services have, characteristics, ingredients, uses or benefits, or approval, that they do not in fact have
- Disparagement of the goods, materials, services, or business of someone else by false or misleading representations
- Engaging in any other conduct that similarly creates a likelihood of confusion or of misunderstanding about the quality of goods, materials, or services
- Causing confusion or misunderstanding as to the source or approval of goods or services
- Using deceptive representations or designations of the geographic source of the goods or service
- Representing that a person has some certification, approval, or connections that he or she does not

While the laws in each state vary, most include an imposition of double or treble compensatory damages and attorney fees.

EXAMPLE

Lewis owned a successful gymnastics instruction business and decided to expand the business and to build a facility. Lewis contracted with Barnett to build the new facility; Barnett was in charge of the management, oversight, and construction of the project.

Lewis testified that he hired Barnett because of Barnett's representations, and he relied on Barnett's representations. Barnett promised Lewis "three times the facility for one-and-a-half times the amount of money" and in a more desirable location. Barnett also led Lewis to believe that the contractual amount of $1.96 million was "more than adequate to build this project." Barnett "guaranteed" that he would finish the project "no matter what." Barnett represented that the building would be completed in six months, that he would have "time to spare," and that the quality of the construction would be "great." Barnett represented to Lewis that he had adequate crews of approximately 20 or more men to "knock it out real quick." Barnett was not on the job on a daily basis, he did not have crews at the job site, and he had fewer men in his crews than he told Lewis he would. After a few months, he walked off the job. The jury determined that Barnett had violated the Deceptive Trade Practices Act.[25]

Fraud

Fraud is an intentional misstatement or failure to disclose information that leads to damage. It is difficult to recover for fraud, and most victims instead proceed with a Deceptive Trade Practices suit, as discussed above. The elements of fraud can be summarized as follows:

1. Misrepresentation of facts or conditions or failure to disclose relevant facts or conditions,
2. With knowledge that the above is false *or* with reckless disregard for the truth
3. Intent to induce the victim to rely on the misrepresentation or failure to disclose

4. The victim justifiably relies upon the misrepresentation or failure to disclose
5. Damages caused by the above

EXAMPLE

Yazd purchased a new home that sank into the unstable soil upon which it was built. Two reports indicated that the site contained moisture-sensitive, collapsible soil. These reports were never revealed to the owners. Woodside, the developer, formulated a plan to dig out the collapsible soil and reduce the grade of several parcels, including the one that eventually became the Yazd parcel. Gordon, an engineer, inspected the area and pronounced the soil fit to support a house.

In a preliminary motion to determine whether or not the matter could proceed to a jury trial, the court said, "In order to prevail on a claim of fraudulent concealment, a plaintiff must prove "(1) that the nondisclosed information is material [the court also used the word "important" elsewhere in the opinion], (2) that the nondisclosed information is known to the party failing to disclose, and (3) that there is a legal duty to communicate (citations omitted)." The court held that the developer had a legal duty to communicate the information and that the first two elements must be presented to a jury for determination. If the jury determined that the information was material or important and that the nondisclosed information was known to Woodside, Woodside would be liable for fraud.[26]

Defenses to fraud include puffery, or seller's talk, and statement of opinion. In addition, the purchase of property "as is" is a defense; that is, no cause of action for fraud will lie if the statements are puffery, seller's talk, opinions, or the property is sold "as is."

EXAMPLE

M&D purchased commercial property from McConkey. The evidence showed that the property had experienced flooding problems for many years. McConkey testified at trial that he had witnessed flooding on the property. M&D never asked about flooding. McConkey refused to prepare a seller's disclosure statement and made this refusal an explicit part of the purchase agreement. He included the following term on the face of the sale agreement, "Owner has never occupied this property. No representations or warranties implied as to condition. Property being sold in 'as is' condition." The court refused to allow the lawsuit for fraud to proceed, stating since McConkey had made no affirmative misrepresentation and the property was sold "as is," no fraud could lie.[27] Note that in many jurisdictions sellers of residential property are required by statute to disclose a lot of information about the property and failure to do so is a violation of that statute rather than fraud.

Nuisance

A **nuisance** arises when an owner of real property uses the property in such a way as to unreasonably interfere with neighbors' use and enjoyment of their own property. The typical remedy in a nuisance lawsuit is an injunction or court order to stop the unreasonable use.

EXAMPLE

Cleland, the owner of a home next to a vacant lot, informed Segars, an experienced homebuilder, that the stakes for a proposed house to be built on the lot were too close to her lot and violated the setback ordinance. Segars responded in a "threatening, intimidating, and demeaning manner." Segars refused to check with the county about the setback. Cleland complained to the county, and a stop order was issued but not until the house was almost completed. Segars testified that he did not know about the setback requirement—despite the fact that Cleland had told him when the stakes were in the ground and before laying the concrete that he was not meeting the setback requirement. The setback requirement was clearly listed on documents, and Segars was an experienced contractor. The county eventually issued Segars a hardship variance and allowed the completion of the project.

Cleland sued Segars for nuisance. The jury awarded her $31,000 in attorney fees, $30,000 in compensatory damages for loss in the value of her home, and $81,742 in punitive damages. The jury also awarded $1 in nominal damages against the purchasers of the home built by Segars.

In the appeal upholding the jury's decision, the appeal court stated, "[A] nuisance is anything that causes hurt, inconvenience, or damage to another....The inconvenience complained of shall not be fanciful, or such as would affect one of fastidious taste, but it shall be such as would affect an ordinary, reasonable [person]."[28]

Rhode Island sued Sherwin for, among other things, public nuisance relating to lead in paint sold and applied in numerous buildings in the state. A public nuisance is a nuisance to the public at large rather than to just specific neighbors. Such actions can be filed by the state government.

The jury was instructed that it could find the existence of a public nuisance based upon "actual present harm or the threat of likely future harm. The threat of likely future harm means harm likely would occur in the future...as a result of a condition which exists today." Evidence presented to the jury included evidence that lead in deteriorated paint and paint on friction surfaces either caused actual harm or threatened to do so. After trial, the jury returned a verdict of public nuisance.[29]

Interference with Contract and Interference with Prospective Business Advantage

These torts occur when the tortfeasor intentionally interferes with the business of another. The tort of wrongful interference with contract occurs when a party outside of a certain contract relationship intentionally induces one of the parties to that contract to breach the contract. "To state a claim for tortious interference, plaintiff must allege: 1) the existence of a valid and enforceable contract; 2) defendant was aware of the contract; 3) defendant intentionally and unjustifiably induced a breach; 4) the breach was caused by defendant's wrongful conduct; and 5) damages."[30]

Tresdale entered into a contract with Actra Construction for the remodeling of his home. However, before Actra Construction could begin the work, Bosra told Tresdale he could do the work for $5,000 less. He suggested that Tresdale tell Actra Construction that he had to back out of the contract because he could not get the financing. Tresdale did this and gave the work to Bosra. Bosra is liable to Actra Construction for intentional interference with a contract.

The second tort, intentional interference with prospective business advantage, does not require that an actual contract exist between two parties. "[T]he following elements have evolved into the tort of intentional or tortious interference with prospective business advantage:

1. the existence of a valid business relationship or a prospective advantage or expectancy sufficiently definite, specific, and capable of acceptance in the sense that there is a reasonable probability of it maturing into a future economic benefit to the plaintiff;

2. knowledge of the relationship, advantage, or expectancy by the defendant;

3. a purposeful intent to interfere with the relationship, advantage, or expectancy;

4. legal causation between the act of interference and the impairment of the relationship, advantage, or expectancy; and

5. actual damages."[31]

EXAMPLE

In this case, the court said that a matter had to proceed to a jury trial to determine if GMP, the engineering firm, had engaged in this tort. The city had retained GMP to provide engineering and design services for a wastewater treatment project to be constructed in the city. GMP prepared the specifications for the project, and these were used by the contractors in bidding the job.

The specifications required a specific system called the "Trojan" system or equal. The specification stated that the "major axis of the UV lamps shall be *horizontal* and parallel to flow. UV systems that operate with lamps placed perpendicular to the flow are unacceptable."

Ultratech, another UV system manufacturer, had a system that contained *perpendicular* lamps. Ultratech's system was rejected by GMP as being nonconforming to the specifications.

Ultratech contended that the specifications were written by GMP, the engineers, in such a way that only the Trojan system, with its horizontal lamps, would be acceptable. One of the principles of GMP had a personal relationship with a principal of Trojan: they were brothers-in-law.

Ultratech sued GMP for intentional interference with a contractual relationship, and GMP moved to have the matter dismissed without going to the jury. The court refused to dismiss the matter and stated, "While a jury may ultimately agree with GMP that it did nothing more than provide truthful advice and information, Ultratech [...has] provided sufficient evidence that GMP's actions were not justified. A reasonable jury could find that GMP wrote the specifications to favor Trojan and that GMP did so for the purpose of harming Ultratech...(rather than for the purpose of providing the City with the most beneficial sewage treatment system)."[32]

EXAMPLE

In the following case, the court did dismiss the matter before trial on the merits. Boyle was the unsuccessful bidder for a public contract to replace the boilers at a community college. Dewberry was the project architect and Kallenberger, the project engineer. Boyle claimed they led Boyle to believe 150-psi boilers were required, but in fact only 15-psi boilers were required. Boyle bid using a 150-psi boiler requirement but alleged that had it known the requirement was only 15 psi, it would have been the low bidder and would have been awarded the job.

Prior to bidding, Boyle called Dewberry and asked why high-pressure boilers were being replaced with low-pressure boilers. Boyle was told, "We always specify Section I

continued

boilers in schools for safety purposes." However, the owner accepted the bid of Donohue which included 15-psi boilers. Boyle complained. The owner investigated the matter and determined that the specification for the 150-psi boilers was an error and the 15-psi boilers were always intended.

Boyle sued the designers for intentional interference with a prospective business advantage. The matter was dismissed, because Boyle had no evidence that the designers acted with a purpose to interfere with Boyle's prospective contract for the boilers.[33]

Trespass to Land

Trespass to land is the invasion of the land of another without permission. It can occur by actual physical invasion or the causing of something, such as runoff or waste, to invade the land of another. It is not necessary to prove injury in a trespass to land claim. However, in the construction industry, these claims usually arise because the construction process has damaged neighboring properties.

EXAMPLE

Sorrow purchased land next to her existing home and wanted to keep it in a natural state. A developer owned several other lots in the area and hired a contractor to clear his lots in preparation for development. The contractor accidentally began to clear Sorrow's lot. The contractor committed a trespass to land and was liable for damages.[34]

EXAMPLE

Justus was the owner of land selected by the state of Virginia for possible construction of a new highway. Her land had to be studied and evaluated, and a state statute gave the state the authority to do this. Mactec Engineering was the subcontractor in charge of the geotechnical studies and underground boring on the property. Mactec's actions caused water to run onto the Justus' property, damaging her water table, the quality of her drinking water, her residence, and her personal property. Boulders and debris were left on her property. Basically, after the geotechnical studies and boring were completed, Justus' residence was no longer habitable.

Justus sued the general contractor in charge of the job for trespass. The court refused to dismiss the claim because, if proved, the facts would support a cause of action for trespass.[35]

Trespass to Personal Property and Conversion

Trespass to personal property is the interference with the owner's right to the use and enjoyment of personal property such as cars or tools. **Conversion** is the wrongful taking of another's personal property. Damages are the value of the property taken.

EXAMPLE

Garruso was employed by Haynes as an architect. As part of her employment, she was supplied with a company laptop computer, drafting table, and other items. Upon leaving her employment at Haynes, Garruso did not return the items to Haynes. She is liable for conversion.

Intentional Infliction of Emotional Distress

Intentional infliction of emotional distress arises from a severe or outrageous act. Merely being rude or crude is not sufficient. An employee could recover for this injury from the employer if the act was intentional, reckless, extreme, or outrageous, and caused the employee severe emotional distress.

EXAMPLE

Lightning worked as a janitor for Roadway, a trucking company. Lightning liked his job and earnestly tried to please his superiors, but he was a slow worker and a marginal performer. Supervisors subjected Lightning to verbal abuse on numerous occasions. For example, Roadway supervisors Mitchell and Darrell stood over Lightning while he cleaned under a truck, and, in the presence of other employees, one of the supervisors stated, "Look at that piece of s___ down there." On another occasion, a supervisor called Lightning into his office and stated, "We pay you really good for the s___ you do, which is nothing. We hate you. You don't belong here." During one discussion, a supervisor tried to hit Lightning but was prevented by other employees in the room. Lightning also received anonymous phone calls at home telling him to quit. Many other similar incidents occurred. Toward the end of his employment, Lightning suffered from a psychotic episode that included manifestations of paranoid delusions. This episode occurred on an evening when managers had "chewed out" Lightning on three separate occasions. Lightning was hospitalized and received treatment.

Lightning was eventually fired and brought a lawsuit for intentional infliction of emotional distress. After a nonjury trial, Lightning was awarded $33,720 in damages for intentional infliction of emotional distress—$25,000 for pain and suffering and $8,720 in medical expenses—and $100,000 in punitive damages, both awards in 1995 dollars. The judgment was upheld on appeal.[36]

EXAMPLE

A defendant sent an e-mail to a third party, pretending to be the plaintiff. The court held that it was not outrageous enough to qualify as intentional infliction of emotional distress.[37]

■ Strict Liability, Product Liability, and Defamation

Strict liability is a category of tort law that places liability for injuries on the tortfeasor without regard to intent or unreasonableness of the conduct. The tortfeasor is liable for merely engaging in the conduct. Even if the tortfeasor takes all reasonable care or even extraordinary care to prevent injury, the tortfeasor is liable. Strict liability exists for the owners of wild animals, blasting, and manufacturers of defective products. Defamation is a form of strict liability. Some jurisdictions have extended strict liability to construction. Some authority exists for holding designers strictly liable for injuries caused by defective designs.[38]

Product Liability

Manufacturers and/or sellers of defective products are liable for the injuries caused by their defective products. The manufacturer and/or seller must provide a fund or insurance coverage to compensate entities injured by any defective products manufactured or sold. This requirement exists even if the manufacturer and/or seller take all reasonable and even extraordinary precautions to prevent the manufacture or sale of defective products. In other words, if a defective product is sold and it causes injury, the manufacturer and/or seller is liable.

Defects generally fall into one of four categories:

- Manufacturing defects. The production process fails at some point and a defective product is produced.
- Design defects. The design of the product does not take into account the prevailing standard of safety as demanded by the society.
- Failure to warn. The product does not warn the user of potential hazards that can cause injury.
- Failure to properly label. The product does not contain required or adequate labels informing the user of the contents.

EXAMPLE

Filipina and Nestor Jimenez purchased a new home in a subdivision. Cobb supplied windows that leaked and damaged the home. The court held that manufacturers of component parts used to construct mass-produced houses, such as Cobb, are strictly liable for damages caused by defective component parts, including damage to other parts of the houses in which the defective products are installed. The evidence proved that the windows were defective and that the defect caused damage to stucco, insulation, framing, drywall, paint, wall coverings, floor coverings, baseboards, and other parts of the home. Cobb was therefore liable for the damage caused by its defective windows.[39]

Product misuse is a limited defense; that is, if the user misuses the product and is injured, the manufacturer and/or seller may not be liable. However, if the misuse is foreseeable by the manufacturer and/or seller, the defense does not apply. In other words, a manufacturer and/or seller must take steps to avoid consumer injury by foreseeable misuse of a product.

The defense of **commonly known dangers** states that manufacturers and/or sellers are not liable for injuries that the user should realize can result from the use of certain products such as knives and guns.

Strict Liability for Building

Most jurisdictions have been reluctant to extend the concepts of strict liability to the construction industry; however, a trend to do so can be seen in the law.[40] If the construction is defective and injury is caused, a cause of action could lie.

EXAMPLE

In this case, the court held that the developer could be held liable under strict liability for life-threatening defects in its duck ponds. If the evidence shows that the developer had the responsibility for producing ponds, and a reasonable homebuyer would not hire its own architect or engineer to evaluate the design of ponds before purchasing the house, a jury could find the developer liable under a product liability theory. [41]

EXAMPLE

Imperial, the contractor, constructed a beach club project that contained defects in the grading, paving, installation of decking, parking structures, tennis courts, and major structural defects in the residential units. In addition, the development was constructed on an eroding cliff. The court held the doctrine of strict liability could lie and refused to dismiss the case.[42]

Defamation

Defamation is the intentional or unintentional telling of a falsehood about another that causes the person injury. Intention is irrelevant, although an intentional defamation can fall under the category of intentional torts and support a punitive damage award. If the falsehood is unintentional and made in the course of a legitimate business decision or interest, the business is excused.

Defamation is divided into two broad categories: slander and libel. **Slander** is spoken defamation or defamation that is not recorded in any way. **Libel** is any form of defamation that is printed, recorded, digitalized, or preserved in any format that allows the defamation to be copied or reproduced. A party is liable for defamation if he makes a false statement that causes injury. The tortfeasor is liable even if the tortfeasor does not realize that the statement is false.

EXAMPLE

After 27 years as an employee of New York Life Insurance Company ("New York Life"), Phyllis Meloff began talking to her supervisors about her lack of advancement in the company despite favorable reviews. She began pointing out that many males had been promoted over her, even males with less experience and knowledge of the work.

After a few months, she was fired for billing seven months of personal commuting expenses to the company's American Express card and failing to reimburse the company. She had prepared a reimbursement check to the company but had not yet turned it in due to the Christmas holidays. It was fairly routine practice, although contrary to company policy, for employees to use the company's credit card for personal expenses and then reimburse the company. She was the only person terminated for this practice. Males who had done the same had not been terminated, although one male employee had had the company credit card taken away for this practice.

Her supervisor sent an e-mail to several managers concerning the termination. The e-mail said, "We found it necessary today to terminate Phyllis Meloff, who used her corporate American express card in a way in which the company was defrauded..." The e-mail was forwarded to over 20 employees.

Phyllis sued, claiming that the e-mail was defamation. A jury found in favor of Phyllis's claim.[43]

MANAGEMENT TIP

If you cannot say something positive someone, do not say it *unless* the information is related to a legitimate business decision or interest. If the information is so related, keep it as private as possible. If someone is terminated and it is necessary to make a general informational statement to other employees, say as little as possible and *never* explain why the person was terminated.

A statement of opinion is not defamation.

EXAMPLE

Grace was awarded a contract to construct the heating, ventilating, and air-conditioning systems for the city. Todd, an engineering company, prepared a report that found Grace responsible for delays on the project and cost overruns. Grace sued Todd for defamation. The court held that no defamation existed, as the report was a matter of opinion.[44]

Because defamation requires the telling of a falsehood, truth is an absolute defense to defamation. No cause of action for defamation can exist for telling the truth about another.

EXAMPLE

Parker was the electrical and lighting design engineer on a major hospital project. He had a brother-in-law in the lighting supply business. Larson was also in the lighting supply business. Larson wrote two letters, one to the general contractor and one to the attorney general, questioning whether the lighting specifications were "rigged" in favor of Parker's brother-in-law.

The court determined that the letters could be actionable defamation per se [defamation per se is discussed below] except that the letters could be substantially true, and it was for the jury to determine if they were substantially true or not. If the allegations in the letters were substantially true, no defamation could exist. The matter was remanded to trial.[45]

In limited circumstances, the telling of the truth about a party can result in tort liability. See the below discussion of invasion of privacy and public disclosure of private facts.

Business Defamation or Defamation Per Se

Some special categories of defamation, called **defamation per se,** do not require the victim to prove damages; that is, the jury may award the victim nominal and/or punitive damages even without proof of actual or compensatory damages. Defamation per se can lie if the tortfeasor makes an untrue statement about another and if the statement:

- disparages a person's ability to engage in their profession;
- accuses a person of crime;
- claims an unmarried person is unchaste;
- claims a person is infected with a sexually transmitted disease.

In the construction industry, the first type of defamation per se is the most likely to occur. Because damages for defamation per se are limited to nominal damages in many jurisdictions, few victims file lawsuits.

EXAMPLE

The president of a company committed defamation per se when she wrote letters accusing an electrical engineer of rigging bids on a construction project so that his brother-in-law would win a lucrative contract. This was not true. The statement implied that the engineer lacked integrity in performing the duties of his office, which amounted to defamation per se.[46]

■ Invasion of Privacy

In recent decades, United States law has begun to protect the privacy of individuals. The protection comes in the form of the recognition of torts for invasion of privacy and statutory protections for employees. Employment law is discussed in Chapter 3.

These torts are still undergoing formation and clarification in the law. Four privacy torts have been recognized to varying degrees by most states:

1. Public disclosure of private facts.
2. The use of a person's name, picture, or other likeness for commercial purposes without permission.
3. Intrusion into personal affairs or areas.
4. False light or publication of information that places a person in a false light.

Public disclosure of private facts is the dissemination of private or personal information about a person without their permission. Even if the information is true, the teller of the private information is subject to liability. This tort can arise when "(1) the fact or facts disclosed must be private in nature; (2) the disclosure must be made to the public ["public" is loosely defined to mean just about any other person]; (3) the disclosure must be one which would be highly offensive to a reasonable person; (4) the fact or facts disclosed cannot be of legitimate concern to the public; and (5) the defendant acted with reckless disregard of the private nature of the fact or facts disclosed."[47]

EXAMPLE

Borquez, a lawyer working for a law firm, was a homosexual but had not told his colleagues or others at the firm. One day, his partner was diagnosed with AIDS. Borquez called a colleague, Ozer, explained the situation, and asked the colleague to cover for him at a deposition and at an arbitration. He asked Ozer to keep his situation private. Ozer said nothing in response. Later, Ozer told several people, including management, that Borquez was gay. Borquez was fired one week later. He sued Ozer for invasion of privacy. The state Supreme Court recognized that the case could support the tort of invasion of privacy and sent the matter back for a jury trial.[48]

MANAGEMENT TIP

Care should be taken to keep information about employees private. Company policies should be in place to protect the files and information about employees. Never disclose private information about an employee to others. Keep personnel records under lock.

In today's world of small digital cameras and recording devices, it is very easy to record a person's picture and record people's acts and words. However, people have a right to privacy that should not be invaded without their permission.

MANAGEMENT TIP

Never use a person's name, picture, or other likeness for commercial purposes without written permission. If a person is not recognizable in a photo, it is acceptable to use the photo, but care should be taken to make sure that no specific person can be identified.

I, _____ hereby grant to _____ (name of business) for no payment, the right and permission to use and reproduce, in whole or in part, photographs, video tapes, digital images, recordings made of me and/or my voice and/or any written extraction of same (hereafter referred to as "the materials"), taken on _____ at _____. I realize that _____ may use these for commercial purposes such as advertising or training.

I understand that the materials will not be modified, except as necessary for production purposes. I understand that the materials may be shown to employees of _____, its agents and principals, the public at large, and specific audiences. I understand that the materials may be shown at conferences, meetings, and may be disseminated on the Web.

I waive any right to inspect or approve any versions of the materials.

I hereby release and discharge _____ from any and all claims and demands arising out of or in connection with the use of the material, including any and all claims for negligence, libel and/or invasion of privacy.

(Signature, date, printed name)

The tort of **intrusion into personal affairs or areas** arises when one party invades the personal space of another or intrudes into an area where that party has no legitimate interest.

MANAGEMENT TIP

Businesses should have a personnel manual or policy that informs employees that desks and workspaces, including computers, are not private places but subject to employer investigation. Employees should be informed that they are not to keep personal information or documents in their desks or on company computers, as the information may be subject to review and evaluation by other employees.

False light is "offensive publicity [that] attributes to the plaintiff' characteristics, conduct, or beliefs that are false, such that the plaintiff is placed before the public in a false position."[49] The tort of false light can result from attributing to another ideas, characteristics, and/or opinions that they do not in fact have. False light is similar to traditional defamation law, but this tort gives protection when the information disseminated may not be false, just skewed.

■ Tort Damages

Tort damages can be divided into five categories: actual, compensatory, punitive, nominal, and injunctive. The first four involve the payment of money. The last, injunctive, is a court order to do or not to do something.

Actual damages are all of the damages resulting from the tortious conduct. No court awards actual damages. Some courts have defined "actual damages" to mean "compensatory damages."

Compensatory damages are the damages that a court will allow the jury to award the injured party. Compensatory damages in tort actions include lost wages, pain and suffering, property damage, and medical care. Compensatory damages that are not foreseeable, or are too remote from the act causing the damage, are not recoverable.

The **economic loss rule** says that a party with only contract damages cannot sue for tort damages. This rule prevents parties from trying to turn a breach-of-contract action into a tort action. If no personal injury has occurred, the parties are limited to their contract causes of action and contract damages, and they cannot sue in tort.

EXAMPLE

Bull sold a house to Alejandre. Later, the buyers determined that the septic system was defective and sued Ms. Bull for fraudulently or negligently misrepresenting the condition of the septic system. The buyers' tort claims of fraud and negligence were dismissed, because the septic system was within the scope of the parties' contract, and the buyers must sue for breach of contract.[50]

Punitive damages are damages above and beyond compensatory damages and are designed to punish the tortfeasor. Punitive damages are only recoverable for intentional torts. The jury may take into account the net worth of the tortfeasor when awarding punitive damages. Punitive damages could be hundreds of thousands or even millions of dollars.

Nominal damages are a small amount of damages such as $1. Nominal damages are most often seen in defamation per se and invasion of privacy actions. Nominal damages signal that a party has won the lawsuit.

An **injunction** is a court order in a tort action ordering one party to do or not to do something. An injunction is a common remedy in a nuisance or trespass action to stop the tortfeasor from the tortious act.

Attorney fees are not usually allowable in tort actions. This concept is commonly referred to as the **American rule.** The alternative is the **British rule.** In Britain, it is common to award the winning party attorney fees. In the United States, attorney fees are not awarded unless provided for by statute or a contract.

■ Warranties

The law of warranties actually includes concepts from both contract and tort law. For this reason warranties are often placed in a separate category. Whether or not a party has breached a warranty and the amount of the resulting damages, if any, is a jury question.

Express Warranties

An **express warranty** is an assertion that a product or service will meet a certain standard and is a contract. Failure of the product or service to meet the standard of an express warranty will result in liability for breach of contract. A warranty does not need to use the words "warranty" or "guarantee." The following forms are waranties.

Neither the final certificate of payment nor any provision in the Documents, nor partial or entire occupancy of the premises by the Owner, shall constitute an acceptance of work not done in accordance with the Contract Documents or relieve the Contractor of liability in respect to any express warranties or responsibility for faulty materials or workmanship. The Contractor shall remedy any defects in the work and pay for any damage to other work resulting therefrom, which shall appear within a period of one year from the date of final acceptance of the work unless a longer period is specified. The Owner will give notice of observed defects with reasonable promptness.[51]

A. Contractor provides a limited warranty on all labor and materials used in this project for a period of two years following substantial completion of all work.

B. **THE EXPRESS WARRANTIES CONTAINED HEREIN ARE IN LIEU OF ALL OTHER WARRANTIES, EXPRESS OR IMPLIED, INCLUDING ANY WARRANTIES OF MERCHANTABILITY, HABITABILITY, OR FITNESS FOR A PARTICULAR USE OR PURPOSE. THIS LIMITED WARRANTY EXCLUDES CONSEQUENTIAL AND INCIDENTAL DAMAGES AND LIMITS THE DURATION OF IMPLIED WARRANTIES TO THE FULLEST EXTENT PERMISSIBLE UNDER STATE AND FEDERAL LAW.**

Owners have begun to demand longer and more comprehensive warranties. These can create long-term liabilities for the contractor, which must be carried on balance sheets and may affect such things as bonding capacity, lines of credit, and ability to take on new work.

EXAMPLE

The State Department of Transportation requires a contractor to guarantee a road for 25 years. All repairs during that time are to be undertaken by the contractor. The contractor must allocate resources for future repairs.

Implied Warranties

Warranties can arise in two ways: (2) voluntarily by the parties through a contract or (2) required by the law and therefore a tort duty. Warranties required by the law are generally called **implied warranties,** because the law will imply the warranty into the transaction whether or not the parties have agreed to the warranty by contract.

Implied warranties related to the construction industry generally fall into three categories:

- Warranty of merchantability
- Warranty of fitness
- Warranty to build in a good an workmanlike manner or warranty of habitability

The warranty of merchantability and the warranty of fitness are implied into sales of goods by the Uniform Commercial Code and therefore apply to such items as fans and air conditioners.

The warranty of merchantability states that a product will be fit for the normal use for which the product is manufactured. If the product does not work for its normal purpose, the warranty has been violated, and the buyer is entitled to repair or replacement.

The warranty of fitness only arises when the seller makes a recommendation to the buyer that the product will perform according to the buyer's needs.

EXAMPLE

The buyer asks for paint that will cover metal. The seller suggested and sold to the buyer Brand X, but Brand X peels off because it is designed only to cover wood. The warranty of fitness arose when the seller made the recommendation of Brand X. The failure of Brand X to perform as represented is a breach of the warranty.

The warranty to build in a good and workmanlike manner or warranty of habitability is a common-law implied warranty and, as such, it may have a different name in any given state. Some jurisdictions and authors separate the two and apply the first to the acts of the contractor and the second to the building.[52] However, the results are similar: protection of a home buyer for faulty construction. This warranty is similar to the warranty of merchantability. The law will imply into a construction contract the obligation of the builder to build in a good and workmanlike manner. Failure to do so will result in liability.

EXAMPLE

A cause of action for violation of an implied warranty exists when wiring is faulty and spliced without junction boxes in violation of the National Electrical Code.[53]

A basement wall caved in, resulting in a wet and damp basement. This was a breach of the implied warranty.[54]

Warranty Limitations

The law in many jurisdictions allows the waiver of implied warranties or the substitution of a lesser warranty for an implied warranty.[55] The Uniform Commercial Code allows sellers of products to limit, but not entirely dispense with, the implied warranties of merchantability and fitness. This is most commonly done by giving a specific time period, such as 90 days, during which the product is warranted. In addition, the seller may limit the damages for breach of an implied warranty to such things as replacement cost of the good or product. This action relieves the seller for such damages as pain and suffering, lost wages, medical bills, and any other damage other than the replacement cost of the good or product.

MANAGEMENT TIP

Businesses should develop express warranties for their products and services, as the failure to do so will result in application of the warranties implied by law. Warranty limitations should be prominently displayed and not buried in documentation. If appropriate, the buyer should be asked to sign or initial all warranties. See discussion of warranty limitations in Chapter 14.

■ Endnotes

1 *K & K Recycling, Inc. v. Alaska Gold Co.*, 80 P.3d 702 (Alaska 2003).

2 *Hydro Investors, Inc. v. Trafalgar Power, Inc.*, 227 F.3d 8 (2d Cir. 2000).

3 *Schiltz v. Cullen-Schiltz & Assoc., Inc.*, 228 N.W.2d 10, 17 (Iowa 1975), *Ambassador Baptist Church v. Seabreeze Heating and Cooling Co.*, 28 Mich App 424, 184 NW2d 568 (1970).

4 Example based on *Balagna v Shawnee County*, 233 Kan 1068, 668 P2d 157 (Ka. 1983).

5 Example based on *Howard v. Usiak*, 172 Vt. 227, 775 A.2d 909 (Vt. 2001).

6 *Columbus v. Smith & Mahoney, PC*, 59 A.D.2d 857, 686 N.Y.S.2d 235 (1999).

7 Example based on *Mudgett v. Marshall*, 574 A.2d 867 (Me. 1990).

8 Example based on *Hutcheson v Eastern Engineering Co.*, 132 Ga App 885, 209 SE2d 680, (1974).

9 Example based on *Brown v. Gamble Constr. Co.* 537 S.W.2d 685 (1976, Mo App).

10 Example based on *Caranna v. Eades*, 466 So. 2d 259 (Fla. Dist. Ct. App. 2d Dist. 1985).

11 *Pierce v. ALSC Architects*, 270 Mont. 97, 890 P.2d 1254 (Mont. 1995).

12 Example based on *Sheek v. Asia Badger, Inc.* 235 F.3d 687 (1st Cir. 2000).

13 Example based on *Lee Lewis Construction, Inc. v. Harrison. No. 99-0793, 2001 Tex. LEXIS 132* (Tex. Dec. 20, 2001).

14 See Emmanuel S. Tipon, *Modern Status of Rules Regarding Tort Liability of Building or Construction Contractor for Injury or Damage to Third Person Occurring after Completion and Acceptance of Work*, 75 A.L.R.5th 413.

15 *Louk v. Isuzu Motors*, 198 W. Va. 250, 479 S.E.2d 911 (W. Va. 1996).

16 *Terry v. New Mexico State Highway Comm.*, 98 N.M. 119, 645 P.2d 1375 (NM 1982).

17 Example based on *Stewart v. Cox*, 55 Cal. 2d 857, 13 Cal. Rptr. 521, 362 P.2d 345 (Ca. 1961).

18 Example based on *Atkinson v. MGM Grand Hotel, Inc.*, 120 Nev. 639, 98 P.3d 678 (Nev. 2004).

19 *Malvicini v. Stratfield Motor Hotel, Inc.*, 206 Conn. 439, 442, 538 A.2d 690 (1988).

20 *Byrne v. Boadle*, 2 H. & C. 722, 159 Eng. Rep. 299 (England 1863).

21 *Rosenblum v. Deerfield Woods Condominium Asso.*, 1991 Conn. Super. Lexis 1656 (Conn. 1991).

22 Example based on *Goedert v. Newcastle Equipment Co., Inc.*, 802 P.2d 157 (Wyo. 1990).

23 Example based on *Andren v. White-Rodgers Co.*, 465 N.W.2d 102, 104-05 (Minn. 1991).

24 *Johnson v. Armfield*, 2003 SD 134, 672 N.W.2d 478, 481 (S.D. 2002) (quoting *Boomsma v. Dakota, Minn. & E.R.R. Corp.*, 2002 SD 106, 651 N.W.2d 238, 245-46 (S.D. 2003)).

25 *Barnett v. Coppell North Texas Court, Ltd.*, 123 S.W.3d 804 (Tex. App. 2003).

26 Example based on *Yazd v. Woodside Homes Corp.*, 2006 UT 47, 143 P.3d 283 (Ut. 2004).

27 Example based on *M&D, Inc. v. W.B. McConkey*, 231 Mich. App. 22, 585 N.W.2d 33 (Mi. App. 1998).

28 *Segars v. Cleland*, 255 Ga. App. 293, 564 S.E.2d 874 (2002).

29 Example based on *Rhode Island v. Lead Industries Association, Inc.*, 2007 R.I. Super. LEXIS 32 (R.I. 2007).

30 *HPI Health Care Servs. Inc. v. Mt. Vernon Hosp., Inc.*, 131 Ill. 2d 145, 545 N.E.2d 672, 676, 137 Ill. Dec. 19 (Ill. 1989). *Weinberg v. Mauch,* 78 Haw. 40, 50, 890 P.2d 277, 287 (Haw. 1995).

31 *Robert's Hawaii School Bus, Inc. v. Laupahoehoe Transp. Co., Inc.,* 91 Haw. 224, 258, 982 P.2d 853, 887 (Haw. 1999).

32 Example based on *Bodell Const. Co. v. Ohio Pacific Tech, Inc.,* 458 F. Supp. 2d 1153 (D. Haw. 2006).

33 Example based on *Boyle Services, Inc. v. Dewberry Design Group, Inc.,* 2001 OK CIV APP 63, (Okla. App. 2001).

34 Example based on *Sorrow v. Hadaway,* 269 Ga. App. 446, 604 S.E.2d 197 (2004).

35 Example based on *Justus v. Kellogg Brown & Root Services, Inc.,* 373 F. Supp. 2d 608 (WD Vir. 2005).

36 Example based on *Lightning v. Roadway Express, Inc.,* 60 F.3d 1551 (11th Cir. 1995).

37 Example based on *Rall v. Hellman,* 726 N.Y.S.2d 629 (N.Y.A.D. 1 Dept. 2001).

38 *Architect Tort Liability in Preparation of Plans and Specifications,* Calif. L Rev 1361 (November, 1967); and *Liability of Design Professionals—The Necessity of Fault,* at 58 Iowa L Rev 1221 (June, 1973).

39 Example based on *Jimenez v. Superior Court of San Diego,* 29 Cal. 4th 473, 58 P.3d 450, 127 Cal. Rptr. 2d 614 (Ca. 2002).

40 See *Recovery, Under Strict Liability in Tort, For Injury or Damage Caused by Defects in Building or Land,* 25 A.L.R.4th 351 (2005).

41 Example based on *Chitkin v. Lincoln Nat'l Ins. Co.,* 879 F Supp 841. (SD Cal, 1995).

42 Example based on *Del Mar Beach Club Owners Asso. v. Imperial Contracting Co.* 123 Cal App 3d 898, 176 Cal Rptr 886, 25 ALR4th 336 (Ca.App. 1981).

43 Example based on *Meloff v. New York Life Ins. Co.,* 240 F.3d 138 (2d Cir. 2000).

44 Example based on *John Grace & Co., Inc. v. Todd Associates, Inc.,* 188 A.D.2d 585, 591 N.Y.S.2d 477; (N.Y. App. 1992).

45 *Parker v. House O'Lite Corp.,* 324 Ill. App. 3d 1014, 756 N.E.2d 286 (Ill. App. 2006).

46 Example based on *Parker v. House O'Lite Corp.,* 324 Ill. App. 3d 1014, 756 N.E.2d 286, 296, 258 Ill. Dec. 304 (Ill. App. Ct. 2001).

47 *Oxer v. Borquez,* 940 P.2d 371 (Co. 1997).

48 Example based on *Oxer v. Borquez,* 940 P.2d 371 (Co. 1997).

49 Frank C. Morris, "E-Mail Communication: The Next Employment Nightmare," ALI/ABA Course of Study: Employment Discrimination and Civil Rights Acting in Federal and State Courts 623, 702 (1995).

50 *Alejandre v. Bull,* 2007 Wash. LEXIS 132 (Wash. 2007).

51 From *City of Kennewick v. Hanford Piping, Inc.,* 558 P.2d 276 (Wash.App. 1977).

52 Wendy B. Davis, *Corrosion by Codification: The Deficiencies in the Statutory Versions of the Implied Warranty of Workmanlike Construction,* 39 Creighton L. Rev. 103 (2005).

53 Example based on *Moxley v. Laramie Builders, Inc.,* 600 P.2d 733, 734 (Wyo. 1979).

[54] Example based on *Henggeler v. Jindra,* 191 Neb. 317, 318, 214 N.W.2d 925, 926 (1974).

[55] *Brevorka v. Wolfe Constr., Inc.,* 573 S.E.2d 656 (N.C. Ct. App. 2002). Jason Hoyt Hayes, *Centex Homes v. Buecher:* Separating the Conjoined Humber Warranties and Providing Disclaimer Requirements for Each, 55 Baylor L. Rev. 1137 (2003).

323

CHAPTER **13**

Sales of Materials and Supplies

■ Introduction

The "law of sales" refers to the special laws adopted to deal with the sales of **goods**, defined as all forms of tangible personal property including specifically manufactured goods, supplies, mobile homes, and materials. Contracts regarding these items are not governed by the common law of contracts but by Article 2 of the state's version of the **Uniform Commercial Code (UCC)**. Every state except Louisiana[1] has adopted, with minor changes, Article 2 of the UCC.

The UCC, as all uniform codes, is developed by a private organization and recommended to state legislatures; no uniform code has any validity until adopted as law by a state. The state legislature of each state is free to adopt all, part, or none of any uniform code developed. The term "UCC" used herein refers to the form of the UCC adopted by any particular jurisdiction.

Major differences exist between the UCC and the common law, as discussed in Chapter 7 and Chapter 8. Contract formation, acceptance of the contract, modifications to the contract, and the battle of the forms are areas with different results depending on whether or not the UCC or the common law controls.

For example, under common law, a contract must contain the essential terms of the contract to be enforceable. Under the UCC, a contract is enforceable even if some of the essential terms are *not* included. Essential terms not included in the contract are automatically added to the contract by the UCC. The following terms will be added to the contract governed by the UCC even if the parties have not included the terms in the contract:

- UCC 2-305(1): the law will add a "reasonable" price.
- UCC 2-310(a): if no time for payment is set, payment is due at the time and place where the buyer receives the goods.
- The UCC recognizes requirements and output contracts. A requirements contract is one in which the buyer agrees to buy and the seller agrees to sell all of the goods required by the buyer. An output contract is one where the buyer agrees to buy all of the goods the seller produces. The parties can limit these contracts to a maximum or minimum amount rather than "all."

- UCC 2-206(1) says a contract can be accepted in any reasonable manner.
- One of the major differences between the UCC and the common law is that under the UCC, modifications to the contract need not be supported by new consideration; see Chapter 8. However, changes must be made in good faith.

The UCC has recently undergone a major revision and contains important changes that may or may not have been adopted by a particular state. Important differences between the UCC and the Revised UCC are outlined below.

The UCC does not apply to the prime contract or any contract for services. It applies to the sales of goods only. In the construction industry, the term "materials and supplies" is used instead of the term "goods." Purchases by the contractor for such things such as lumber, paint, windows, drywall, fans, garage door openers, and the like are governed by the UCC rather than the common law of contracts discussed elsewhere in this book. Service contracts such as excavation and grading are not covered by the UCC.

EXAMPLE

Triple H Construction entered into a construction contract with Hunters' Run Stables to erect a horse barn and riding arena on the Hunters' Run property. The contract contained a warranty on the roof. After the roof collapsed, the owner sued the contractor and the supplier of the roof for breach of UCC warranties. The contract between the owner and Triple H is not governed by the UCC because it is primarily for construction services and not materials. The contract between the contractor and the supplier is governed by the UCC. The owner therefore cannot get UCC warranties on the work, only common law warranties.[2] If the contractor were to sue the supplier of the roofing system, the matter would be governed by the UCC.

A contract containing a combination of service and goods, such as a subcontract to install drywall, is looked at to determine the primary purpose of the contract. If the primary purpose is a service, then the contract is not governed by the UCC. If the primary purpose is the sale of a good and the service is secondary, then the transaction is governed by the UCC.

EXAMPLE

Pittsley contracted with Houser for the purchase and installation of carpet. The cost of the carpet was $4,000, and the cost of installation was $700. Pittsley gave Houser a downpayment of $1,000. Following the installation, some seams were visible, gaps appeared, and the carpet did not lie flat. Houser attempted to fix the problems by stretching the carpet, but this did not work. Pittsley sued Houser for return of the down payment and rescission of the contract. It was held this contract was primarily a sale of goods, rather than a service contract, and was subject to the UCC.[3]

Meyers, the contractor, entered into a contract with Henderson, the owner, to install overhead doors at Henderson's plant. The disassembled doors were purchased by Meyers and installed by Henderson at the site. It was held the contract for the doors was primarily for the sale of goods and thus governed by the UCC.[4]

The UCC covers sales by both merchants and nonmerchants. A **merchant** is one who is normally engaged in the selling of goods of the type in the transaction. In the construction industry, a merchant would normally be called a "supplier." Mobile homes are covered by the UCC, and the seller of a mobile home may have a different title.

As with the common law, the UCC will not uphold a contract or part of a sales contract that is unconscionable.[5] In addition, the UCC implies into every contract the duty of good faith and fair dealing.[6] Any party's violation of the duty of good faith and fair dealing is a breach of the contract allowing the other party to recover damages.

■ Formation of a Sales Contract

Contracts for the sale of materials and supplies do not ordinarily fail for indefiniteness. As long as the subject matter of the sale, such as nails or air conditioners, and the quantity is given, and it appears that the parties intended to make contract, a contract will be formed.[7] In the construction industry, the quantity is not always available; these contracts are considered requirements contracts. A **requirements contract** is one that obligates the buyer to purchase its requirements for a certain job from the seller. This type of contract is enforced, but the parties must operate in good faith, and no disproportionate amount may be demanded at the agreed-upon unit price.

EXAMPLE

Lower received a verbal bid from Century for the concrete on a school expansion project. The job specifications included an estimate of concrete needed for completion, although because of the nature of the project, the quantities were only estimates. After being awarded the job, Lower sent a purchase order to Century containing the location of the project and the price to be paid per cubic yard, and it called for the concrete to be delivered "as called for." The purchase order did not contain any specific amount of concrete. Century estimated that about 5,500 cubic yards would be necessary to complete the project, and this later proved to be correct.

Century delivered several loads of concrete, but Lower was not satisfied with it. An independent testing laboratory determined that the preliminary loads of concrete did not meet the strength standards required by the school district. Century protested the

continued

results of the tests and obtained independent testing, at its own expense, which showed the concrete was within the required standards. Lower terminated Century's contract, but it did not tell Century—Century found out when its loaded trucks were turned away from the site.

Lower said that no contract existed between it and Century for more than the amount delivered pursuant to UCC 2-201 which states, "...the contract is not enforceable under this paragraph beyond the quantity of goods shown in the writing." Century claimed that the contract was a requirements contract, which is recognized by in UCC 2-306.

The appeal court held that a contract did exist. The jury had to decide if the contract was a requirements contract or not and whether or not there was a breach.[8]

A **firm offer**[9] is an offer by a supplier which cannot be revoked if it (1) states that it will be kept open, (2) is in writing, and (3) is signed by the supplier. A bid by a supplier may be considered a firm offer if these requirements are met.[10]

EXAMPLE

Santini, a supplier of air conditioners, sends the following to YingLing Construction, "We hereby offer you 10 Airzone model #234-ABC combination heater-air conditioners for $1,000 per unit. This offer will remain open for 30 days. Signed, *Santini.*" After 10 days, Santini calls the contractor and saying the offer is no longer open. YingLing Construction purchases the same air conditioner from another supplier at $1,200. Santini is liable to YingLing Construction for $200/air conditioner or $2,000.

The above example illustrates a major difference between the common law and the UCC. Under common law, offers are normally revocable at any time prior to acceptance unless consideration has been given to the supplier to keep the offer open. In non-UCC contracts, the doctrine of promissory estoppel *may* come into play to prevent injustice, such as when the general contractor relies upon the subcontractor's bid (see Chapter 2).

MANAGEMENT TIP

Contractors should request suppliers include the term "firm offer" in their quotations to make sure that the offer is binding for the given time period. While the term "firm offer" is not specifically required by the statute, it is better to have it.

Acceptance[11] of an offer[12] is easily accomplished under the UCC. An offer can be accepted in any reasonable manner.[13] Another major difference between the common law of contracts and the UCC is that under the UCC, the acceptance need not mirror the offer but can have different or additional terms. In other words, the contract is still formed at the moment of acceptance, but the terms of the contract may be different from those in the offer.

Additional or Different Terms

A major difference between the UCC and the common law of contracts is in the area of acceptance of the offer. If the supplier has made a bid, the bid is the offer, and the purchase order is the acceptance. If the general contractor places an order or sends a purchase order, that is the offer, and the invoice or the shipment of the materials is the acceptance.

Under the common law, the acceptance must mirror the offer; that is, if the response to an offer adds any terms to the offer or modifies the offer in any way, the response to the offer is *not* an acceptance but a counteroffer. No contract is formed at that point (see Chapter 7).

Under the UCC, however, the acceptance need *not* mirror the offer and may contain additional, different, and even conflicting terms. The contract is formed even though the acceptance does not mirror the offer. This has resulted in what is commonly called the battle of the forms. The **battle of the forms** occurs when the parties' forms contain additional, different, or conflicting clauses. The issue then arises: what are the terms of the contract?

If the acceptance contains additional terms, the additional terms become part of the contract *unless* they are material terms.[14] The issue then switches to whether or not any given term is material. Binding arbitration and warranty disclaimers have been found to be material terms. This means attempts to include these terms in the acceptance are ineffective. These terms must appear in the offer in order to be binding.

329

EXAMPLE

Shimono sends a bid (offer) to Nemec Construction to supply the lumber on a particular job. The bid does not provide for binding arbitration of disputes. Nemec is eventually awarded the job and sends a purchase order (acceptance) to Shimono. The purchase order contains a provision for binding arbitration and a 10-year warranty. The contract is formed between Shimono and Nemec. However, the provision for binding arbitration and the 10-year warranty are not included in the contract. The contract will contain the implied warranty of merchantability.

Parties often try to craft language in their documentation stating that only *their* documentation controls and no additional, different, or contradictory terms of the other party have any legal validity. This language is of no legal effect. If either the offer or the acceptance contains language stating that *its* terms must control, then no contract can be formed on the paperwork. At this point, neither party is required to perform.

Kang, the contractor, sends Bart, the supplier, a purchase order for drywall. The purchase order contained provisions for binding arbitration and specified that the drywall was to have a five-year warranty. It also contained language that the acceptance could not contain any additional or different terms. The seller sent an invoice disclaiming all warranties. No contract is formed; Kang need not purchase the drywall from Bart. Bart need not sell the drywall to Kang.

In the above situation, it is common for the parties to complete the sale anyway even though no contract has been formed on the paperwork. If the materials or supplies are shipped and accepted, a contract is formed though not on the paperwork, and *neither* party's provisions control.[15] The contract between the parties contains those provisions that exist in both the offer and the acceptance *and* all other terms are supplied by the UCC's default terms.

ETP sent a purchase order (offer) to OxyChem for plastic liners. OxyChem then sent an invoice (acceptance) to ETP. On the back of OxyChem's invoice, in light blue print 1/16-inch high, OxyChem disclaimed all warranties and limited ETP's damages. Oxy-Chem's invoices also stated that the contract was "subject to and expressly conditioned" upon ETP's assent to the terms and conditions printed on OxyChem's invoice. The invoice also expressly rejected the terms in ETP's purchase order. ETP never assented to the terms and conditions on OxyChem's invoices. OxyChem shipped the liners anyway.

When the plastic liners failed, ETP sued OxyChem for breach of warranty. Oxy-Chem claimed that its warranty limitations controlled and the contract contained *only* the terms in its invoice. OxyChem was incorrect. No contract was formed on the paperwork. The contract arose under UCC 2-207(3), that is, by the acts of the parties. The terms of the contract included those terms upon which both the invoice and the purchase order agreed *and* any terms supplied by the UCC. Since the UCC implies a warranty of fitness and merchantability, those warranties were in the contract.[16]

Conflicting Terms

If the terms of the offer and the acceptance conflict, neither party is obligated to perform. However, if the sale is completed, a contract is formed by the acts of the parties and not on the paperwork. The conflicting terms cancel each other out and are not part of the contract. If the UCC has a term covering that situation, the UCC term controls.

Steiny purchased an electric switchboard with circuit breakers from Cutler and installed it on a construction project. The switchboard was defective and exploded, causing injury to the project. Steiny sued Cutler for breach of warranty. Steiny's purchase order (offer) contained a requirement that the product was to include broad warranties. Cutler's invoice (acceptance) disclaimed all warranties. Under UCC 2-207, the conflicting warranty provisions dropped out of the contract, and the UCC warranty provision took its place. The warranty of merchantability was therefore implied into the contract and Steiny could sue for breach of the implied warranty.[17]

"Open communication and reasonableness are the keys to avoiding conflicts in the battle of the forms. If you want some specific term in your contract, be sure to inform the buyer or supplier. Do not try to bury it in fine print on the form. Draw the other side's attention to it. Do not try to be sneaky. Remember that every contract has an implied term of good faith. "**'Good faith'**...means honesty in fact and the observance of reasonable commercial standards of fair dealing." (UCC 1-304)

Problems occur when additional terms in receipts or invoices come as a surprise or are materially different from what a party expects. For example, provisions limiting damages, warranties, and/or requiring arbitration may come as a surprise to one party.

Requirement of a Writing or Statute of Frauds

The UCC requires sales over $500 to be evidenced by a writing signed by the party to be charged. The Revised UCC requires sales over $5,000 to be evidenced by a record. Both the terms **writing** and **record** have been broadly defined to include e-mail and facsimile documents. The word **signed** is also broadly defined and includes a printed, stamped, written name such as on a letterhead. Initials or a thumbprint are acceptable. "No catalog of possible authentications can be complete, and the court must use common sense and commercial experience in passing upon these matters."[18]

Even if there is no writing/record signed by the party to be charged, a contract is formed upon shipment and acceptance of the materials and supplies. A signed writing/record is only needed to recover damages if one of the parties refuses to perform. Once the parties have performed the contract, no writing is necessary to prove the contract exists.

EXAMPLE

Rentenbach, the general contractor, sent a purchase order (offer) to Home Lumber to supply ready-mix concrete for a building project. Home Lumber did not sign the purchase order or send an invoice. It did, however, ship the concrete. Later, when a dispute arose between Rentenbach and Home Lumber with regard to the quality of the concrete, Home Lumber claimed that no contract had been entered into between the parties. However, Home Lumber's shipment of the concrete evidenced the contract.[19] Though issues may have existed as to the exact terms of the contract, a contract existed.

Another important exception to the writing requirement exists. When the parties are business people, *either* party to the transaction can send a **confirmation memorandum,** which is a memo confirming or verifying the agreement that is signed or authenticated by *either* of the two parties to the oral agreement. As long as the other party does not object to the confirmation memorandum within 10 days, it represents the agreement of the parties. At least one state has said that an e-mail can satisfy the requirement.[20]

EXAMPLE

Louisiana-Pacific orally agreed to buy a large quantity of cedar shakes from GPL. GPL filled out and signed order confirmation forms that contained the stated price and quantity (200 truckloads) and sent these to Louisiana-Pacific. These forms contained a "sign and return clause" asking Louisiana-Pacific to sign and return one copy. Louisiana-Pacific never signed and returned a copy, but GPL shipped the product. Louisiana-Pacific accepted delivery of 13 truckloads of shakes but then stated that it need not purchase the remaining quantity because it had never signed and returned the form, and therefore the contract failed for lack of a writing signed by the entity to be charged.

The court held that the order confirmation form, even though unsigned by Louisiana-Pacific, was sufficient to satisfy the merchant exception because it unambiguously identified the parties to the oral contract and the prices and quantities of the goods being sold, it was signed by one of the parties, sent to the other party, and no objection was made within 10 days of receipt.[21]

EXAMPLE

In the following case, no memo satisfying the exception existed, and therefore, no contract was formed. Clarion and Phelon were in negotiations for Phelon's purchase of certain component parts (counterweights) from Clarion. Clarion sent a fax to Phelon on July 16 that read, "Regarding the 19873-00-A counterweight...please send me an e-mail... confirming that [we, Clarion] will also be supplying these parts..." Later, a dispute arose between Phelon and Clarion, and Clarion sued Phelon for breach of the contract to buy the counterweights.

Clarion claimed that the fax complied with the merchant's exception, but the court said, "In order to be a 'confirmation' under *Section 36-2-201(2)*, [the North Carolina statute corresponding to UCC 2-201(2)], a writing must at least 'indicate that a binding or completed transaction has been made.'...By requiring the buyer to take further action in order to signal acceptance..., [the seller] indicate[s] to the buyer...that the terms quoted were still subject to acceptance or rejection rather than representing a memorialization of an oral contract[;]...[a] true confirmation requires no response." The fax was not a confirmation of the contract's existence but part of a continuing negotiation.[22]

MANAGEMENT TIP

Contractors should always send a supplier a signed confirmation memorandum including the terms of the oral agreement. The confirmation memo should state, at a minimum, the price, time, and place of performance.

Other exceptions to the requirement of a writing include the following:

- Specially manufactured goods.[23]

EXAMPLE

Schultz is in the business of designing, developing, and producing custom electronic components per customer specification. It designed a product called the Head End System for Nynex and received a purchase order for 10 systems at a cost of $100,000. After receipt of these 10 systems, Nynex asked Schultz to modify the components and instructed Schultz that Nynex's management had approved an order for an additional 90 of the modified Head End Systems at an aggregate price of $800,000. Nynex required 30 of the systems be rushed to completion and also indicated that production was to be commenced on another 60 systems.

No purchase order was prepared for the additional 90 modified systems, and later, Nynex refused to purchase any but the original 10 systems. Schultz sued Nynex for breach of contract for failing to purchase the other 90 systems.

Nynex claimed that it was only obligated to purchase the number of systems evidenced by the writing. However, because the goods were specially manufactured, no writing was needed. The matter must proceed to a jury trial to determine if a contract existed.[24]

- Receipt and acceptance of goods

EXAMPLE

Gerner, the seller, and Vasby, the buyer, entered into an oral contract to sell goods valued at over $500. The seller delivered the goods to the buyer, and the buyer tendered a check to the seller for $10,000. The seller claimed that the amount agreed upon in the oral contract was $12,000 and sued the buyer. The court held that a contract did exist, because the goods had been accepted. The only issue then was a factual issue: was the agreed-upon amount $10,000 or $12,000? After testimony at a bench trial, the judge decided that the more credible witness was Vasby, and therefore the amount owed was $10,000.[25]

- Payment of goods
- Admission of existence of contract

■ Modifications to the Sales Contract

Another major area of difference between the UCC and the common law of contracts is that a sales contract may be modified by the parties without additional consideration.[26] Under common law. modifications to contracts are considered *new* contracts and must be supported by additional consideration. Under the UCC, as long as the modification is voluntarily entered into by both parties, it will be valid.[27] The issue then shifts to the voluntariness of the modification rather than the existence of additional consideration. Parties must act in good faith in modifying the contract, and a failure to act in good faith can show lack of voluntariness.

Attempts to require all modifications to the UCC contract be in writing suffer from the same problem in enforcement as common-law contracts. The UCC has attempted to clarify what happens in this situation. If the buyer is *not* a merchant, such as a purchaser of a mobile home, the writing requirement must be separately signed by the buyer. The problem is always that this provision, like any provision in a contract, can be waived by acts inconsistent with the provision. It is impossible to draft a contract provision that can *only* be waived by a writing.[28]

Parson gave Monroc, a road contractor, a price quotation for ½-inch and ¾-inch bituminous surface course mixes and a seal coat mix. Monroc was the successful bidder and was awarded the project. Monroc sent a purchase order to Parson. Parson modified the purchase order by increasing the price and sent it back to Monroc. Monroc made no objection to the increased price in the modified purchase order.

The project was delayed for approximately one year. Parson testified that Monroc contacted Parson the next year and asked if Parson would supply the materials at the same price as in the purchase order. Parson said no, that price had expired. Parson then testified that the parties orally agreed to a higher price, and Parson sent a confirmatory memo to Monroc containing the elements of the oral agreement, including the new prices. Monroc never objected to this confirmatory memo.

Monroc claimed that it never orally agreed to the higher prices stated in the confirmatory memo but instead told Parson it would pay only according to the *original* purchase order. Monroc claimed that it did not respond to Parson's confirmatory memo containing the higher prices because it was relying upon the contract provision that stated, "Claims for extras positively will not be allowed unless ordered in writing and signed by Monroc."

The court held that the contract had been modified to include the higher prices as indicated in Parson's confirmatory memo sent one year after the original purchase order had been sent. If Monroc had objected to the confirmatory memorandum with the new prices, it should have done so. Language in Monroc's purchase order did not overcome the UCC requirement that a merchant must object to a confirmatory memo within 10 days.[29]

■ Interpretation of the Sales Contract

The common law of contract interpretation and the law of the interpretation under the UCC are very similar (see Chapter 9).

American Carpet, the seller, sold carpet to World Carpet, the buyer. The contract was oral but was confirmed by letters, invoices, and other memoranda. None of the writings contained provisions for advertising credit, but both parties agreed that it was the custom of the industry to include such a credit. The only disagreement was the *amount* of the credit and *how* the credit was to be calculated.

The court relied upon the testimony of Smith as to the custom in the carpet industry regarding advertising credits and ruled in favor of the buyer's version of the amount of the credit and how it was to be calculated. Under the UCC, industry custom can be used to explain or supplement the agreement.[30]

> ## EXAMPLE
>
> Metric entered into a contract with NASA for the construction of its space station processing facility at the Kennedy Space Center in Florida. The contract, drafted by NASA, required that "new lamps shall be installed immediately prior to completion" of the facility. Metric claimed that this meant replacement of broken and defective lamps, and NASA claimed that it meant total replacement of all lamps. On appeal, the court held that the contract contained a latent ambiguity. Since latent ambiguities are resolved against the drafter of the document, NASA lost, and the contractor's interpretation prevailed.[31]

■ Risk of Loss and Transfer of Title

The risk of loss and title to conforming materials and supplies passes to the buyer when the supplier delivers the goods to the carrier for shipment[32] unless the parties have agreed otherwise. This means that the buyer must have insurance to cover the materials and supplies in transit.

> ## EXAMPLE
>
> Jordan bought custom-made windows from Windows, Inc. During shipment, the windows were broken and many of the frames twisted. Jordan ordered another set of windows from Windows, Inc., which arrived without damage. Jordan did not pay for either shipment.
>
> Risk of loss passed to Jordan when Windows, Inc. transferred the windows to the carrier, and Jordan had to pay Windows, Inc. for both sets of windows.[33]

> ## MANAGEMENT TIP
>
> Contractors and purchasers of materials and supplies should have insurance coverage for materials and supplies in transit or clarify that the title does not pass until the materials and supplies are received by the buyer.

The risk of loss and transfer of title of nonconforming materials and supplies does not pass to the buyer.[34]

> ## EXAMPLE
>
> A contractor orders 10 white garage doors, Model #Ab-110. The supplier ships 10 tan garage doors, Model #Ab-120. The garage doors are damaged in transit. The loss falls on the supplier, not the contractor because the goods are nonconforming.

Sellers of materials and supplies should have insurance coverage for materials and supplies in transit to cover losses to nonconforming goods.

If a contractor accepts the materials or supplies, and later discovers a defect, the acceptance can be revoked, and the risk of loss passes back to the supplier to the extent that the contractor's insurance does not cover the loss.[35]

If the material or supplies are to be installed by the seller, risk of loss stays with the seller prior to the installation.

EXAMPLE

Thomas purchased a pool heater from Sun Kissed, and Sun Kissed was to install the pool heater. The heater was delivered to Thomas's home and left in the driveway. It was too heavy for Thomas to move, and she called Sun Kissed, but Sun Kissed did not return her calls. The pool heater eventually disappeared. Sun Kissed was liable for the pool heater, as title had not transferred to Thomas because it had not been installed.[36]

337

■ Nonconforming Goods

Upon receipt of the goods, except for COD goods, the buyer has a duty to inspect the goods within a reasonable time and determine if they conform. If they do not conform, the buyer can reject all of the goods, some of the goods, or none of the goods. If the time for performance has not yet passed, the seller is automatically given a chance to cure, this is to supply conforming goods, within the contract time period.[37] If the time of performance is past, the buyer need not afford the seller the opportunity to cure.

EXAMPLE

Davis purchased and paid for a mobile home from Colonial. Shortly after the mobile home was installed, Davis called Colonial with numerous complaints: the cabinets were out of line and would not shut; the door of the refrigerator was about ¼-inch higher than the top of the body, thus preventing a seal; the windows would not close; the front door would not shut; a large gap existed between the paneling and the framing in the back; at one point, corrugated metal exterior was bent out of shape and pulled away from the frame; the "I" beam running under the chassis was completely warped; the floors were buckled; the rafters were warped and bent out of shape; water poured into the unit when it rained; and the hot water heater did not work.

continued

Davis called and complained. Colonial installed a new hot water heater and put a substance on the roof that stopped about 50% of the leaking. On several occasions, Colonial employees said they could not say when other repairs could be made. About three months after the purchase, Davis sent a letter to Colonial stating that she no longer wanted the mobile home. She did live in it on occasion and, shortly thereafter, sued for return of the purchase price.

The court held that Davis had provided the seller with a valid UCC revocation or rejection of the goods and that living in the home on occasion did not amount to acceptance. She had rightfully rejected, or revoked her acceptance of, nonconforming goods under UCC 2-602(b), (c), and 2-608(3). At that point, Davis had "no further obligations to purchase or accept any mobile home from defendant, whether the original unit repaired or a replacement." Davis was entitled to return of the purchase price and her consequential damages.[38]

Remedies

The buyer's remedies include what are called "incidental" damages such as transportation costs and costs required to care for rejected goods. Consequential damages are also allowable. Any physical injury resulting from a breach of warranty is also recoverable.[39] Remedies can be limited or expanded by contract.[40]

The statute of limitations in the UCC is four years unless the breach is only discovered in the fourth year. In that event, the statute of limitations is extended to five years.[41] States may vary the statue of limitation for actions.

Express and Implied Warranties

An express warranty[42] arises under the UCC when the seller makes any affirmation or promise relating to the goods, describes the goods, or supplies a sample or model.

The words "warrant" or "guarantee" are not necessary to form an express warranty.

EXAMPLE

The Jensens' bought a mobile home from Seigel. Advertising materials and the owner's manual described the mobile home as "well-built, thoroughly inspected, in compliance with state, local, and federal building codes and industry standards...trouble-free." The sales contract contained a disclaimer of all warranties, express or implied. However, this disclaimer could not negate the express warranties given in the advertising materials and the owner's manual.[43]

In addition to express warranties, the UCC implies several warranties into sales contracts:

- Warranty of title: the seller warrants it has title to the goods and the right to transfer title to the buyer.

- Warranty against encumbrances: the seller warrants the goods are not encumbered by a lien or other security interest.
- Warranty against infringement: the seller warrants the goods are free of the rightful claim of any third party for copyright, patent, or trademark infringement.

The warranties of particular interest to contractors are as follows:

- Warranty of merchantability or fitness for normal use.

This warranty arises in the sale of all goods (unless disclaimed, see below) and states that the goods purchased will be fit for the normal purpose for which goods of that kind are normally used. The UCC says that merchantable goods

"(a) pass without objection in the trade under the contract description; and

(b) in the case of fungible goods, are of fair average quality within the description; and

(c) are fit for the ordinary purposes for which such goods are used; and

(d) run, within the variations permitted by the agreement, of even kind, quality, and quantity within each unit and among all units involved; and

(e) are adequately contained, packaged, and labeled as the agreement may require; and

(f) conform to the promises or affirmations of fact made on the container or label, if any."[44]

EXAMPLE

Coldiron purchased a mobile home from Vintage. Within six months, it developed holes in the windows, doors, and roof. The electrical, plumbing, and insulation systems repeatedly failed. The home breached the implied warranty of merchantability since it was unfit for residential use.[45]

EXAMPLE

Oasis purchased room air-conditioning units from Home Gas that proved unsatisfactory and had to be replaced. Oasis sued Home Gas for breach of the implied warranty of merchantability. The warranty between the parties contained a limitation of damages provision that Oasis could not recover for any special, indirect or consequential damages or "for any damages arising from use" of the units. Home Gas claimed that this limitation of damages provision disclaimed the implied warranty of merchantability. The court did not agree and stated the implied warranty of merchantability could not be disclaimed unless done so in a conspicuous manner and containing the word "merchantability." Oasis was allowed to proceed with its claim for determination of whether or not the units were merchantable.[46]

- Warranty of fitness for a particular purpose.

This warranty arises when the seller of materials or supplies recommends to the buyer that a certain material or supply will perform to the buyer's specification. If the material or supply does not perform as indicated by the seller, the seller has breached this warranty.

EXAMPLE

Scientific manufactured a catalytic coating product called Spraylock that was recommended as a sealer for surfaces made of metal, concrete, or wood. The product information indicated a the user would be able to cover 6,000 square feet per day. Delkamp purchased the Spraylock but had problems with the product and could not reach the application level of 6,000 square feet per day. The issue of whether or not the product violated the warranty of fitness for a particular purpose was presented to the jury, which determined that the implied warranty of fitness had been violated.[47]

EXAMPLE

Morris, was a general contractor who purchased a set of plans for a spec house. He gave the plans to Atkins, an employee of Lee's Home Center, Inc. and they discussed the use of 16-inch I-joists for support of the garage. Atkins had supplied Morris with similar advice in the past. However, Lee's Home Center did not have a sufficient number of the 16-inch I-joists on hand. While Morris was in the room, Atkins called Lumberman's Wholesale Distributors, the manufacturer of 16-inch and 14-inch I-joists. At this point, the parties disagreed on what happened. Morris claimed that after the conversation with Lumberman's Wholesale Distributors, Atkins recommended that the 14-inch I-joists were suitable. Atkins testified that he did not recommend the 14-inch I-joists.

After using the 14-inch I-joists, the kitchen, bedroom, and upstairs hallway began to sag. To remedy the problem, Morris was required to jack up the second story of the house and insert steel plates to provide the structural strength necessary to carry the weight of the upper stories of the house.

This matter must be presented to the jury for determination of the factual issue of whether or not Lee's Home Center, Inc., through its agent Atkins, had given Morris a warranty of fitness.[48]

■ Warranty Limitations

Sellers are free to limit or disclaim warranties.

EXAMPLE

Conestoga Mall purchased a roofing system from Seal Dry. Seal Dry provided Conestoga with technical and performance information and the warranty. The warranty said, "Roofing materials will provide watertight protection for a period of ten (10) years..." The warranty excluded damage and leaks in the roof resulting from "natural disasters, including but not limited to earthquakes, lightning, tornados, gales, hurricanes, fires, etc."

After a hailstorm, the Seal Dry roofs were damaged, and Conestoga's insurance company sought reimbursement from Seal Dry for the damage. The court held that both Conestoga and Seal Dry were experienced parties, the express warranty limited Seal Dry's liability, and Seal Dry was not liable for hail damage.[49]

UCC 2-316 requires that any attempt to exclude or modify an implied warranty of merchantability must specifically mention the word "merchantability" and be conspicuous.

FORM

SELLER HEREBY DISCLAIMS ALL WARRANTIES INCLUDING THE WARRANTY OF MERCHANTABILITY.

MANAGEMENT TIP

Sellers of materials and supplies should carefully review all warranties. Buyers should also carefully review all warranties.

■ Endnotes

[1] Louisiana has modified its commercial code to resemble Article 2. See James W. Bowers, *Incomplete Law: On the Inevitable Inadequacies of Codes, Civil and Commercial*, 62 La. L. Rev. 1229 (2002).

[2] Example based on *Hunter's Run Stables, Inc. v. Triple H, Inc.*, 938 F.Supp 166 (WD NY 1996).

[3] Example based on *Pittsley v. Houser*, 875 P.2d 232 (Ida. App. 1994).

[4] Example based on *Meyers. v. Henderson Constru. Co.*, 370 A.2d 547 (NJ Sup. 1977).

[5] UCC § 2-302.

[6] This implied covenant has been adopted by the common law and now exists in all contracts.

[7] UCC § 2-204(3).

[8] Example based on *Century Ready-Mix Co. v. Lower & Co.*, 770 P.2d 692, 697 (Wyo. 1989).

[9] UCC § 2-205.

[10] *Rich Products Corp. v. Kemutec, Inc.*, 66 F.Supp.2d 937 (ED Wa 1999).

[11] UCC 2-207(1).

[12] UCC 2-207 (2)(b).

[13] UCC § 2-206(1).

[14] UCC 2-207 (2)(b).

[15] UCC 2-207(3).

[16] Example based on *Coastal & Native Plant Specialties, Inc. v. Engineered Textile Products, Inc.*, 139 F. Supp. 2d 1326 (ND Fla. 2001).

[17] Example based on *Steiny & Co., Inc. v. California Elec. Supply Co.*, (Cal.App. 2000) 79 Cal App 4th 285, 294, 93 Cal Rptr 2d 920.

[18] UCC 1-201, Official Comment 39.

[19] Example based on *A46 Home Lumber Co. v. Appalachian Regional Hospitals, Inc.*, 722 S.W.2d 912, 3 U.C.C. Rep. Serv. 2d 494 (Ky. Ct. App. 1987).

[20] *Bazak Inter. Corp. v. Tarrant Apparel Group*, 378 F. Supp. 2d 377 (SD NY 2005).

[21] Example based on *GPL Treatment, Ltd. v. Louisiana-Pacific Corp.*, 914 P.2d 682 (Or. 1996).

[22] *R.E. Phelon Company, Inc. v. Clarion Sintered Metals, Inc.*, 2006 U.S. Dist. Lexis 53219, 60 U.C.C. Rep. Serv. 2d (Callaghan) 951 (S.C. 2006).

[23] UCC 2-201.

[24] Example based on *R.M. Schultz & Associates, Inc. v. Nynex Computer Services Co.*, 1994 U.S. Dist. Lexis 4509 (ND Ill. 1994).

[25] Example based on *Gerner v. Vasby*, 75 Wis. 2d 660; 250 N.W.2d 319, 21 U.C.C. Rep. Serv. (Callaghan) 44 (Wis. 1977).

[26] UCC 2-209.

27 In some jurisdictions, the requirement of consideration is disappearing, and the common law is absorbing this concept of the UCC. In other words, the modification is upheld as long as it is in good faith and voluntary.

28 UCC 2-209(5).

29 Example based on *Monroc, Inc. v. Jack B. Parson Constr. Co.*, 604 P.2d 901, U.C.C. Rep. Serv. (Callaghan) 18 (Uta. 1979).

30 Example based on *American Carpet Sales, Inc. v. World Carpets, Inc.*, 477 So. 2d 974, 42 U.C.C. Rep. Serv. (Callaghan) 1197 (Ala. App. 1985).

31 Example based on *Metric Constructors, Inc. v. National Aeronautics & Space Admin.*, 169 F.3d 747 (Fed. Cir. 1999).

32 UCC § 2-401(2)(a).

33 Example based on *Windows, Inc. v. Jordan Panel System Corp.*, 177 F.3d 114 (2d Cir. 1999).

34 UCC § 2-510(2).

35 UCC § 2-510(2).

36 Example based on *In re: Thomas*, 182 BR 347 (SD Fla. 1995).

37 UCC 2-508.

38 Example based on *Davis v. Colonial Mobile Homes, Inc.*, 28 N.C. App. 13, 220 S.E.2d 802 (NC App. 1975).

39 UCC 2-711.

40 UCC 2-719.

41 UCC § 2–725.

42 UCC 2-313.

43 Example based on *Jensen v. Seigel Mobile Home Group*, 105 Idaho 189, 668 P.2d 65, 35 U.C.C. Rep. Serv. (Callaghan) 804 (Ida. 1983).

44 UCC 2-314.

45 Example based on *Vintage Homes, Inc. v. Coldiron*, 585 S.W.2d 886 (Tex. Ct. App. 1979).

46 Example based on *Admiral Oasis Hotel Corp. v. Home Gas Industries, Inc.*, 68 Ill. App. 2d 297, N.E.2d 282, 3 U.C.C. Rep. Serv. (Callaghan) 531 (Ill. App. 1965).

47 *Scientific Application, Inc. v. Delkamp*, 303 N.W.2d 71, 30 U.C.C. Rep. Serv. (Callaghan) 1256 (N.D. 1981).

48 Example based on *Lee's Home Center, Inc. v. Morris*, 2006 Tenn. App. Lexis 412 (Tenn. App. 2005).

49 Example based on *J.G. Conestoga LLC, v. Seal Dry/USA, Inc.*, 2000 U.S. Dist. Lexis 1850 (DC Neb. 2000).

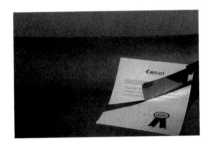

CHAPTER **14**

Killer Clauses

■ Introduction

In an attempt to control and allocate risks in the construction industry, parties have developed many clauses that have important and far-reaching effects. All parties to construction contracts should be familiar with these clauses and what they attempt to do. Not all jurisdictions will uphold these clauses in all circumstances, and statutes and case law may have limited these clauses in certain circumstances. If one of these clauses produces an onerous or unconscionable result, it is unlikely a the court will uphold the clause. In addition, since every contract contains an implied term of good faith, all such clauses should be used only in good faith.

On the other hand, particularly in commercial construction where the parties are experienced, the law tends to uphold these clauses. Merely suffering damage because of a clause is not an onerous or unconscionable result if a party has taken on the risk.

■ Dispute Resolution: Arbitration and Mediation Requirements

In an attempt to avoid the costs and time associated with litigation, it is common in the industry to have clauses requiring binding arbitration or mediation. The advantages of these processes are that they are faster, more efficient, and generally private.

These contract-required processes should not be confused with court-ordered arbitration or court-ordered mediation. In order to reduce the caseload of the courts, and because these processes are so effective, many courts have adopted processes that order parties to court-ordered arbitration or court-ordered mediation. However, because of due process protections guaranteed by the U.S. Constitution, parties are entitled to proceed to trial if they so desire.

Arbitration

Arbitration is similar to litigation in that the dispute is presented to a third party, called an arbitrator, to resolve the dispute. Some arbitrations may require a panel of arbitrators. Each party presents his case to the arbitrator, and the arbitrator then decides who wins and the damages. The arbitrator's decision is called an **award.** The winning party may file the arbitration award in a court in order to obtain a judgment. This allows the winning party to enforce the arbitration through government processes should the losing party not voluntarily pay the award.

No one can be compelled to arbitrate; he must have agreed to it in a contract.

EXAMPLE

A contractor filed a motion with the court asking it to compel the homeowners association to arbitrate certain construction defects claims. The construction contracts between the contractor and the individual owners contained a requirement for binding arbitration. The court did not compel arbitration, because the homeowners association was not a party to the construction contracts.[1]

Although parties can agree to arbitrate at any time, after a dispute arises it is likely one party will not want to arbitrate, if only to take advantage of the time delay associated with litigation. For this reason, the agreement to arbitrate is usually part of the original contract between the parties. If one of the parties to the mandatory arbitration clause in the contract does not wish to arbitrate, the other party can compel arbitration through a court order.

FORM

Any and all claims, disputes, and other matters in question arising out of, or relating to, this Subcontract, or the breach thereof, shall be submitted to arbitration in accordance with the Construction Industry Arbitration Rules of the American Arbitration Association. The award rendered by the arbitrators shall be final, and judgment may be entered upon it in accordance with applicable law in any court having jurisdiction thereof.

AIA 201-1997

§ 4.6.1. Any Claim arising out of or related to the Contract, except Claims relating to aesthetic effect and except those waived as provided for in Subparagraphs 4.3.10, 9.10.4 and 9.10.5, shall, after decision by the Architect or 30 days after submission of the Claim to the Architect, be subject to arbitration. Prior to arbitration, the parties shall endeavor to resolve disputes by mediation in accordance with the previsions of Paragraph 4.5.

§ 4.6.2.

Claims not resolved by mediation shall be decided by arbitration...in accordance with the Construction Industry Arbitration Rules of the American Arbitration Association...

Judicial Review of Arbitration Awards

Courts will seldom overturn or even review an arbitration award. Unlike judgments, which can be reviewed by appeal courts for legal correctness, no process exists to review an arbitration award for legal correctness. No process exists to make sure that the arbitrators apply the law or uphold the contract.[2] Arbitrators are not required to justify their award or write any type of memorandum explaining the law or facts depended on to reach the award. Arbitrator's decisions are not published and are not subject to review and critique by others in the field. For these reasons, arbitrators may feel less pressure to follow the law. In truth, arbitrators may not even know the applicable law.

Courts do not routinely review arbitrator's awards. An arbitrator's award is entitled to a "special degree of deference on judicial review," and "every presumption is in favor of the validity of the award."[3] Judicial review of awards is "among the narrowest known to the law."[4] The "court will set [the arbitrator's] decision aside only in very unusual circumstances."[5] "We must underscore at the outset the limited scope of review that courts are permitted to exercise over arbitral decisions."[6] "In short, upon judicial review, the question is "whether the arbitrator did his job—not whether he did it well, correctly, or reasonably, but simply whether he did it."[7]

Some jurisdictions will only overturn awards based on corruption, fraud, misconduct, partiality, or failure to follow proper procedures. Not included on the list is misapplication or disregard of the law.

EXAMPLE

Keane entered into a subcontract with Signal. The subcontract required Keane personnel to provide "scarce and unusual" services related to Signal's prime contract with the Federal Highway Administration's unusual needs. The subcontract also contained a "no-hire provision" that basically stated neither party would solicit or engage the services of any employee of the other for one year after the subcontract ended. The contract contained an arbitration clause.

Signal terminated the subcontract because Keane did not submit its invoices to Signal on a timely basis. After termination of the subcontract, Signal hired 22 key Keane employees.

Keane initiated an arbitration proceeding against Signal, alleging that Signal had wrongfully terminated its subcontract and breached it by hiring 22 of Keane's former employees contrary to the no-hire provision in the contract. In addition, Keane alleged that Signal had committed civil conspiracy and sought treble damages. After arbitration an award was entered in Keane's favor for breach of contract and treble damages. The total amount of the award was $6,883,029 and attorney's fees. One arbitrator on the panel objected to the treble damage award.

Signal asked the court of Virginia to review the award, claiming that the award was arbitrary and irrational. The case proceeded through the appeal process to the Virginia Supreme Court, which also refused to overturn the award. The court said a Virginia statute[8] allows vacating an award only in the following circumstances:

"Upon application of a party, the court shall vacate an award where:

1. The award was procured by corruption, fraud or other undue means;
2. There was evident partiality by an arbitrator appointed as a neutral, corruption in any of the arbitrators, or misconduct prejudicing the rights of any party;
3. The arbitrators exceeded their powers;
4. The arbitrators refused to postpone the hearing upon sufficient cause being shown therefore...
5. There was no arbitration agreement..."

The court said, "Essentially, Signal argues that the arbitrators exceeded their powers because they purportedly applied the *wrong legal standard* in the resolution of the contract claim. *We express no opinion regarding the correctness of the arbitrators' legal analysis.* The issue before this Court is not whether the arbitrators' conclusions were legally correct, but rather whether the arbitrators had the power to resolve the parties' contractual claims....We hold that the arbitrators did not exceed their powers because the issues that they resolved were within the scope of the powers conferred upon the arbitrators by the subcontract....Even though courts in other jurisdictions have vacated arbitration awards when there has been a 'manifest disregard of the law,' we refuse to adopt that standard in this case because to do so would require that this Court add words to *Code § 8.01-581.010*, which enumerates the bases on which a court shall vacate an arbitration award....We conclude that Signal's arguments lack merit and, therefore, we will affirm the judgment of the circuit court." [9]

The second set of jurisdictions include the federal government,[10] and in these jurisdictions, a court will vacate an arbitration award for manifest disregard of the law if

(1) the error was obvious and capable of being readily and instantly perceived by the average person qualified to serve as an arbitrator; (2) the arbitration panel appreciated the existence of a clearly governing legal principle but decided to ignore it; and (3) the governing law alleged to have been ignored by the arbitration panel is well defined, explicit, and clearly applicable.[11]

EXAMPLE

Demaria, the contractor, and Word of Faith Church entered into a contract to construct a new sanctuary. The church terminated the contract prior to completion. The parties submitted the matter to binding arbitration, and after the hearing, the arbitrator awarded the contractor approximately $2 million in damages. The church filed an appeal of the award. The court upheld the award except for approximately $150,000 relating to certain "fees" awarded the contractor. In a specific change order, the contractor had waived those same fees, but the arbitrator awarded them to the contractor anyway. All other parts of the award were upheld.[12]

Even in jurisdictions that will review the arbitrator's award for legal correctness, it is extremely rare for a court to review or overturn an award, and a great deal of deference is given to these awards. "A court may set aside an arbitration award only if it clearly appears on the face of the award or in the reasons for the decision that the arbitrator made an error of law and that, but for that error, a substantially different award must be made. In an arbitration arising from a contract dispute, the arbitrator is bound to render an award that comports with the terms of the parties' contract."[13]

One-sided Clauses

In a typical example of how parties attempt to use the law to their own advantage many attempt to include one-sides clauses. A contractor had this clause in the subcontract.

FORM

LIENS AND CLAIMS: ...At Contractor's sole option any and all claims, disputes and other matters in question arising out of, or relating to, this Subcontract, or the breach thereof, shall be submitted to arbitration in accordance with the Construction Industry Arbitration Rules of the American Arbitration Association....The award rendered by the arbitrators shall be final, and Judgment may be entered upon it in accordance with applicable law in any court having jurisdiction thereof. Subcontractor expressly acknowledges and agrees that arbitration shall be at Contractor's option and waives any right to require claims disputes and other matters in question arising out of, or relating to, this Subcontract, or the breach thereof, be submitted to arbitration.

This clause obligated only the subcontractor to binding arbitration and not the contractor. Although the court upheld the clause,[14] this result may have been an anomaly. Many courts might find this clause to be onerous or unconscionable. Similar clauses in employment contracts[15] have not been upheld.

Mediation

Although less common, some construction contracts contain a requirement to mediate claims before filing arbitration or litigation. Mediation is different from arbitration or litigation. In mediation, a trained person or persons, the mediator(s), *helps* the parties reach their own resolution of the dispute. The mediator does not decide the dispute, does not interpret the contract, does not apply the law, and does not resolve factual issues. These issues may be discussed during the mediation, but no resolution of any legal or factual issue is made, and the parties are encouraged to resolve the dispute despite the contractual, legal, and/or factual issues that may exist. If the parties are able to resolve the matter, an agreement is drawn up, usually on the spot, and the parties sign it. The mediation agreement is enforceable under contract law. Since mediation is never binding, the parties are free to file litigation or arbitration should it prove ineffectual.

FORM

Prior to arbitration of any issue or claim, the parties agree to mediate any such claim or issue.

■ Clauses Related to Damages

While the law would not enforce a clause totally preventing the payment of damages upon breach of a contract, the law will allow the parties to limit recoverable damages or eliminate certain types of damages. The most common damage limitation clauses are no damage except time extension, also called the "no damages for delay" clause, the waiver of consequential damages clause, and the liquidated damages clause.

No Damages for Delay

This clause limits the contractor to a time extension in the event of owner-caused delay or delay from outside forces such as acts of God. The contractor is not entitled to money damages, only a time extension. In a subcontract, this clause limits the subcontractor to the same in the event of contractor-caused delay. These clauses are most effective in eliminating arguments for relatively minor sums of money that may be due because of owner-caused delay; that is, these clauses prevent the contractor from collecting damages for reasonable delays or the types of owner-caused delays that are normal on a project. If the delay is major or has been deliberately caused by the owner, the clause is not likely to be enforced.

An extension of time shall be the sole remedy under the Contract for any delay caused by any reason or occurrence. The Contractor acknowledges such extension of time to be its sole remedy hereunder and agrees to make no claim for damages of any sort for delay in the performance of the Contract for any reason, including but not limited to delay occasioned by any act or failure to act of the Owner....[16]

In the event Subcontractor is obstructed or delayed in its performance of its work by Contractor or Owner, Subcontractor will be entitled to a reasonable extension of time. It is agreed that the extension of time will be Subcontractor's sole and exclusive remedy for such obstruction or delay, and that in no event will the Subcontractor be entitled to recover monetary damages of any type from Contractor or Owner for any such obstruction or delay.

These clauses are not favored in the law, and many exceptions to enforceability exist.[17] Some states have passed statutes to prevent the enforceability of these clauses for unreasonable or willful owner-caused delay.

EXAMPLE

Any provision contained in any public construction contract that purports to waive, release, or extinguish the rights of a contractor to recover costs or damages for unreasonable delay in performing such contract, either on his behalf or on the behalf of his subcontractor, if and to the extent the delay is caused by acts or omissions of the public body, its agents or employees and due to causes within their control, shall be void and unenforceable as against public policy.[18]

Exceptions to Enforceability of No Damage for Delay Clauses

While most courts will enforce these clauses, they tend to be strictly construed. "Most courts begin with a presumption that these clauses are valid so long as they are clearly drafted. As noted above, however, courts will generally construe no damages for delay clauses strictly, because they are exculpatory in nature and because the consequences of enforcement can be very severe. The result of this careful judicial scrutiny has been the development of a considerable list of widely recognized exceptions under which a no damages for delay clause will *not* be enforced."[19]

The exceptions to enforceability to the no damages for delay clauses have been seen as necessary because, in some circumstances, the clause will conflict with the implied covenant of good faith. "Four of the five proposed exceptions [to the enforceability of a no damages for delay clause] relate directly to and are logical extensions of the implied covenant of good faith and fair dealing: (1) willful concealment of foreseeable circumstances that impact timely performance, (2) delays so unreasonable in length as to amount to project abandonment, (3) delays caused by the other party's bad faith or fraud, and (4) delays caused by the other party's active interference."[20]

The exceptions to enforceability are discussed in more detail below.

Exception for Active Interference

Some courts have found that owner-caused delays are not delays contemplated by the parties.

EXAMPLE

A contractor entered into a subcontract for spray fireproofing on structural steel. However, the contractor allowed other subcontractors to attach pipes, ducts, and conduits to the structural steel. The subcontractor was hampered in applying the fireproofing and had to build scaffolding to apply it instead of standing on the structural steel itself. In addition, the roofing was not completed and caused water damage, requiring the work to be redone. The subcontractor sued the contractor for delay damages and was able to collect monetary damages, because the court held that the delay was not the type contemplated by the parties at the time the contract was entered into and therefore the subcontractor was allowed to recover.[21]

Exception for Fraud, Bad Faith, Wrongful Conduct, or Willful Conduct

These exceptions all relate to owner conduct that intentionally delays a project or at least is close to being intentional.

EXAMPLE

Triple R Paving entered into a contract with the government to install a bridge. During the process, Triple R recommended an alteration in part of the government's design, which the government approved. Triple R redesigned the bridge but did not check or verify any element of the horizontal geometry, and it did not check horizontal sight distances as supplied in the government's original design, as no part of the horizontal sight

continued

distance was impacted by the alteration. The government did agree at that time to check the horizontal sight distances but did not. Unknown to Triple R, but known to the government engineers, the government's original design did not meet horizontal sight distance standards. After construction began, the problem was noticed and caused delay in the completion of the project. Two other government-caused delays also occurred. The jury awarded damages to Triple R. The appeal court refused to overturn the jury's decision, stating, "However, we find that the facts surrounding the delay which resulted from the horizontal sight distance design flaw were sufficient to allow a jury to decide the question of fraud, bad faith, or active interference."[22]

Exception for Unreasonable Delays

As a general statement, the law requires parties to act reasonably toward one another. If the actions of the delay-causing party have been unreasonable, the clause may be ineffective.

EXAMPLE

Gray was one of the contractors on a project to build several schools. The contract contained a no damage for delay clause. The court refused to enforce the clause because of the school district's actions. The school district caused delay to the project by failing to timely obtain easements for electrical and drain sewer installations, failing to adequately supervise and coordinate the work of the various contractors, failing to prepare coordinated construction schedules and drawings, terminating the general contractor, terminating its own construction manager, and the owner's decision to hire 30 subcontractors instead of replacing the general contractor.[23]

Requirement of Notice as a Prerequisite to Damages

Even if an exception to the no damages for delay clause exists, contractors must still comply with all notice and procedural requirements of the contract related to claims for delay.

EXAMPLE

The contract between the contractor and the owner contained a no damages for delay clause and a requirement that the contractor submit a written request for delay damages to the owner within 72 hours of the claimed delay. The school district caused many delays on the project. However, for the first few months, the contractor never sent any written requests for additional time or damages. The contractor's delay claims that arose prior to its sending notice to the school district were dismissed; the contractor could only collect damages for owner-caused delays that occurred after he had sent the required notice.[24]

Marriott Corporation began construction of an elaborate resort complex in Orlando, Florida. It was to contain 14 different building segments, including a 28-story guest tower, numerous convention rooms, restaurants, ballrooms, and several outbuildings for swimming pools, golf courses, and tennis courts. At the time, it was the largest building in terms of square footage in the state of Florida. Dasta was one of the contractors on the project, and it bid and won a contract for the exterior skin work on the 28-story guest tower.

Because of the complexity of the project, the fast-track building approach, and the fact that the plans were only partially complete, it was clear in the contract documents that Marriott had absolute authority to modify the construction schedule, while Dasta was obligated to abide by Marriott's instructions and all modifications in the schedule. In addition, the project was a lump-sum bid project and contained a no damages for delay clause that gave the contractor only a time extension for owner-caused delays.

The court said, "Presumably, contractors involved with the Resort project made provision for the risks presented by such contingencies in two ways: first, by setting their bid prices high enough to absorb the 'additional costs' they might incur in performing the work (including the costs Dasta seeks to recover in this case); and, second, by negotiating contract terms and conditions designed to minimize such costs. In the present case, the parties thoroughly addressed and allocated the risks of delay inherent in the Resort project."

Problems started from day one for Dasta. Its team arrived on the project site only to learn that the project was five months behind schedule. It could not begin work until Marriot had attached the safety net required by OSHA standards to catch debris falling from work being done on the tower's upper levels. Marriot also failed to provide vertical transportation in the tower. Some of the concrete work to which Dasta was supposed to attach exterior coatings was defective and had to be redone. The fast-track approach resulted in frequent and significant modifications in the plans and progress schedule.

Despite many problems, Dasta actually undertook greater responsibility for the project, entering into change orders for additional work to, for example, correct other contractors' work. At one point, Marriot accelerated the work, and Dasta continued to perform. Eventually however, Dasta was unable to perform and ran out of money. Dasta presented Marriott with an efficiency claim, and a lawsuit resulted.

Dasta's claims for delay damages were denied, not because the no damage for delay clause prevented them, but because Dasta never made a demand for time extension. "Dasta's failure to comply with the contractually provided measures for relief bars Dasta from recovering for its delay, impact, and inefficiency damages."[25]

MANAGEMENT TIP

Owners should always have a no damages for delay clause and a clause requiring notice of delays in their contracts with contractors. Contractors should have the same in their subcontracts.

Parties should always give notice of delays.

Waiver of Consequential Damages

This type of clause limits a party or parties from collecting certain types of damages such as lost profit or overhead. These clauses are generally enforced even if the owner-caused delay is unreasonable. If the owner caused delay amounts to active interference or intentional conduct, the clause may not be effective.

FORM

The Contractor shall not be entitled to payment or compensation of any kind from the Owner for direct, indirect, or impact damages arising because of any hindrance or delay from any cause whatsoever.

FORM

AIA 201-1997 § 4.3.10. Claims for Consequential Damages. The Contractor and Owner waive Claims against each other for consequential damages arising out of or relating to this Contract. This mutual waiver includes:

1. Damages incurred by the Owner for rental expenses, for losses of use, income, profit, financing, business, and reputation, and for loss of management or employee productivity or of the services of such person; and

2. Damages incurred by the Contractor for principle office expenses including the compensation of personnel stationed there, for losses of financing, business, and reputation, and for loss of profit except anticipated profit arising directly from the Work.

EXAMPLE

An owner and a contractor enter into a contract to build a casino. The contract contains a waiver of consequential damages including lost profits. The contractor is delayed one month in the construction of a casino. The owner can produce evidence that casinos of this type and in this location can be expected to produce $500,000 in profit in the first month. The contractor is not liable to owner for the $500,000 in lost profit damages.

Liquidated Damages

A liquidated damage provision is a provision giving the damaged party a sum certain, usually calculated per day of delay. For example, the contractor may be required to pay the owner $200 per day for failure to complete the project on time. These clauses are upheld as long as the amount is reasonable, the actual damages are difficult to prove, and the clause is not a penalty. "Accordingly this court has long regarded provisions for liquidated damages as prima facie valid on the assumption that the parties in naming a liquidated sum intended it to be a fair compensation for an injury caused by a breach of contract and not a penalty for nonperformance."[26]

EXAMPLE

The contractor and the Department of Transportation (DOT) entered into a road construction contract that contained a liquidated damages provision. The contractor was late in turning over the project, and the DOT attempted to sue for certain actual damages it had suffered. This provision fixed both the maximum and minimum sum that could be collected for a breach of the contract. The provision was upheld. The DOT's claim was dismissed.[27]

EXAMPLE

The Kellys signed a purchase agreement to purchase residential real estate from the Marxs and gave $17,750 or 5% of the total price as a deposit. The contract contained the following liquidated damage clause:

If the BUYER shall fail to fulfill the BUYER'S agreements herein, all deposits made hereunder by the BUYER shall be retained by the SELLER as liquidated damages.

The Kellys never purchased the home, because they could not sell the home they were then living in. The Marxs eventually sold the house for slightly more than the Kellys had agreed to pay.

The court upheld the liquidated damage clause. It concluded, "A liquidated damages clause in a purchase and sale agreement will be enforced where, at the time the agreement was made, potential damages were difficult to determine and the clause was a reasonable forecast of damages expected to occur in the event of a breach."[28]

This type of clause should be considered for its ease of enforcement. However, the amount of liquidated damages must be carefully determined. If the amount is so high as to be a penalty, the court will not enforce it, and the damaged party will be forced to prove compensatory damages in order to recover (see Chapter 7).

■ Delay and Force Majeure

The law allows an extension of time to perform the contract for many circumstances including changes, acts of the owner or its agents, acts of God, and similar events. Many contracts contain a list of events causing a delay to accrue. This type of clause is called a **force majeure** clause by some jurisdictions.

FORM

357

AIA 201-1997 § 8.3.1. If the Contractor is delayed at any time in the commencement or progress of the Work by an act or neglect of the Owner or Architect, or of an employee of either, or of a separate contractor employed by the Owner, or by changes ordered in the Work, or by labor disputes, fire, unusual delay in deliveries, unavoidable casualties or other causes beyond the Contractor's control, or by delay authorized by the Owner pending mediation and arbitration, or by other causes which the Architect determines may justify delay, then the Contract Time shall be extended by Change Order for such reasonable time as the Architect may determine.

Force majeure or an event of force majeure means any cause beyond the control of the Seller or of Idaho Power which, despite the exercise of due diligence, such Party is unable to prevent or overcome, including but not limited to an act of God, fire, flood, explosion, strike, sabotage, an act of the public enemy, civil or military authority, court orders, laws or regulations, insurrection or riot, an act of the elements, or lack of precipitation resulting in reduced water flows for power production purposes. If either party is rendered wholly or in part unable to perform its obligations under this Agreement because of an event of force majeure, both parties shall be excused from whatever performance is affected by the event of force majeure.[29]

This clause will not excuse a contractor that has breached the contract.

EXAMPLE

Cogeneration and Idaho Power entered into an agreement whereby Cogeneration would construct a power plant and have it on line by January 1, 1996. Idaho Power would purchase the power generated by the plant. Cogeneration's performance was secured by two payments: one for $250,000, which was paid, and one for $1,874,800, which was never paid. The agreement between the parties contained the force majeure clause above.

Cogeneration was unable to renew Idaho Department of Environmental Quality and Army Corps of Engineers certificates because of its failure to meet certain criteria. Because it was unable to meet the criteria, it was unable to obtain financing and pay Idaho Power the second security installment. Idaho Power notified Cogeneration that it was in default of the agreement. Cogeneration invoked the force majeure provision, asserting that the Department of Environmental Quality and the Army Corps of Engineers had unilaterally and arbitrarily revoked the required certificates. Cogeneration argued that because these entities were civil authorities, and acts of civil authorities triggered the clause, Cogeneration was excused from its required performance: giving the security.

The trial court determined, and the appeal court upheld, that no force majeure had occurred to prevent Cogeneration from posting the security. Cogeneration's obligation to post the security was not stopped by the acts of the government agencies. Cogeneration could not post the security, because its lender would not advance it the money. Its lender would not advance the money because Cogeneration could not get the required environmental certificates, but the government entities were not preventing it from getting funding. Failure to pay the second security payment was a breach of the contract, allowing the other party to terminate it.[30]

358

Image courtesy © istockphoto.com/EricHood

■ Insurance Clauses

These clauses outline which party is responsible for insurance coverage and are enforced by the courts.

Management needs to think carefully about insurance costs and have internal processes to manage and coordinate coverages. It may be cheaper and easier to obtain one policy of general liability coverage for the project rather than spend money on the management time it costs to oversee the insurances of the several parties.

■ Indemnity Clauses

Indemnity clauses shift the responsibility for injuries to one of the parties, usually the general contractor. However, it is becoming more common to see these clauses in subcontracts, and this means the risk is shifted to the subcontractor.

Three types of indemnity clauses exist: the limited form, the intermediate form, and the broad form. The limited form is just a restatement of the law and will be upheld by the courts. The intermediate form shifts the responsibility to one of the liable parties, usually the contractor, and this clause will be upheld as long as it is clear. The broad form shifts the liability to a nonliable party, and many courts refuse to uphold this clause. The better alternative to the broad form indemnity clause is the additional insured clause, which the law will uphold.

Indemnity clauses are coupled with a waiver of subrogation clause (discussed below), which prevents the parties or their insurance companies from suing each other.

Limited Form

The law requires a liable party to be responsible for the injuries or damages he causes. In the event that several parties, such as the owner, general contractor, and subcontractor, contribute to one injury, each is responsible for his portion of the damages. This clause merely restates the legal liability parties have anyway and adds nothing to a party's legal liability.

FORM

Subcontractor shall at all times indemnify, defend, and hold harmless General Contractor and Owner from all loss and damage, lawsuits, arbitrations, mechanic's liens, legal actions of any kind, attorney's fees, and/or costs caused or contributed by, or claimed to be caused or contributed to, by any act, omission, fault, and/or negligence, whether passive or active, of Subcontractor or its agents or employees, in connection with the Work but only to the extent caused by the Subcontractor or its agents or employees.

A school hires a contractor to do remodeling. The prime contract and the subcontract contain a limited form indemnity clause. The school is to remain open during the remodeling. The contractor subcontracts the replacement of some drywall. The subcontractor places a rope and a sign around the area under construction. However, while the subcontractor is on a break outside, a first grader goes under the rope and is injured. Assume the damages are $10,000. Depending on the facts, the school might be partially negligent for failing to supervise the student, and the general contractor and the subcontractor may be partially negligent for using an inadequate barrier given that the work is being done in a school setting with young children present.

A court or arbitrator would have to determine the percentage of liability of each of the parties, and each would be liable for the student's injuries in proportion to their liability. For example, assuming that the liability of the school was found to be 50% for negligent supervision, the liability of the contractor was 25% for failing to provide an adequate barrier, and the liability of the subcontractor was 25% for failing to provide an adequate barrier, each would be responsible for the following damages under the law or under the clauses in the contracts:

- School: $5,000
- Contractor: $2,500
- Subcontractor: $2,500.

MANAGEMENT TIP

The limited form clause can be inserted in any contract.

Intermediate Form

This clause shifts the liability for injuries to the contractor as long as the contractor or his agents is at least partially at fault for the injury. If the clause is in the subcontract, it shifts the liability to the subcontractor if the subcontractor is partially at fault for the injury.

To the fullest extent permitted by law, Contractor shall indemnify and hold harmless Owner, its consultants and agents and employees of any of them from and against claims, damages, losses, and expenses, including but not limited to attorney's fees, arising out of or resulting from performance of the Work, provided that such claim, damage, loss or expense is attributable to bodily injury, sickness, disease. or death, or to injury to or destruction of tangible property (other than the work itself) including loss of use resulting therefrom, **whether caused in whole or in part by negligent acts or omissions of the Contractor** or anyone directly or indirectly employed by them or anyone for whose acts they may be liable, regardless of whether or not such claim, damage, loss or expense is caused in part by a party indemnified hereunder (emphasis added).

EXAMPLE

Assume the facts as in the example above except that the prime contract and the subcontract contain an intermediate form indemnity clause. With this clause, the liability of the parties is as follows:

- School: $0
- Contractor: $0
- Subcontractor: $10,000.

MANAGEMENT TIP

This clause may reduce the costs of construction if one party is responsible for the insurance. However, the party that is accepting the liability is also accepting the risk of a claim, which could increase its premiums for years to come. Whichever party accepts the liability must inform its insurance company of this increased liability. Failure to do so will mean the party is liable but without insurance coverage. The insurance company is not obligated to absorb the liability if it has not been informed that its insured has taken on additional risk.

Broad Form

This clause attempts to shift the liability onto a party whether or not that party was negligent. These clauses are not favored in the law, and many jurisdictions will not enforce them.

To the fullest extent permitted by law, Contractor shall indemnify and hold harmless owner, its consultants and agents and employees of any of them from and against claims, damages, losses and expenses, including but not limited to attorney's fees, arising out of or resulting from performance of the Work, whether or not such claim, damage, loss, or expense is attributable to bodily injury, sickness, disease, or death, or to injury to or destruction of tangible property (other than the work itself) including loss of use resulting therefrom, **whether or not caused in whole or in part by negligent acts or omissions of the Contractor** or anyone directly or indirectly employed by them or anyone for whose acts they may be liable, and regardless of whether or not such claim, damage, loss or expense is caused in whole or in part by a party indemnified hereunder (emphasis added).

EXAMPLE

A school hires a contractor to do remodeling of the school in the summer when school is not scheduled and students are not allowed in the building. As part of the contract agreement, the school takes on the responsibility for keeping the building locked and secure so students and others do not enter. However, one teacher decides that it would be a good idea to show some students the work in progress. She uses her key on a Sunday afternoon, when neither the contractor nor any of the subcontractors are on the site, to admit herself and some students. A student is injured on the construction site. The student's damages are $10,000.

Assuming that the liability of the school is 100% in this situation, the above clause in the prime contract would have the following effect if a court were to enforce it:

- School: $0
- Contractor: $10,000

MANAGEMENT TIP

This clause should not be used. The owner must weigh the possibility that the only contractors who would accept this clause in a contract would be those who have little alternative and are desperate for the work. These are not likely to be high-quality contractors. The general contractor has the same problem in connection with subcontracts.

The better approach is to have one of the parties add the other parties as additional insureds on their policy.

Additional Insured

While the broad form indemnity clause is not favored and many jurisdictions will not enforce it, a clause requiring a contractor or subcontractor to name others as additional insureds on their policy is acceptable.

FORM

Before commencing work on the project, Contractor and its Subcontractors of every tier will supply to Owner duly issued Certificates of Insurance, naming Contractor as an "additional insured," showing in force the following insurance for comprehensive general liability, automobile liability, and Worker's Compensation:

Comprehensive general liability with limits of not less than $_____ per occurrence;

Automobile liability in comprehensive form with coverage for owned, hired, and nonowned automobiles; and

Worker's Compensation insurance in statutory form.

All insurance binders must contain a clause indicating that certificate holders and additional insured will be given a minimum of 10 days' written notice prior to cancellation of the insurance. Contractor must furnish the insurance binder referred to above as an express condition precedent to the Owner's duty to make any progress payments to Contractor pursuant to this Agreement.

Contractor's insurance shall be the primary insurance and Owner's insurance shall be called on to contribute to a loss caused in whole or part by the negligence of Subcontractor.

◼ Release and Waiver Clause

A waiver is the knowing relinquishment of a known right. Waivers are extremely powerful and are upheld by the law.

Entergy solicited bids for, among other things, removing debris from and cleaning of 13 "spillway cells" in the dam that had become clogged with debris. Contractors, including Hall, were invited to inspect the cells. Hall inspected only five of the cells. Entergy invited bids on a time and materials basis; this would have allowed the contractors to be compensated for removing the actual amount of the debris in the spillway cells. However, Hall decided to submit a lump-sum bid and was eventually awarded the project.

Naturally, many of the cells contained a much greater volume of debris than originally anticipated, including, of all things, a small railroad car. The cost for cleaning the cells far exceeded Hall's the lump-sum bid, and it sued for additional compensation. Its claim was denied, however, because it had waived all claims for additional compensation in the contract:

Contractor waives all claims for additional compensation beyond that allowed in this Agreement unless the claim is expressly authorized..."[31]

In the construction industry, these clauses are often part of a comprehensive insurance plan that allocates the liability to one party and their insurance. When this is the case, the law will uphold the clauses. These clauses are usually coupled with subrogation clauses.

FORM

Contractor and Subcontractors fully release, waive, and hold harmless Owner from any liabilities and/or injuries that may occur during the course of the project, to the Contractor, the Subcontractors, their employees, agents, and assigns. In no way does this provision affect the absolute duty of Contractor and every Subcontractor to provide Worker's Compensation insurance coverage as required by law.

EXAMPLE

Sponaugle was the electrical subcontractor for the prime contractor, Hunt, on the construction of a 306,000-square-foot indoor arena. The project fell behind schedule but was essentially completed on time due to, according to Sonaugle, acceleration of its work by the prime contractor. Sponaugle claimed that it incurred additional costs when Hunt required it to work overtime, in a random and inefficient manner, and in congestion with other subcontractors.

The subcontract required Sponaugle to submit a release to Hunt with its payment applications. The release contained the following language:

> In addition, for and in consideration of the amounts and sums received, [Sponaugle] hereby waives, releases, and relinquishes any and all claims, rights, or causes of action whatsoever arising out of or in the course of the work performed on [the project], contract, or event transpiring prior to the date hereof, excepting the right to receive payment for work performed and properly completed and retainage, if any, after the date of the above-mentioned payment application or invoices.

In June 2002, Sponaugle began to exchange letters and invoices with Hunt for amounts it claimed were due for its acceleration of the project. Hunt claimed that Sponaugle was supplying insufficient labor to complete the project on time.

In January 2003, Sponaugle returned to Hunt a signed release of lien with its payment application. The court dismissed Sponaugle's claim on a motion for summary judgment, stating, "We conclude that the release Sponaugle submitted with its payment applications bars the instant action seeking recovery for Sponaugle's extra work allegedly caused by Hunt's delay and interference. In pertinent part, the release, provides that Sponaugle 'waives any and all claims, rights or causes of action whatsoever arising out of or in the course of the work performed on [the project]...prior to the date' of the particular payment application. This language is broad, but the aspect relevant here is that it is plain, clear, and unambiguous and waives any and all claims arising out of work performed on the project before the date of the particular payment application. This means that Sponaugle certainly waived its right to recover for any extra work when it submitted the January 20, 2003 release."[32]

The court rejected Sponaugle's arguments that the releases were not intended to cover additional or extra work. One of Sponaugle's arguments was industry practice. Sponaugle's copresident submitted an affidavit stating, "Consistent with my 17 years of experience in the industry...releases submitted with monthly payment applications are not construed as general releases barring subsequent requests for payment." The court said, "We reject this argument, because industry custom or practice can only be used for an ambiguous contract...and the instant release is not ambiguous. We therefore conclude that the releases Plaintiff executed throughout the project, culminating in the January 20, 2003 release, bars any claim for extra work supposedly caused by Hunt's delay and requires us to enter summary judgment against Plaintiff [subcontractor]."

Waiver of Subrogation

This clause prevents the parties and/or their insurance companies from suing each other if one party causes property damage covered by insurance. This clause is necessary because contracts for insurance usually contain a provision allowing the insurance company to attempt to recoup any proceeds it pays to its insured from a third party that caused the damage or injury. Parties signing such an agreement must notify their insurance company that they have done so.

EXAMPLE

Chang and Pacquete are in an automobile accident. Chang's health insurance pays for Chang's injuries. The company believes that Pacquete is liable for Chang's injuries because Pacquete was driving unreasonably. Chang's insurance company can sue Pacquete, in Chang's name, for negligence and seek reimbursement for the health benefits paid to Chang.

The waiver of subrogation clause is an attempt to force one insurance policy to cover the project and prevent that insurance company from suing others on the job who may have contributed to the damage. Waiver-of-subrogation clauses are usually upheld because they avoid litigation.[33] Unqualified waivers waive the right to all claims including warranty and strict liability claims.[34]

FORM

AIA 201-1997 §11.4.7. Waivers of Subrogation. The Owner and Contractor waive all rights against (1) each other and any of their subcontractors sub-subcontractors, agents, and employees, each of the other, and (2) the Architect...for damages caused by fire or other causes of loss to the extent covered by property insurance obtained pursuant to this Paragraph 11.4 or other property insurance applicable to the Work, except such rights as they have to proceeds of such insurance held by the Owner as fiduciary...the polices shall provide such waivers of subrogation by endorsement or otherwise...

Waiver of Subrogation—To the extent that a loss is covered by insurance in force, and recovery is made for such loss, the Owner, Contractor, Subcontractors, their agents, employees and assigns, hereby mutually release each other from liability and waive all rights of subrogation and all rights of recovery against each other for any loss insured against under their respective policies (including extended coverage), no matter how caused. Contractor shall require all Subcontractors to similarly waive their rights of subrogation in each of their respective construction contracts with respect to the work.

EXAMPLE

A church entered into a contract with Knowles to remove lead-based paint and do other construction work on an historic church building. During the process, Knowles was removing the paint with chemicals and a paint scraper. He left the area for a smoking break but was still too close to the work area. The fumes ignited, and the entire church was destroyed. Reliance, the church's insurance, paid for the destroyed church.

The construction contract contained the following clause:

"The Owner and Contractor waive all rights against each other, separate contractors, and all other subcontractors for damages caused by fire or other perils to the extent covered by Builder's Risk or any other property insurance, except such rights as they may have to the proceeds of such insurance." The only fire insurance on the project was the church's policy.

Reliance sued Knowles to recoup the damages it paid to the church and claimed that Knowles was negligent for smoking too close to the site. Reliance also sued the manufacturer of the paint scraper.

Reliance claimed that the subrogation clause was not applicable for several reasons, all of which were rejected by the court. The reasons claimed by Reliance but rejected by the court included the following: Knowles misrepresented his understanding of the standards applicable to historic restoration; Knowles' act in smoking was gross negligence, which voided the contract; Knowles' act in smoking was willful and wanton misconduct, which voided the contract; waivers of subrogation are void as against public policy; and the clause should not prevent Reliance from suing the paint stripper manufacturer under either strict liability or breach of warranty theories.[35] The court held that the waiver of subrogation clause waived Reliance's claims against both the contractor and the manufacturer of the paint scraper. The case was dismissed.

MANAGEMENT TIP

A waiver of subrogation clause should be included in all contracts and subcontracts.

■ Termination for Convenience

The termination for convenience clause allows the owner to terminate the contract for any reason or no reason. It eliminates the owner's liability for wrongful termination or breach-of-contract damages to the contractor for wrongful termination. This is a very powerful clause, and many courts have even used the concept of "constructive termination for convenience"; that is, if the owner terminates the contract but does not do so based on this clause, the court will assume that the termination was done pursuant to this clause. This relieves the owner of damages for wrongful breach. If the owner terminates for convenience, it must pay for work done prior to the termination.

FORM

Owner may terminate performance of work under this contract in whole or, from time to time, in part if the Owner or Architect determines that a termination is in the Owner's interest. Upon termination, Owner will not be liable for payment for Work not performed, nor for prospective profits on Work not performed, nor for any other consequential damages, and in no event shall such payment be greater than the Contract price.

MANAGEMENT TIP

Owners should always have this clause in their contracts, and contractors should always have this clause in their subcontracts.

Any reason may be too broad, as the law requires good faith in all contract actions, including terminating for convenience,[36] but as long as the termination is not arbitrary, capricious, or an abuse of discretion, the termination for convenience will probably be upheld. On the other hand, see the example below where the owner terminated the contract in bad faith but was still liable to the contractor only for limited damages.

Conomos entered into a contract with the owner, Sun, for painting of industrial piping at a refinery. The Steel Structures Painting Council has set standards for different types of surface preparation. The least costly surface preparation is washing (SP-1), followed by hand-tool cleaning (SP-2), followed by the use of scrapers and light use of a power tool called a "needle gun" (SP-3), followed by eight other levels of surface preparation, each more exacting than the last. SP-11, the final standard, required leaving a paint-free, shiny surface.

The more exacting the type of surface preparation the more costly it is to perform. However, the more exacting the type of surface preparation the better adherence of the paint and thus a more durable paint job. The standard required of Conomos in the contract was SP-3. Some subjectivity existed in determining whether or not Conomos had performed according to the specification.

Problems developed within a few days of Conomos' start of performance. Sun's inspector continually found Conomos' preparation unacceptable and demanded a higher quality of surface preparation than Conomos believed was called for by an SP-3 specification. The owner eventually terminated the contract for convenience.

Conomos sued for breach of contract. After a bench trial, the trial court determined that Sun's inspector was attempting to get a SP-4 level of surface preparation by claiming Conomos' preparation was unacceptable. This amounted to a breach of Sun's duty of good faith. Conomos was awarded damages and reasonable attorney fees.

In addition to the termination for convenience clause, the contract contained the following clause limiting Conomos' damages, in the event of a termination for convenience clause, to the contract price.

> Contractor [Conomos] waives any claim for damages, including loss of anticipated profits....However, provided that Contractor is not in default of any of its obligations hereunder, Owner [Sun] agrees that Contractor shall be paid an amount which, when added to all previously paid installments, will equal the sum of all costs properly incurred by Contractor prior to the date of cancellation, plus any earned profit on such incurred costs, but in no event shall such amount be greater than the Contract price.

The trial court held that Sun's bad faith breach nullified the above provision and therefore Conomos' damages were not limited to the contract price, even though the damages awarded to Conomos were less than the contract price. This determination was overturned on appeal, and the appeal court said that the limitation of damages to the contract price still applied even if the termination for convenience was in bad faith. It said that if the breach were to rise to the level of fraud or unconscionability, the outcome might be different.[37]

■ Unforeseen Site Conditions

An unforeseen or Type 2 differing site condition is a condition at the site that is unusual or unexpected. The law places the risk of unforeseen or Type 2 differing site conditions on the contractor. However, most contractors are unwilling to sign a contract unless the owner has agreed to absorb the risk in an unforeseen site condition clause. This clause is often coupled with a differing or Type 1 site condition clause that puts the risk of a differing site condition on the owner. However, since the law already does that, the differing or Type 1 clause is unnecessary.

If conditions are encountered at the site which are unknown physical conditions of an unusual nature, and which differ materially from those ordinarily found to exist and generally recognized as inherent in construction activities of the character provided for in the contract documents...the contractor shall be entitled to a reasonable addition to the contract sum and/or the contract time to complete the work.

MANAGEMENT TIP

Contractors and subcontractors should always have this clause in the contract. Failure to include this clause substantially increases risk, and therefore contractors and subcontractors must include a large sum to cover the risk. For owners who routinely buy construction services, this clause will eliminate this sum, and the overall costs of all projects will be lower.

■ Noncompete and Nonsolicitation Clauses

Noncompete clauses prevent previous employees from competing with their former employers. These clauses are upheld only if the duration and geographical restrictions are reasonable.[38] For example, an agreement that the former employee would never engage in the practice of architecture would not be upheld. However, an agreement that the employee would not do so within 50 miles of the employer's office and for a period of five years probably would be. As with all of the clauses in this section, the states vary widely on the enforceability of these types of agreements depending on the circumstances.

Employee agrees that Employee shall not, for a period of three (3) years immediately following the termination of this agreement, directly or indirectly, solicit or take away, or attempt to solicit or take away, any Clients of Employer, either for Employee's own benefit or for the benefit of any other person or entity.

If a firm is working with another similar firm, such as two prime contractors working in a joint venture, the firms may wish to include a clause that prevents the other from soliciting the employees of either. This is called a nonsolicitation clause.

FORM

Nonsolicitation. Firm A and Firm B agree that they, their agents, employees, or assigns shall not, for a period of three (3) years immediately following the termination of this agreement, directly or indirectly, solicit or take away, or attempt to solicit or take away, any Employees of the other, either for their own benefit or for the benefit of any other person or entity.

FORM

Firm B agrees that, except for those contacts made in the ordinary course of business, it will not initiate contact with any officer, director, or employee of Firm A or any customers, clients, or accounts of Firm A regarding Firm A's business or operation, except with the express permission of Firm A. Additionally, Firm B agrees not to solicit or entice, other than normal employment discussions not initiated by Firm B, for employment any of the current employees of Firm A for purposes of hiring such employee during the period of this Agreement and for a period of one (1) year thereafter, without the prior written consent of Firm A. Firm A agrees to the same nonsolicitation clause as above.

FORM

During the period of performance of this subcontract, and for a period of one (1) year thereafter, neither party shall solicit or engage the services of any employee of the other party engaged in performance of work related to this subcontract, without express written notification to and acceptance by the other party.[39]

The subcontract between Computech and Stan required Computech personnel to provide "scarce and unusual" services related to Stan's prime contract with the State Environmental Administration's unusual needs. The subcontract also contained a "no-hire provision," used in the example form immediately above, which basically stated that Stan would not solicit or engage the services of any employee of the other for one year after the subcontract ended.

Stan terminated the subcontract because Computech did not submit its invoices to Stan on a timely basis. This happened repeatedly, and Stan attempted to get Computech to submit the invoices on time. After termination of the subcontract, Stan hired 22 key Computech employees.

The arbitration panel awarded Computech damages, including punitive damages in the amount of treble the actual damages, because Stan terminated the contract, solicited, and hired Computech employees.[40]

■ Warranty Limitations

The law will imply warranties into both construction contracts and contracts for the sales of goods. Warranties can be limited in commercial construction and also for goods annexed to the construction. In order to limit an implied warranty, the contractor or supplier must usually supply an alternative express warranty such as a one-year replacement warranty.

Most states have statutes or case law covering warranties for new home construction. These warranties often cannot be limited by a contract provision.

The express warranties contained herein are in lieu of all other warranties, express or implied, including any warranties of merchantability, habitability, or fitness for a particular use or purpose. This limited warranty excludes consequential and incidental damages and limits the duration of implied warranties to the fullest extent permissible under state and federal law.

Modern sued to recover consequential damages for the alleged breach by Rapid for the design and installation of a furnace at Modern's Blue Island, Illinois, facility. The contract between the parties contained the following limitation:

> If any defects in material and workmanship of the product develop within the applicable warranty period, Seller shall supply replacement/repair parts FOB. Blue Island, freight prepaid. This shall be Buyer's exclusive remedy for Seller's liability hereunder.

This warranty limitation was upheld, and Modern was unable to sue for consequential damages.[41]

Frey Dairy entered into a contract for the purchase and installation of grain silos manufactured by Harvestore. After installation, Frey Dairy claimed that the silos leaked; the food stored in them became wet and when eaten by its cows caused them injury. Frey sued for injury to its herd.

The purchase order contained the following "WARRANTY OF MANUFACTURER AND SELLER" clause:

> If within the time limits specified below, any product sold under this purchase order, or any part thereof, shall prove to be defective in material or workmanship upon examination by the Manufacturer, the Manufacturer will supply an identical or substantially similar replacement part FOB the Manufacturer's factory, or the Manufacturer, at its option, will repair or allow credit for such part.
>
> NO OTHER WARRANTY, EITHER EXPRESS OR IMPLIED AND INCLUDING A WARRANTY OF MERCHANTABILITY AND FITNESS FOR A PARTICULAR PURPOSE, HAS BEEN OR WILL BE MADE BY OR IN BEHALF OF THE MANUFACTURER OR BY OPERATION OF LAW WITH RESPECT TO THE EQUIPMENT AND ACCESSORIES OR THEIR INSTALLATION, USE, OPERATION, REPLACEMENT OR REPAIR. THE MANUFACTURER SHALL NOT BE LIABLE BY VIRTUE OF THIS WARRANTY, OR OTHERWISE, FOR ANY SPECIAL OR CONSEQUENTIAL LOSS OR DAMAGE (INCLUDING BUT NOT LIMITED TO THOSE RESULTING FROM THE CONDITION OR QUALITY OF ANY CROP OR MATERIAL STORED IN THE STRUCTURE) RESULTING FROM THE USE OR LOSS OF THE USE OF EQUIPMENT AND ACCESSORIES.
>
> IRRESPECTIVE OF ANY STATUTE, THE BUYER RECOGNIZES THAT THE EXPRESS WARRANTY SET FORTH ABOVE IS THE EXCLUSIVE REMEDY TO WHICH HE IS ENTITLED, AND HE WAIVES ALL OTHER REMEDIES, STATUTORY OR OTHERWISE.

Frey Dairy could not recover for damages to its herd, as those damages had been expressly waived by the above clause.[42]

■ Time Is of the Essence Clause

This clause attempts to hold the contractor's failure to complete a contract by the date called for in the contract as a material breach. "The cases are legion which stand for the proposition that a material breach of contract gives rise to election by the nonbreaching party: the latter may either terminate the contract or perform and later sue for breach damages."[43]

This clause has lost its importantance in recent times, not because time is or is not of the essence but because the law is not controlled by the clause; that is, time may be of the essence with or without the clause. "However, the existence of a contract deadline itself establishes that time is of the essence."[44] Since the clause is usually boilerplate, the court may refuse to uphold it in light of conflicting clauses or acts of the owner contrary to time being of the essence.

EXAMPLE

On a major construction project, the masonry subcontract contained a "time is of the essence" clause. After several delays caused by the owner and the contractor, the masonry subcontractor terminated the masonry subcontract. The court held that the masonry subcontractor had wrongfully terminated the subcontract because other clauses in the contract suggested the time is of the essence clause was only for the benefit of the contractor, not the subcontractor. Another clause in the contract said that the subcontractor's only choice when it was delayed by the architect was to make a claim in writing. "In light of these other provisions, we conclude that the phrase "time-is-of-the-essence" as used in this subcontract did not give to [the subcontractor] a right to rescind on the basis of any delay, however slight and from whatever source, in the readiness of the site for its work."[45]

■ Pay-If-Paid Clause or Pay-When-Paid Clause

A pay-if-paid clause or pay-when-paid clause transfers the risk of owner failure to pay onto the subcontractor. Under this clause, if the owner does not pay the general contractor, the general contractor has no obligation to pay the subcontractor.

Subcontractor to be paid progress payments after General Contractor has received payment from Owner for project work that has been successfully completed by Subcontractor. Receipt of payment from Owner is a condition precedent to paying Subcontractor under this Agreement.

Receipt of payment by the Contractor from the Owner for the Subcontractor Work is a condition precedent to payment by the Contractor to the Subcontractor. The Subcontractor hereby acknowledges that it relies on the credit of the Owner, not the Contractor, for payment of Subcontract Work. Progress payments received from the Owner for the Subcontractor for satisfactory performance of the Subcontract Work shall be made no later than seven (7) days after receipt by the Contractor of payment from the Owner for the Subcontract Work.

Many jurisdictions will enforce these clauses,[46] but many refuse to do so or have interpreted the clause as a "pay within a reasonable time" clause, discussed below.[47] New York[48] and California[49] have refused to uphold the clauses because they violate public policy. In North Carolina[50] and Wisconsin,[51] they are void by statute. Other states protecting subcontractors are Maryland, Illinois, and Missouri. Federal projects are controlled by the Millar Act, and the contractor is obligated to the subcontractor despite a pay-if-paid clause.[52]

■ Pay within a Reasonable Time

Many courts and statutes refuse to uphold pay-if-paid or pay-when-paid clauses. In these jurisdictions, no matter what the clause actually says, it is interpreted to mean that the contractor must pay the subcontractor within a reasonable time whether or not the contractor is paid by the owner.[53]

Printz was the general contractor on a casino construction project. Prior to the completion of the job, the owner went bankrupt and did not pay Printz. Printz did not pay Main Electric, a subcontractor, because of a pay-when-paid clause in the subcontract. "In this case, we interpret the payment provisions of a construction contract to require a general contractor to pay a subcontractor even though the owner has failed to pay the general contractor. This provision is referred to as a 'pay-when-paid' clause. We hold that the 'when' of this clause is not a contingency but rather means that payment may be delayed. We decline to find that this clause is a 'pay-if-paid' clause that excuses the general contractor's obligation to the subcontractor if the owner does not pay."[54] Whatever the above means or does not mean to a normal person, the court said that the contractor must pay the subcontractor even if the contractor is not paid.

■ Flow Down or Incorporation by Reference

The flow-down provision in a subcontract obligates the subcontractor to all of the provisions in the general contract. This clause has been referred to as a "conduit" through which all of the obligations in the prime contract flow down to the subcontractor. For example, if the general contract has an arbitration clause, the clause automatically flows down into the subcontract. Although these clauses are always extremely broadly written, no court has held that they require the subcontractor to perform the prime contract should the prime contractor fail to do so. The scope of the work of the subcontract remains the same.

FORM

The Subcontractor agrees to be bound to the Contractor by the terms of the prime contract documents and assume toward the Contractor all of the obligations and responsibilities that the Contractor by aforesaid documents assumes toward the Owner.

Issues usually revolve around insurance and arbitration requirements. In one case involving the ability of the flow-down clause to obligate the subcontractor's insurer to defend the general contractor, the court said, "This case requires us to deconstruct various construction contracts, insurance agreements, and other related documents. In particular, we must decide whether an insurer for a subcontractor on a real estate construction project agreed to provide liability insurance and a defense for the general contractor with respect to a lawsuit alleging that the general contractor was negligent. After hacking our way through a dense thicket of so-called incorporation-by-reference, flow down, and additional-insured provisions contained in the pertinent documents, *we answer this question in the negative*" (emphasis added)."[55]

FORM

SUBCONTRACTOR agrees to fully perform and to assume all obligations and liabilities of the CONTRACTOR under the General Contract for the WORK as related to the scope of the WORK of this SUBCONTRACT.

Failure to limit the flow down to the work can mean the many provisions in the prime contract control the subcontractor. A delay damages provision was held to pass through to the subcontractor.[56] Notice provisions were incorporated and flowed down.[57] A provision requiring binding arbitration flowed down to the subcontractor.[58] On the other hand, several courts have refused to flow down the contractor's liability for injuries to the subcontractor.

EXAMPLE

Acciavatti, the general contractor, entered into a contract with Super Fresh Grocery to do repair work at a store. The store was to remain open during the project. The prime contract contained an intermediate form indemnity clause (discussed above) stating the general contractor would "assume entire responsibility and liability for any and all damage or injury of any kind...caused by...the execution of the work provided for in this Contract..." as long as the general contractor was at least partially responsible for the damage.

Acciavatti entered into an electrical subcontract with Goldsmith, and the subcontract contained the provision that the "[prime] contract documents form a part of this Subcontract and are as fully a part of this Subcontract as if attached to this agreement and as if herein set forth at length."

Bernotas, a patron at the store, fell into a hole in the floor in the construction area. All three of the parties settled with Bernotas, each paying $200,000 in damages. The owner, general contractor, and subcontractor then sued each other, each claiming that they were not responsible for the damages.

continued

Chapter 14

The trial court determined that Super Fresh was not solely negligent, as both Accia-vatti and Goldsmith failed to provide a safe work area. This finding triggered the inter-mediate form indemnification clause in the prime contract: Super Fresh, the owner, was not obligated for Bernotas' injuries, only the general contractor was. The general con-tractor then claimed that the flow-down provision obligated the subcontractor to absorb the contractor's liability to Super Fresh. The state Supreme Court held no, as although many provisions in a prime contract do flow-down to the subcontractor, an indemnity clause would not unless the contract expressly stated this. Therefore, the contractor was responsible for his own damages and Super Fresh's damages ($400,000). The sub-contractor was responsible for his portion of the damages ($200,000).[59]

■ Release or Waiver of Lien

A release of lien prevents the contractor or supplier from exercising their statutory lien rights. The following form also releases the owner from all liability for monies due. Releases of lien should not be signed without receiving payment. It is not uncommon for contractors or owners to demand that the subcontractor or material supplier sign these before payment is made.

FORM

In consideration of the amounts stated below, receipt of which is hereby acknowledged, _____ (contractor, subcontractor, supplier), the potential lienor, doing business in the county of _____, city of _____, state of _____, having performed work on and/or supplied materials to the property legally described as _____, commonly known as _____, does hereby release, waive, and discharge the owner of this property from any and every liability for moneys owed arising out of the performance of the work and/or for materials supplied pursuant to our contract or invoice dated _____.

The consideration received by the potential lienor for this release is: _____ (signed)

EXAMPLE

Gandric was the prime contractor on 1,000-bed hospital construction project. Storos was the plumbing and mechanical subcontractor. The subcontract required Storos to sign a release of all liens prior to partial payments, and Storos signed such a release for work performed prior to January 31, 1997.

On April 30, 1997, Gandric was terminated from the project. Gandric informed Storos that it was going to challenge the termination and that Storos's was suspended, but that the subcontracts remained in full force and effect until further notice. Gandric and the owner resolved the dispute, and Gandric continued work on the project some months later. However, Storos never returned to work, and another subcontractor was hired to finish the job. Gandric sued Storos for damages, and Storos cross-complained, claiming that Gandric had not in fact paid the entire amount due Storos on its various requisitions prior to January 31, 1997. The court said, "There is no need to address this argument, however, because the lien waiver and release—executed *after* payment was made—was sufficient to waive all claims arising prior to January 31, 1997."[60]

The release, which stated "we [Storos] hereby release Gandric and the owner from any further liability in regard to payment for all materials, labor, and services furnished by us through the period stated above [1/31/97]," did exactly what it said: it released the general contractor from liability.

MANAGEMENT TIP

Once signed, these documents are effective, and subcontractors and suppliers should exercise caution in signing a form before payment is received.

■ Endnotes

[1] *R.J. Griffin & Co. v. Beach Club II Homeowners Asso., Inc.*, 384 F.3d 157 (4th Cir. 2004).

[2] "Without doubt, arbitration of certain commercial disputes, and even of certain commercial insurance disputes, is desirable. When the law governing an issue is well settled, or where predominantly factual questions are at issue, arbitration offers a confidential, potentially expeditious method of resolving disputes. *When legally unresolved issues may arise, however, arbitration of insurance disputes is a trap for the unwary. Because many major disputes under commercial insurance policies tend to raise such issues, commercial policyholders would be well advised to think twice before purchasing insurance policies that provide for mandatory, binding arbitration. They must weigh the advantages of confidentiality, cost saving, and decision-maker expertise in certain respects, on the one hand, against the disadvantages of potentially lawless resolution of the dispute without recource to correction by higher authority. And because binding arbitration impedes the production of publicly known legal rules, the courts themselves should be more neutral toward arbitration than they have tended recently to be. The strong presumption in favor or arbitrating disputes that at least arguably belongs in courts of law should be abolished (emphasis added)."* Kenneth S. Abraham, The Lawlessness of Arbitration, 9 Conn. Ins. L.J. 355 (2002/2003).

[3] *Upshur Coals Corp. v. United Mine Workers, Dist. 31*, 933 F.2d 225, 228 (4th Cir. 1991).

[4] *Richmond, Fredericksburg & Potomac R.R. Co. v. Transp. Commc'ns Int'l Union.* 973 F.2d 276, 278 (4th Cir. 1992), quoting *Union Pac. R.R. Co. v. Sheehan*, 439 U.S. 89, 91, 99 S. Ct. 399, 58 L. Ed. 2d 354 (1978).

[5] *First Options of Chicago, Inc. v. Kaplan*, 514 U.S. 938, 944, 115 S. Ct. 1920, 131 L. Ed. 2d 985 (1995).

[6] *Remmey v. PaineWebber. Inc.*, 32 F.3d 143, 146 (4th Cir. 1994).

[7] *TC Arrowpoint, LP v. Choate Constr. Co.*, 2006 U.S. Dist. Lexis 2881 (U.S. Dist. NC, 2006) quoting *Mountaineer Gas Co. v. Oil, Chem. & Atomic Workers Int'l Union*, 76 F.3d 606, 608 (4th Cir. 1996).

[8] Virginia Uniform Arbitration Act *Code § 8.01-581.010.*

[9] Clause from *Signal Corp. v. Keane Federal Systems, Inc.*, 265 Va. 38, 574 S.E.2d 253 (Vir. 2003).

[10] Federal Arbitration Act, Section 10(a)(4).

[11] *Merrill Lynch, Pierce, Fenner & Smith, Inc. v. Bobker*, 808 F.2d 930, 933-34 (2d Cir. 1986).

[12] *Demaria Building Co., Inc. v. Word of Faith International Christian Center*, 2005 Mich. App. Lexis 1686 (2005) (unpublished opinion).

[13] See *Gordon Sel-Way, Inc v. Spence Bros, Inc*, 438 Mich. 488, 497, 475 N.W.2d 704 (1991), *DAIIE v. Gavin*, 416 Mich. 407, 428-429, 443, 331 N.W.2d 418 (1982), *Dohanyos v. Detrex Corp.* (after Remand), 217 Mich. App. 171, 176; 550 N.W.2d 608 (1996).

[14] *Dustrol, Inc. v. Champagne-Webber, Inc.*, 2001 U.S. Dist. Lexis 16593 (U.S. Dist. ND Tx. 2001).

[15] *Hooters v. Phillips*, 39 F. Supp. 2d 582 (D.S.C. 1998).

381

[16] *Blake Construction Co., Inc. v. Upper Occoquan Sewage Authority*, 266 Va. 564, 587 S.E. 2d 711 (Vir. 2003).

[17] See *E.C. Ernst, Inc. v. Manhattan Construction Co. of Texas*, 551 F.2d 1026, 1029 (5th Cir. 1977), pet. for reh. granted in part and denied in part; pet. for reh. en banc denied, 559 F.2d 268 (5th Cir. 1977), cert. denied, 434 U.S. 1067, 98 S. Ct. 1246, 55 L. Ed. 2d 769 (1978).

[18] Va. Code § 2.2-4335(A).

[19] Beattie, Carl S., *Apportioning the Risk of Delay in Construction Projects: A Proposed Alternative to the Inadequate "No Damages For Delay" Clause*, 46 Wm and Mary L. Rev. 1857 (March 2005).

[20] *J.A. Jones Construction Co. v. Lehrer Mcgovern Bovis, Inc.*, 89 P.3d 1009 (Nev. 2004). See also *U.S. v. Metric Constructors, Inc.*, 325 S.C. 129, 480 S.E.2d 447 (S.C. 1997).

[21] Example based on *Blake Constr. Co. v. C.J. Coakley Co.*, 431 A.2d 569, 576 (D.C. App. 1981).

[22] Example based on *Triple R Paving, Inc. v. Broward County*, 774 So. 2d 50 (Fla. 4th DCA 2000).

[23] Example based on *Clifford R. Gray, Inc. v. City School District of Albany*, 277 A.D.2d 843, 716 N.Y.S.2d 795 (N.Y. App. 2000).

[24] Example based on *Clifford R. Gray, Inc. v. City School District of Albany*, 277 A.D.2d 843, 716 N.Y.S.2d 795 (N.Y. App. 2000).

[25] Example based on *Marriott Corp. v. Dasta Constr. Co.*, 26 F.3d 1057 (11th Cir. 1994).

[26] *Gorco Construction Co. v. Stein*, 256 Minn. 476, 99 N.W.2d 69 (Minn. 1959).

[27] Example based on *Fortune Bridge Co. v. Department of Transportation*, 242 Ga. 531 (Ga. 1978).

[28] Example based on *Kelly v. Marx*, 428 Mass. 877, 705 N.E.2d 1114 (Ma. 1999).

[29] Clause from *Idaho Power Co. v. Cogeneration, Inc.*, 134 Idaho 738; 9 P.3d 1204 (Ida. 2000).

[30] Example based on *Idaho Power Co. v. Cogeneration, Inc.*, 134 Idaho 738, 9 P.3d 1204 (Ida. 2000).

[31] Example based on *Hall Contracting Corp. v. Entergy Services, Inc.*, 309 F.3d 468 (8th Cir. 2002).

[32] Example based on *G.R. Sponaugle & Sons, Inc. v. Hunt Constr. Group, Inc.*, 366 F. Supp. 2d 236 (M.D. Penn. 2004).

[33] *Reliance Nat'l Indem. v. Knowles Indus. Servs. Corp.*, 868 A.2d 220, 225-27 (Me. 2005), *Willis Realty Assocs. v. Cimino Constr. Co.*, 623 A.2d 1287, 1288 (Me. 1993).

[34] *Bastian v. Wausau Homes, Inc.*, 635 F. Supp. 201, 204-05 (N.D. Ill. 1986, *Town of Silverton v. Phoenix Heat Source Sys. Inc.*, 948 P.2d 9, 13–14 (Colo. Ct. App. 1997), *Village of Rosemont v. Lentin Lumber Co.*, 144 Ill. App. 3d 651, 494 N.E.2d 592, 601, 98 Ill. Dec. 470 (Ill. App. Ct. 1986).

[35] *Reliance National Indemnity v. Knowles Industrial Services, Corp.*, 868 A.2d 220 (Me. 2005).

[36] *Krygoski Constr. Co. v. United States*, 94 F.3d 1537, 1541 (Fed. Cir. 1996).

[37] Example based on *John B. Conomos, Inc. v. Sun Company, Inc.*, 831 A.2d 696, 2003 Pa Super 310 (Pa. Sup. 2003).

[38] Sela Stroud, *Non-Compete Agreements: Weighing the Interests of Profession and Firm*, 53 Ala. L. Rev. 1023 (2002).

[39] Clause from *Signal Corp. v. Keane Federal Systems, Inc.*, 265 Va. 38, 574 S.E.2d 253 (Vir. 2003).

[40] Clause from *Signal Corp. v. Keane Federal Systems, Inc.*, 265 Va. 38, 574 S.E.2d 253 (Vir. 2003).

[41] Example based on *Modern Drop Forge Co. v. Rapid Technologies, Inc.*, 1996 U.S. App. Lexis 20097 (7th Cir. 1996).

[42] Example based on *Frey Dairy V. A.O. Smith Harvestore Products, Inc.*, 886 F.2d 128 (6th Cir. 1989).

[43] *John W. Johnson, Inc. v. Basic Construction Co.*, 292 F. Supp. 300 (D.D.C.1968), aff'd. 139 U.S. App. D.C. 85, 429 F.2d 764 (D.C.Cir.1970).

[44] *Empire Energy Management Systems, Inc. v. Roche*, 362 F.3d 1343 (Fed.Cir. 2004).

[45] Example based on *Vermont Marble Co. v. Baltimore Contractors, Inc.*, 520 F. Supp. 922 (D. D.C. 1981).

[46] *Christman Co. v. Anthony S. Brown Development Co., Inc.*, 210 Mich.App. 416 (1994).

[47] See Corbin on Contracts, § 30.15 *Promises to Pay Money Out of Funds Yet to Be Acquired,* and § 38.8 *Aleatory Contracts for Compensation of Agents and Subcontractors,* Matthew Bender & Company, Inc. (2006).

[48] *West-Fair Electrical Contractors v. Aetna Casualty & Surety Co.*, 661 N.E.2d 967 (N.Y. 1995).

[49] *Wm. R. Clarke Corp. v. Safeco Ins. Co.*, 15 Cal. 4th 882, 938 P.2d 372, 374 (Cal. 1997).

[50] N.C. Gen. Stat. § 22C-2 (2003).

[51] Wisconsin Stat. § 779.135.

[52] See *U.S. ex rel. T.M.S. Mechanical Contractors, Inc. v. Millers Mut. Fire Ins. Co.*, 942 F.2d 946 (5th Cir. 1991).

[53] Margie Alsbrook, *Contracting Away an Honest Day's Pay: An Examination of Conditional Payment Clauses in Construction Contracts*, 58 Ark. L. Rev. 353 (2005).

[54] *Main Electric, Ltd v. Printz Services Corp.*, 980 P.2d 522, 1999 Colo. J. C.A.R. 1353 (Co. 1999).

[55] *A.F. Lusi Construction, Inc. v. Peerless Ins. Co.*, 847 A.2d 254 (R.I. 2004).

[56] *Indus. Indem. Co. v. Wick Constr. Co.*, 680 P.2d 1100, 1106 (Alaska 1984).

[57] *Sime Constr. Co. v. Washington Pub. Power Supply Sys.*, 28 Wn. App. 10, 621 P.2d 1299, 1302 (Wash. Ct. App. 1980).

[58] *Turner Constr. Co. v. Midwest Curtainwalls, Inc.*, 187 Ill. App. 3d 417, 543 N.E.2d 249, 250–52, 135 Ill. Dec. 14 (Ill. App. Ct.1989).

[59] Example based on *Bernotas v. Super Fresh Food Markets, Inc.*, 581 Pa. 12; 863 A.2d 478; 2004 Pa. LEXIS 3238 (Penn. 2004).

[60] Example based on *Morganti National, Inc. v. Petri Mechanical Co.*, 2004 U.S. Dist. Lexis 8659 (D.Conn. 2004).

Notice of Commencement (Michigan Law)

Prepared by, recording requested by and return to: Name (designee):	
Address:	
City:	
State: Zip:	
Phone:	Above this line for Official Use Only

Notice of Commencement - M.S.A. § 570.1108a

To lien claimants and subsequent purchasers:

Take notice that work is about to commence on an improvement to the real property described in this instrument. A person having a construction lien may preserve the lien by providing a notice of furnishing to the above named designee and the general contractor, if any, and by timely recording a claim of lien, in accordance with law.

A person having a construction lien arising by virtue of work performed on this improvement should refer to the name of the owner or lessee and the legal description appearing in this notice. A person subsequently acquiring an interest in the land described is not required to be named in a claim of lien.

A copy of this notice with an attached form for Notice of furnishing may be obtained upon making a written request by certified mail to the owner or lessee; the designee; or the person with whom you have contracted.

SWORN STATEMENT
State of Michigan, County of _____

_____(deponent), being sworn, states the following:

The name and address of the fee owner of the real property is:

The name, address, and capacity of the person contracting for the improvement, if different from the above, is:

The name and address of the general contractor, if any:

_____
Signature of Deponent

Subscribed and sworn to before me this the ____ day of _____, 20____.
_____ Notary Public, _____ County, MI.

_____
Notary Public

My Commission Expires: _____

Instructions to Owners: Notice of Commencement - M.S.A. § 570.1108a

NON-RESIDENTIAL PROPERTY OWNERS or LESSEES:
The law requires you to file this Notice of Commencement at the office of the register of deeds in the county where the real property to be improved is located before the start of the first actual physical improvement.

You are also required to post this document and a blank Notice of Furnishing in a conspicuous place on the property.

WARNING TO THE HOMEOWNER:
You, the homeowner are required to send a completed Notice of Commencement and blank Notice of Furnishing to an entity supplying labor or materials to your construction project within ten days of the request if:

- an entity supplying labor or materials has requested the Notice of Commencement in writing and mailed the request by certified mail; and
- has attached a blank Notice of Commencement; and
- has attached a blank Notice of Furnishing.

If you do not live on the site you are also required to post the completed Notice of Commencement and blank Notice of Furnishing in a conspicuous place after the above request has been made.

You, the Owner, are not required by law, but you should:
1. Even if you live at the site you should complete and post a copy of this form at the place where the improvement is being done.
2. Make and keep a copy of this form for your records.

FAILURE TO SEND THE COMPLETED NOTICE OF COMMENCEMENT AND BLANK NOTICE OF FURNISHING OR TO POST THE SAME INCREASES THE TIME DURING WHICH ENTITIES SUPPLYING MATERIAL AND LABOR TO YOUR PROJECT CAN FILE A NOTICE OF FURNISHING. YOU CAN ALSO BE REQUIRED TO PAY THE COSTS INCURRED TO OBTAIN THIS INFORMATION.

APPENDIX  **B**

Notice of Furnishing
(Michigan Law)

NOTICE OF FURNISHING

To:

> (*Instructions*: Fill in name and address of the designee, owner, or lessee from Notice of Commencement in above space.)

Please take notice that the undersigned is furnishing to:

> (*Instructions*: Fill in name and address of the entity you have a contract with or are employed by, in the above space.)

certain labor or material for or in connection with the improvements to the real property described in the Notice of Commencement. A copy of the Notice of Commencement is either attached to this Notice of Furnishing or is recorded on page_____, records _____ (name of county).

> (*Instructions*: Attach copy of Notice of Commencement or fill in blanks with information from recorded document above.)

Description of type of work:

> (*Instructions*: Describe the type of work in the above space.)

WARNING TO OWNER: THIS NOTICE IS REQUIRED BY THE MICHIGAN CONSTRUCTION LIEN ACT. IF YOU HAVE QUESTIONS ABOUT YOUR RIGHTS AND DUTIES UNDER THIS ACT, YOU SHOULD CONTACT AN ATTORNEY TO PROTECT YOU FROM THE POSSIBILITY OF PAYING TWICE FOR THE IMPROVEMENTS TO YOUR PROPERTY.

_____   _____ (date)
Signature of party supplying this Notice of Furnishing

Name, address, and capacity of party *supplying* this Notice of Furnishing:

If different than above, the name, address, and capacity of the party *signing* this Notice of Furnishing:

APPENDIX **C**

Claim of Lien

CLAIM OF LIEN

Notice is hereby given that on the _____ day of _____, 20 _____,

(name)

(address)
first provided labor or material for an improvement to

_____,
(legal description of real property from notice of commencement)
the (owner) (lessee) of which property is _____

The last day of providing the labor or material was the _____ day of
_____, 20 _____

TO BE COMPLETED BY A LIEN CLAIMANT WHO IS A CONTRACTOR,
SUBCONTRACTOR, OR SUPPLIER:
The lien claimant's contract amount, including extras, is $_____.The lien
claimant has received payment thereon in the total amount of
$_____ and therefore claims a construction lien upon the
above-described real property in the amount of $_____.

TO BE COMPLETED BY A LIEN CLAIMANT WHO IS A LABORER:
The lien claimant's hourly rate, including fringe benefits and withholdings, is
$_____.
There is due and owing to or on behalf of the laborer the sum of
$_____ for which the laborer claims a construction lien upon the
above-described real property.

_____
Signature of lien claimant, agent, or attorney

Address of party signing this claim of lien:

Date:

State of Michigan, County of _____

Subscribed and sworn to before me this _____ day of _____, 20 _____,

_____
Signature of Notary Public
My commission expires:

Standard General Conditions of the Construction Contract

Reprinted by permission of the Engineers Joint Contract Documents Committee (EJCDC). For more information about the EJCDC, please visit www.nspe.org.

STANDARD
GENERAL CONDITIONS
OF THE
CONSTRUCTION CONTRACT

Prepared by

ENGINEERS JOINT CONTRACT DOCUMENTS COMMITTEE

and

Issued and Published Jointly By

PROFESSIONAL ENGINEERS IN PRIVATE PRACTICE
a practice division of the
NATIONAL SOCIETY OF PROFESSIONAL ENGINEERS

AMERICAN COUNCIL OF ENGINEERING COMPANIES

AMERICAN SOCIETY OF CIVIL ENGINEERS

This document has been approved and endorsed by

The Associated General Contractors of America

Knowledge for Creating
and Sustaining
the Built Environment

Construction Specifications Institute

Copyright ©2002

National Society of Professional Engineers
1420 King Street, Alexandria, VA 22314

American Council of Engineering Companies
1015 15th Street, N.W., Washington, DC 20005

American Society of Civil Engineers
1801 Alexander Bell Drive, Reston, VA 20191-4400

These General Conditions have been prepared for use with the Suggested Forms of Agreement Between Owner and Contractor Nos. C-520 or C-525 (2002 Editions). Their provisions are interrelated and a change in one may necessitate a change in the other. Comments concerning their usage are contained in the EJCDC Construction Documents, General and Instructions (No. C-001) (2002 Edition). For guidance in the preparation of Supplementary Conditions, see Guide to the Preparation of Supplementary Conditions (No. C-800) (2002 Edition).

TABLE OF CONTENTS

Standard General Conditions of the Construction Contract

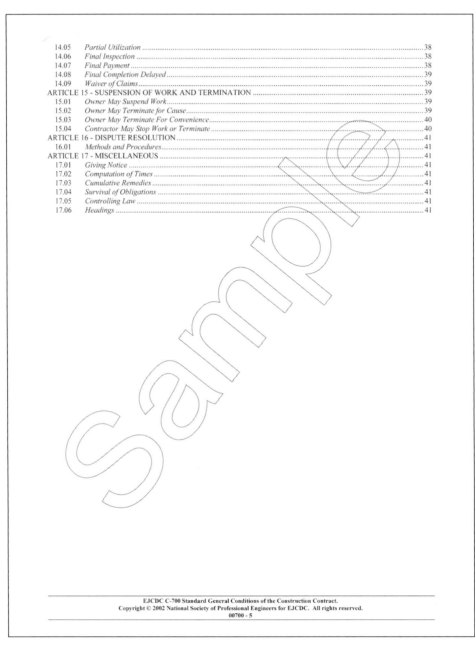

Appendix D

GENERAL CONDITIONS

ARTICLE 1 - DEFINITIONS AND TERMINOLOGY

1.01 Defined Terms

A. Wherever used in the Bidding Requirements or Contract Documents and printed with initial capital letters, the terms listed below will have the meanings indicated which are applicable to both the singular and plural thereof. In addition to terms specifically defined, terms with initial capital letters in the Contract Documents include references to identified articles and paragraphs, and the titles of other documents or forms.

1. *Addenda*--Written or graphic instruments issued prior to the opening of Bids which clarify, correct, or change the Bidding Requirements or the proposed Contract Documents.

2. *Agreement*--The written instrument which is evidence of the agreement between Owner and Contractor covering the Work.

3. *Application for Payment*--The form acceptable to Engineer which is to be used by Contractor during the course of the Work in requesting progress or final payments and which is to be accompanied by such supporting documentation as is required by the Contract Documents.

4. *Asbestos*--Any material that contains more than one percent asbestos and is friable or is releasing asbestos fibers into the air above current action levels established by the United States Occupational Safety and Health Administration.

5. *Bid*--The offer or proposal of a Bidder submitted on the prescribed form setting forth the prices for the Work to be performed.

6. *Bidder*--The individual or entity who submits a Bid directly to Owner.

7. *Bidding Documents*--The Bidding Requirements and the proposed Contract Documents (including all Addenda).

8. *Bidding Requirements*--The Advertisement or Invitation to Bid, Instructions to Bidders, bid security of acceptable form, if any, and the Bid Form with any supplements.

9. *Change Order*--A document recommended by Engineer which is signed by Contractor and Owner and authorizes an addition, deletion, or revision in the Work or an adjustment in the Contract Price or the Contract Times, issued on or after the Effective Date of the Agreement.

10. *Claim*--A demand or assertion by Owner or Contractor seeking an adjustment of Contract Price or Contract Times, or both, or other relief with respect to the terms of the Contract. A demand for money or services by a third party is not a Claim.

11. *Contract*--The entire and integrated written agreement between the Owner and Contractor concerning the Work. The Contract supersedes prior negotiations, representations, or agreements, whether written or oral.

12. *Contract Documents*-- Those items so designated in the Agreement. Only printed or hard copies of the items listed in the Agreement are Contract Documents. Approved Shop Drawings, other Contractor's submittals, and the reports and drawings of subsurface and physical conditions are not Contract Documents.

13. *Contract Price*--The moneys payable by Owner to Contractor for completion of the Work in accordance with the Contract Documents as stated in the Agreement (subject to the provisions of Paragraph 11.03 in the case of Unit Price Work).

14. *Contract Times*--The number of days or the dates stated in the Agreement to: (i) achieve Milestones, if any, (ii) achieve Substantial Completion; and (iii) complete the Work so that it is ready for final payment as evidenced by Engineer's written recommendation of final payment.

15. *Contractor*--The individual or entity with whom Owner has entered into the Agreement.

16. *Cost of the Work*--See Paragraph 11.01.A for definition.

17. *Drawings*--That part of the Contract Documents prepared or approved by Engineer which graphically shows the scope, extent, and character of the Work to be performed by Contractor. Shop Drawings and other Contractor submittals are not Drawings as so defined.

18. *Effective Date of the Agreement*--The date indicated in the Agreement on which it becomes effective, but if no such date is indicated, it means the date on which the Agreement is signed and delivered by the last of the two parties to sign and deliver.

19. *Engineer*--The individual or entity named as such in the Agreement.

20. *Field Order*--A written order issued by Engineer which requires minor changes in the Work but which does not involve a change in the Contract Price or the Contract Times.

21. *General Requirements*--Sections of Division 1 of the Specifications. The General Requirements pertain to all sections of the Specifications.

22. *Hazardous Environmental Condition*--The presence at the Site of Asbestos, PCBs, Petroleum, Hazardous Waste, or Radioactive Material in such quantities or circumstances that may present a substantial danger to persons or property exposed thereto in connection with the Work.

23. *Hazardous Waste*--The term Hazardous Waste shall have the meaning provided in Section 1004 of the Solid Waste Disposal Act (42 USC Section 6903) as amended from time to time.

24. *Laws and Regulations; Laws or Regulations*--Any and all applicable laws, rules, regulations, ordinances, codes, and orders of any and all governmental bodies, agencies, authorities, and courts having jurisdiction.

25. *Liens*--Charges, security interests, or encumbrances upon Project funds, real property, or personal property.

26. *Milestone*--A principal event specified in the Contract Documents relating to an intermediate completion date or time prior to Substantial Completion of all the Work.

27. *Notice of Award*--The written notice by Owner to the Successful Bidder stating that upon timely compliance by the Successful Bidder with the conditions precedent listed therein, Owner will sign and deliver the Agreement.

28. *Notice to Proceed*--A written notice given by Owner to Contractor fixing the date on which the Contract Times will commence to run and on which Contractor shall start to perform the Work under the Contract Documents.

29. *Owner*--The individual or entity with whom Contractor has entered into the Agreement and for whom the Work is to be performed.

30. *PCBs*--Polychlorinated biphenyls.

31. *Petroleum*--Petroleum, including crude oil or any fraction thereof which is liquid at standard conditions of temperature and pressure (60 degrees Fahrenheit and 14.7 pounds per square inch absolute), such as oil, petroleum, fuel oil, oil sludge, oil refuse, gasoline, kerosene, and oil mixed with other non-Hazardous Waste and crude oils.

32. *Progress Schedule*--A schedule, prepared and maintained by Contractor, describing the sequence and duration of the activities comprising the Contractor's plan to accomplish the Work within the Contract Times.

33. *Project*--The total construction of which the Work to be performed under the Contract Documents may be the whole, or a part.

34. *Project Manual*--The bound documentary information prepared for bidding and constructing the Work. A listing of the contents of the Project Manual, which may be bound in one or more volumes, is contained in the table(s) of contents.

35. *Radioactive Material*--Source, special nuclear, or byproduct material as defined by the Atomic Energy Act of 1954 (42 USC Section 2011 et seq.) as amended from time to time.

36. *Related Entity* -- An officer, director, partner, employee, agent, consultant, or subcontractor.

37. *Resident Project Representative*--The authorized representative of Engineer who may be assigned to the Site or any part thereof.

38. *Samples*--Physical examples of materials, equipment, or workmanship that are representative of some portion of the Work and which establish the standards by which such portion of the Work will be judged.

39. *Schedule of Submittals*--A schedule, prepared and maintained by Contractor, of required submittals and the time requirements to support scheduled performance of related construction activities.

40. *Schedule of Values*--A schedule, prepared and maintained by Contractor, allocating portions of the Contract Price to various portions of the Work and used as the basis for reviewing Contractor's Applications for Payment.

41. *Shop Drawings*--All drawings, diagrams, illustrations, schedules, and other data or information which are specifically prepared or assembled by or for Contractor and submitted by Contractor to illustrate some portion of the Work.

42. *Site*--Lands or areas indicated in the Contract Documents as being furnished by Owner upon which the Work is to be performed, including rights-of-way and easements for access thereto, and such other lands furnished by Owner which are designated for the use of Contractor.

43. *Specifications*--That part of the Contract Documents consisting of written requirements for materials, equipment, systems, standards and workmanship as applied to the Work, and certain

Appendix D

administrative requirements and procedural matters applicable thereto.

44. *Subcontractor*--An individual or entity having a direct contract with Contractor or with any other Subcontractor for the performance of a part of the Work at the Site.

45. *Substantial Completion*--The time at which the Work (or a specified part thereof) has progressed to the point where, in the opinion of Engineer, the Work (or a specified part thereof) is sufficiently complete, in accordance with the Contract Documents, so that the Work (or a specified part thereof) can be utilized for the purposes for which it is intended. The terms "substantially complete" and "substantially completed" as applied to all or part of the Work refer to Substantial Completion thereof.

46. *Successful Bidder*--The Bidder submitting a responsive Bid to whom Owner makes an award.

47. *Supplementary Conditions*--That part of the Contract Documents which amends or supplements these General Conditions.

48. *Supplier*--A manufacturer, fabricator, supplier, distributor, materialman, or vendor having a direct contract with Contractor or with any Subcontractor to furnish materials or equipment to be incorporated in the Work by Contractor or any Subcontractor.

49. *Underground Facilities*--All underground pipelines, conduits, ducts, cables, wires, manholes, vaults, tanks, tunnels, or other such facilities or attachments, and any encasements containing such facilities, including those that convey electricity, gases, steam, liquid petroleum products, telephone or other communications, cable television, water, wastewater, storm water, other liquids or chemicals, or traffic or other control systems.

50. *Unit Price Work*--Work to be paid for on the basis of unit prices.

51. *Work*--The entire construction or the various separately identifiable parts thereof required to be provided under the Contract Documents. Work includes and is the result of performing or providing all labor, services, and documentation necessary to produce such construction, and furnishing, installing, and incorporating all materials and equipment into such construction, all as required by the Contract Documents.

52. *Work Change Directive*--A written statement to Contractor issued on or after the Effective Date of the Agreement and signed by Owner and recommended by Engineer ordering an addition, deletion, or revision in the Work, or responding to differing or unforeseen subsurface or physical conditions under which the Work is to be performed or to emergencies. A Work Change Directive will not change the Contract Price or the Contract Times

but is evidence that the parties expect that the change ordered or documented by a Work Change Directive will be incorporated in a subsequently issued Change Order following negotiations by the parties as to its effect, if any, on the Contract Price or Contract Times.

1.02   *Terminology*

A. The following words or terms are not defined but, when used in the Bidding Requirements or Contract Documents, have the following meaning.

B. *Intent of Certain Terms or Adjectives*

1. The Contract Documents include the terms "as allowed," "as approved," "as ordered", "as directed" or terms of like effect or import to authorize an exercise of professional judgment by Engineer. In addition, the adjectives "reasonable," "suitable," "acceptable," "proper," "satisfactory," or adjectives of like effect or import are used to describe an action or determination of Engineer as to the Work. It is intended that such exercise of professional judgment, action or determination will be solely to evaluate, in general, the Work for compliance with the requirements of and information in the Contract Documents and conformance with the design concept of the completed Project as a functioning whole as shown or indicated in the Contract Documents (unless there is a specific statement indicating otherwise). The use of any such term or adjective is not intended to and shall not be effective to assign to Engineer any duty or authority to supervise or direct the performance of the Work or any duty or authority to undertake responsibility contrary to the provisions of Paragraph 9.09 or any other provision of the Contract Documents.

C. *Day*

1.   The word "day" means a calendar day of 24 hours measured from midnight to the next midnight.

D. *Defective*

1. The word "defective," when modifying the word "Work," refers to Work that is unsatisfactory, faulty, or deficient in that it:

a. does not conform to the Contract Documents, or

b. does not meet the requirements of any applicable inspection, reference standard, test, or approval referred to in the Contract Documents, or

c. has been damaged prior to Engineer's recommendation of final payment (unless responsibility for the protection thereof has been assumed by Owner at Substantial Completion in accordance with Paragraph 14.04 or 14.05).

E. *Furnish, Install, Perform, Provide*

1. The word "furnish," when used in connection with services, materials, or equipment, shall mean to supply and deliver said services, materials, or equipment to the Site (or some other specified location) ready for use or installation and in usable or operable condition.

2. The word "install," when used in connection with services, materials, or equipment, shall mean to put into use or place in final position said services, materials, or equipment complete and ready for intended use.

3. The words "perform" or "provide," when used in connection with services, materials, or equipment, shall mean to furnish and install said services, materials, or equipment complete and ready for intended use.

4. When "furnish," "install," "perform," or "provide" is not used in connection with services, materials, or equipment in a context clearly requiring an obligation of Contractor, "provide" is implied.

F. Unless stated otherwise in the Contract Documents, words or phrases which have a well-known technical or construction industry or trade meaning are used in the Contract Documents in accordance with such recognized meaning.

ARTICLE 2 - PRELIMINARY MATTERS

2.01 *Delivery of Bonds and Evidence of Insurance*

A. When Contractor delivers the executed counterparts of the Agreement to Owner, Contractor shall also deliver to Owner such bonds as Contractor may be required to furnish.

B. *Evidence of Insurance:* Before any Work at the Site is started, Contractor and Owner shall each deliver to the other, with copies to each additional insured identified in the Supplementary Conditions, certificates of insurance (and other evidence of insurance which either of them or any additional insured may reasonably request) which Contractor and Owner respectively are required to purchase and maintain in accordance with Article 5.

2.02 *Copies of Documents*

A. Owner shall furnish to Contractor up to ten printed or hard copies of the Drawings and Project Manual. Additional copies will be furnished upon request at the cost of reproduction.

2.03 *Commencement of Contract Times; Notice to Proceed*

A. The Contract Times will commence to run on the thirtieth day after the Effective Date of the Agreement

or, if a Notice to Proceed is given, on the day indicated in the Notice to Proceed. A Notice to Proceed may be given at any time within 30 days after the Effective Date of the Agreement. In no event will the Contract Times commence to run later than the sixtieth day after the day of Bid opening or the thirtieth day after the Effective Date of the Agreement, whichever date is earlier.

2.04 *Starting the Work*

A. Contractor shall start to perform the Work on the date when the Contract Times commence to run. No Work shall be done at the Site prior to the date on which the Contract Times commence to run.

2.05 *Before Starting Construction*

A. *Preliminary Schedules:* Within 10 days after the Effective Date of the Agreement (unless otherwise specified in the General Requirements), Contractor shall submit to Engineer for timely review:

1. a preliminary Progress Schedule; indicating the times (numbers of days or dates) for starting and completing the various stages of the Work, including any Milestones specified in the Contract Documents;

2. a preliminary Schedule of Submittals; and

3. a preliminary Schedule of Values for all of the Work which includes quantities and prices of items which when added together equal the Contract Price and subdivides the Work into component parts in sufficient detail to serve as the basis for progress payments during performance of the Work. Such prices will include an appropriate amount of overhead and profit applicable to each item of Work.

2.06 *Preconstruction Conference*

A. Before any Work at the Site is started, a conference attended by Owner, Contractor, Engineer, and others as appropriate will be held to establish a working understanding among the parties as to the Work and to discuss the schedules referred to in Paragraph 2.05.A, procedures for handling Shop Drawings and other submittals, processing Applications for Payment, and maintaining required records.

2.07 *Initial Acceptance of Schedules*

A. At least 10 days before submission of the first Application for Payment a conference attended by Contractor, Engineer, and others as appropriate will be held to review for acceptability to Engineer as provided below the schedules submitted in accordance with Paragraph 2.05.A. Contractor shall have an additional 10 days to make corrections and adjustments and to complete and resubmit the schedules. No progress payment shall be made to Contractor until acceptable schedules are submitted to Engineer.

1. The Progress Schedule will be acceptable to Engineer if it provides an orderly progression of the Work to completion within the Contract Times. Such acceptance will not impose on Engineer responsibility for the Progress Schedule, for sequencing, scheduling, or progress of the Work nor interfere with or relieve Contractor from Contractor's full responsibility therefor.

2. Contractor's Schedule of Submittals will be acceptable to Engineer if it provides a workable arrangement for reviewing and processing the required submittals.

3. Contractor's Schedule of Values will be acceptable to Engineer as to form and substance if it provides a reasonable allocation of the Contract Price to component parts of the Work.

ARTICLE 3 - CONTRACT DOCUMENTS: INTENT, AMENDING, REUSE

3.01   Intent

A. The Contract Documents are complementary; what is required by one is as binding as if required by all.

B. It is the intent of the Contract Documents to describe a functionally complete Project (or part thereof) to be constructed in accordance with the Contract Documents. Any labor, documentation, services, materials, or equipment that may reasonably be inferred from the Contract Documents or from prevailing custom or trade usage as being required to produce the intended result will be provided whether or not specifically called for at no additional cost to Owner.

C. Clarifications and interpretations of the Contract Documents shall be issued by Engineer as provided in Article 9.

3.02   Reference Standards

A. Standards, Specifications, Codes, Laws, and Regulations

1. Reference to standards, specifications, manuals, or codes of any technical society, organization, or association, or to Laws or Regulations, whether such reference be specific or by implication, shall mean the standard, specification, manual, code, or Laws or Regulations in effect at the time of opening of Bids (or on the Effective Date of the Agreement if there were no Bids), except as may be otherwise specifically stated in the Contract Documents.

2. No provision of any such standard, specification, manual or code, or any instruction of a Supplier shall be effective to change the duties or responsibilities of Owner, Contractor, or Engineer, or any of their subcontractors, consultants, agents, or employees from those set forth in the Contract Documents. No such provision or instruction shall be effective to assign to Owner, or Engineer, or any of, their Related Entities, any duty or authority to supervise or direct the performance of the Work or any duty or authority to undertake responsibility inconsistent with the provisions of the Contract Documents.

3.03   Reporting and Resolving Discrepancies

A. Reporting Discrepancies

1. Contractor's Review of Contract Documents Before Starting Work: Before undertaking each part of the Work, Contractor shall carefully study and compare the Contract Documents and check and verify pertinent figures therein and all applicable field measurements. Contractor shall promptly report in writing to Engineer any conflict, error, ambiguity, or discrepancy which Contractor may discover and shall obtain a written interpretation or clarification from Engineer before proceeding with any Work affected thereby.

2. Contractor's Review of Contract Documents During Performance of Work: If, during the performance of the Work, Contractor discovers any conflict, error, ambiguity, or discrepancy within the Contract Documents or between the Contract Documents and any provision of any Law or Regulation applicable to the performance of the Work or of any standard, specification, manual or code, or of any instruction of any Supplier, Contractor shall promptly report it to Engineer in writing. Contractor shall not proceed with the Work affected thereby (except in an emergency as required by Paragraph 6.16.A) until an amendment or supplement to the Contract Documents has been issued by one of the methods indicated in Paragraph 3.04.

3. Contractor shall not be liable to Owner or Engineer for failure to report any conflict, error, ambiguity, or discrepancy in the Contract Documents unless Contractor knew or reasonably should have known thereof.

B. Resolving Discrepancies

1. Except as may be otherwise specifically stated in the Contract Documents, the provisions of the Contract Documents shall take precedence in resolving any conflict, error, ambiguity, or discrepancy between the provisions of the Contract Documents and:

a. the provisions of any standard, specification, manual, code, or instruction (whether or not specifically incorporated by reference in the Contract Documents); or

b. the provisions of any Laws or Regulations applicable to the performance of the Work (unless such an interpretation of the provisions of the Contract Documents would result in violation of such Law or Regulation).

3.04 *Amending and Supplementing Contract Documents*

A. The Contract Documents may be amended to provide for additions, deletions, and revisions in the Work or to modify the terms and conditions thereof by either a Change Order or a Work Change Directive.

B. The requirements of the Contract Documents may be supplemented, and minor variations and deviations in the Work may be authorized, by one or more of the following ways:

1. A Field Order;

2. Engineer's approval of a Shop Drawing or Sample; (Subject to the provisions of Paragraph 6.17.D.3); or

3. Engineer's written interpretation or clarification.

3.05 *Reuse of Documents*

A. Contractor and any Subcontractor or Supplier or other individual or entity performing or furnishing all of the Work under a direct or indirect contract with Contractor, shall not:

1. have or acquire any title to or ownership rights in any of the Drawings, Specifications, or other documents (or copies of any thereof) prepared by or bearing the seal of Engineer or Engineer's consultants, including electronic media editions; or

2. reuse any of such Drawings, Specifications, other documents, or copies thereof on extensions of the Project or any other project without written consent of Owner and Engineer and specific written verification or adaption by Engineer.

B. The prohibition of this Paragraph 3.05 will survive final payment, or termination of the Contract. Nothing herein shall preclude Contractor from retaining copies of the Contract Documents for record purposes.

3.06 *Electronic Data*

A. Copies of data furnished by Owner or Engineer to Contractor or Contractor to Owner or Engineer that may be relied upon are limited to the printed copies (also known as hard copies). Files in electronic media format of text, data, graphics, or other types are furnished only for the convenience of the receiving party. Any conclusion or information obtained or derived from such electronic files will be at the user's sole risk. If there is a discrepancy between the electronic files and the hard copies, the hard copies govern.

B. Because data stored in electronic media format can deteriorate or be modified inadvertently or otherwise without authorization of the data's creator, the party receiving electronic files agrees that it will perform acceptance tests or procedures within 60 days, after which the receiving party shall be deemed to have accepted the data thus transferred. Any errors detected within the 60-day acceptance period will be corrected by the transferring party..

C. When transferring documents in electronic media format, the transferring party makes no representations as to long term compatibility, usability, or readability of documents resulting from the use of software application packages, operating systems, or computer hardware differing from those used by the data's creator.

ARTICLE 4 - AVAILABILITY OF LANDS; SUBSURFACE AND PHYSICAL CONDITIONS; HAZARDOUS ENVIRONMENTAL CONDITIONS; REFERENCE POINTS

4.01 *Availability of Lands*

A. Owner shall furnish the Site. Owner shall notify Contractor of any encumbrances or restrictions not of general application but specifically related to use of the Site with which Contractor must comply in performing the Work. Owner will obtain in a timely manner and pay for easements for permanent structures or permanent changes in existing facilities. If Contractor and Owner are unable to agree on entitlement to or on the amount or extent, if any, of any adjustment in the Contract Price or Contract Times, or both, as a result of any delay in Owner's furnishing the Site or a part thereof, Contractor may make a Claim therefor as provided in Paragraph 10.05.

B. Upon reasonable written request, Owner shall furnish Contractor with a current statement of record legal title and legal description of the lands upon which the Work is to be performed and Owner's interest therein as necessary for giving notice of or filing a mechanic's or construction lien against such lands in accordance with applicable Laws and Regulations.

C. Contractor shall provide for all additional lands and access thereto that may be required for temporary construction facilities or storage of materials and equipment.

4.02   *Subsurface and Physical Conditions*

A. *Reports and Drawings:* The Supplementary Conditions identify:

1. those reports of explorations and tests of subsurface conditions at or contiguous to the Site that Engineer has used in preparing the Contract Documents; and

2. those drawings of physical conditions in or relating to existing surface or subsurface structures at or contiguous to the Site (except Underground Facilities) that Engineer has used in preparing the Contract Documents.

B. *Limited Reliance by Contractor on Technical Data Authorized:* Contractor may rely upon the general accuracy of the "technical data" contained in such reports and drawings, but such reports and drawings are not Contract Documents. Such "technical data" is identified in the Supplementary Conditions. Except for such reliance on such "technical data," Contractor may not rely upon or make any claim against Owner or Engineer, or any of their Related Entities with respect to:

1. the completeness of such reports and drawings for Contractor's purposes, including, but not limited to, any aspects of the means, methods, techniques, sequences, and procedures of construction to be employed by Contractor, and safety precautions and programs incident thereto; or

2. other data, interpretations, opinions, and information contained in such reports or shown or indicated in such drawings; or

3. any Contractor interpretation of or conclusion drawn from any "technical data" or any such other data, interpretations, opinions, or information.

4.03   *Differing Subsurface or Physical Conditions*

A. *Notice:* If Contractor believes that any subsurface or physical condition at or contiguous to the Site that is uncovered or revealed either:

1. is of such a nature as to establish that any "technical data" on which Contractor is entitled to rely as provided in Paragraph 4.02 is materially inaccurate; or

2. is of such a nature as to require a change in the Contract Documents; or

3. differs materially from that shown or indicated in the Contract Documents; or

4. is of an unusual nature, and differs materially from conditions ordinarily encountered and generally recognized as inherent in work of the character provided for in the Contract Documents;

then Contractor shall, promptly after becoming aware thereof and before further disturbing the subsurface or physical conditions or performing any Work in connection therewith (except in an emergency as required by Paragraph 6.16.A), notify Owner and Engineer in writing about such condition. Contractor shall not further disturb such condition or perform any Work in connection therewith (except as aforesaid) until receipt of written order to do so.

B. *Engineer's Review:* After receipt of written notice as required by Paragraph 4.03.A, Engineer will promptly review the pertinent condition, determine the necessity of Owner's obtaining additional exploration or tests with respect thereto, and advise Owner in writing (with a copy to Contractor) of Engineer's findings and conclusions.

C. Possible Price and Times Adjustments

1. The Contract Price or the Contract Times, or both, will be equitably adjusted to the extent that the existence of such differing subsurface or physical condition causes an increase or decrease in Contractor's cost of, or time required for, performance of the Work; subject, however, to the following:

a. such condition must meet any one or more of the categories described in Paragraph 4.03.A; and

b. with respect to Work that is paid for on a Unit Price Basis, any adjustment in Contract Price will be subject to the provisions of Paragraphs 9.07 and 11.03.

2. Contractor shall not be entitled to any adjustment in the Contract Price or Contract Times if:

a. Contractor knew of the existence of such conditions at the time Contractor made a final commitment to Owner with respect to Contract Price and Contract Times by the submission of a Bid or becoming bound under a negotiated contract; or

b. the existence of such condition could reasonably have been discovered or revealed as a result of any examination, investigation, exploration, test, or study of the Site and contiguous areas required by the Bidding Requirements or Contract Documents to be conducted by or for Contractor prior to Contractor's making such final commitment; or

c. Contractor failed to give the written notice as required by Paragraph 4.03.A.

3. If Owner and Contractor are unable to agree on entitlement to or on the amount or extent, if any, of any adjustment in the Contract Price or Contract Times, or both, a Claim may be made therefor as provided in Paragraph 10.05. However, Owner and Engineer, and any of their Related Entities shall not be liable to Contractor for any claims, costs, losses, or damages (including but not limited to all fees and charges of engineers, architects, attorneys, and other professionals and all court or arbitration or other dispute resolution costs) sustained by Contractor on or in connection with any other project or anticipated project.

4.04   *Underground Facilities*

A. *Shown or Indicated:* The information and data shown or indicated in the Contract Documents with respect to existing Underground Facilities at or contiguous to the Site is based on information and data furnished to Owner or Engineer by the owners of such Underground Facilities, including Owner, or by others. Unless it is otherwise expressly provided in the Supplementary Conditions:

1. Owner and Engineer shall not be responsible for the accuracy or completeness of any such information or data; and

2. the cost of all of the following will be included in the Contract Price, and Contractor shall have full responsibility for:

a. reviewing and checking all such information and data,

b. locating all Underground Facilities shown or indicated in the Contract Documents,

c. coordination of the Work with the owners of such Underground Facilities, including Owner, during construction, and

d. the safety and protection of all such Underground Facilities and repairing any damage thereto resulting from the Work.

B. *Not Shown or Indicated*

1. If an Underground Facility is uncovered or revealed at or contiguous to the Site which was not shown or indicated, or not shown or indicated with reasonable accuracy in the Contract Documents, Contractor shall, promptly after becoming aware thereof and before further disturbing conditions affected thereby or performing any Work in connection therewith (except in an emergency as required by Paragraph 6.16.A), identify the owner of such Underground Facility and give written notice to that owner and to Owner and Engineer. Engineer will

promptly review the Underground Facility and determine the extent, if any, to which a change is required in the Contract Documents to reflect and document the consequences of the existence or location of the Underground Facility. During such time, Contractor shall be responsible for the safety and protection of such Underground Facility.

2. If Engineer concludes that a change in the Contract Documents is required, a Work Change Directive or a Change Order will be issued to reflect and document such consequences. An equitable adjustment shall be made in the Contract Price or Contract Times, or both, to the extent that they are attributable to the existence or location of any Underground Facility that was not shown or indicated or not shown or indicated with reasonable accuracy in the Contract Documents and that Contractor did not know of and could not reasonably have been expected to be aware of or to have anticipated. If Owner and Contractor are unable to agree on entitlement to or on the amount or extent, if any, of any such adjustment in Contract Price or Contract Times, Owner or Contractor may make a Claim therefor as provided in Paragraph 10.05.

4.05   *Reference Points*

A. Owner shall provide engineering surveys to establish reference points for construction which in Engineer's judgment are necessary to enable Contractor to proceed with the Work. Contractor shall be responsible for laying out the Work, shall protect and preserve the established reference points and property monuments, and shall make no changes or relocations without the prior written approval of Owner. Contractor shall report to Engineer whenever any reference point or property monument is lost or destroyed or requires relocation because of necessary changes in grades or locations, and shall be responsible for the accurate replacement or relocation of such reference points or property monuments by professionally qualified personnel.

4.06   *Hazardous Environmental Condition at Site*

A. *Reports and Drawings:* Reference is made to the Supplementary Conditions for the identification of those reports and drawings relating to a Hazardous Environmental Condition identified at the Site, if any, that have been utilized by the Engineer in the preparation of the Contract Documents.

B. *Limited Reliance by Contractor on Technical Data Authorized:* Contractor may rely upon the general accuracy of the "technical data" contained in such reports and drawings, but such reports and drawings are not Contract Documents. Such "technical data" is identified in the Supplementary Conditions. Except for such reliance on such "technical data," Contractor may not rely upon or make any claim against Owner or Engineer, or any of their Related Entities with respect to:

1. the completeness of such reports and drawings for Contractor's purposes, including, but not limited to, any aspects of the means, methods, techniques, sequences and procedures of construction to be employed by Contractor and safety precautions and programs incident thereto; or

2. other data, interpretations, opinions and information contained in such reports or shown or indicated in such drawings; or

3. any Contractor interpretation of or conclusion drawn from any "technical data" or any such other data, interpretations, opinions or information.

C. Contractor shall not be responsible for any Hazardous Environmental Condition uncovered or revealed at the Site which was not shown or indicated in Drawings or Specifications or identified in the Contract Documents to be within the scope of the Work. Contractor shall be responsible for a Hazardous Environmental Condition created with any materials brought to the Site by Contractor, Subcontractors, Suppliers, or anyone else for whom Contractor is responsible.

D. If Contractor encounters a Hazardous Environmental Condition or if Contractor or anyone for whom Contractor is responsible creates a Hazardous Environmental Condition, Contractor shall immediately: (i) secure or otherwise isolate such condition; (ii) stop all Work in connection with such condition and in any area affected thereby (except in an emergency as required by Paragraph 6.16.A); and (iii) notify Owner and Engineer (and promptly thereafter confirm such notice in writing). Owner shall promptly consult with Engineer concerning the necessity for Owner to retain a qualified expert to evaluate such condition or take corrective action, if any.

E. Contractor shall not be required to resume Work in connection with such condition or in any affected area until after Owner has obtained any required permits related thereto and delivered to Contractor written notice: (i) specifying that such condition and any affected area is or has been rendered safe for the resumption of Work; or (ii) specifying any special conditions under which such Work may be resumed safely. If Owner and Contractor cannot agree as to entitlement to or on the amount or extent, if any, of any adjustment in Contract Price or Contract Times, or both, as a result of such Work stoppage or such special conditions under which Work is agreed to be resumed by Contractor, either party may make a Claim therefor as provided in Paragraph 10.05.

F. If after receipt of such written notice Contractor does not agree to resume such Work based on a reasonable belief it is unsafe, or does not agree to resume such Work under such special conditions, then Owner may order the portion of the Work that is in the area affected by such condition to be deleted from the Work. If Owner and Contractor cannot agree as to entitlement to or on the amount or extent, if any, of an adjustment in Contract Price or Contract Times as a result of deleting such portion of the Work, then either party may make a Claim therefor as provided in Paragraph 10.05. Owner may have such deleted portion of the Work performed by Owner's own forces or others in accordance with Article 7.

G. To the fullest extent permitted by Laws and Regulations, Owner shall indemnify and hold harmless Contractor, Subcontractors, and Engineer, and the officers, directors, partners, employees, agents, consultants, and subcontractors of each and any of them from and against all claims, costs, losses, and damages (including but not limited to all fees and charges of engineers, architects, attorneys, and other professionals and all court or arbitration or other dispute resolution costs) arising out of or relating to a Hazardous Environmental Condition, provided that such Hazardous Environmental Condition: (i) was not shown or indicated in the Drawings or Specifications or identified in the Contract Documents to be included within the scope of the Work, and (ii) was not created by Contractor or by anyone for whom Contractor is responsible. Nothing in this Paragraph 4.06. G shall obligate Owner to indemnify any individual or entity from and against the consequences of that individual's or entity's own negligence.

H. To the fullest extent permitted by Laws and Regulations, Contractor shall indemnify and hold harmless Owner and Engineer, and the officers, directors, partners, employees, agents, consultants, and subcontractors of each and any of them from and against all claims, costs, losses, and damages (including but not limited to all fees and charges of engineers, architects, attorneys, and other professionals and all court or arbitration or other dispute resolution costs) arising out of or relating to a Hazardous Environmental Condition created by Contractor or by anyone for whom Contractor is responsible. Nothing in this Paragraph 4.06.H shall obligate Contractor to indemnify any individual or entity from and against the consequences of that individual's or entity's own negligence.

I. The provisions of Paragraphs 4.02, 4.03, and 4.04 do not apply to a Hazardous Environmental Condition uncovered or revealed at the Site.

ARTICLE 5 - BONDS AND INSURANCE

5.01 *Performance, Payment, and Other Bonds*

A. Contractor shall furnish performance and payment bonds, each in an amount at least equal to the Contract Price as security for the faithful performance and payment of all of Contractor's obligations under the Contract Documents. These bonds shall remain in effect until one year after the date when final payment becomes due or until completion of the correction period specified in Paragraph 13.07, whichever is later, except as provided otherwise by Laws or Regulations or by the Contract Documents. Contractor shall also furnish such other bonds as are required by the Contract Documents.

B. All bonds shall be in the form prescribed by the Contract Documents except as provided otherwise by Laws or Regulations, and shall be executed by such sureties as are named in the current list of "Companies Holding Certificates of Authority as Acceptable Sureties on Federal Bonds and as Acceptable Reinsuring Companies" as published in Circular 570 (amended) by the Financial Management Service, Surety Bond Branch, U.S. Department of the Treasury. All bonds signed by an agent must be accompanied by a certified copy of the agent's authority to act.

C. If the surety on any bond furnished by Contractor is declared bankrupt or becomes insolvent or its right to do business is terminated in any state where any part of the Project is located or it ceases to meet the requirements of Paragraph 5.01.B, Contractor shall promptly notify Owner and Engineer and shall, within 20 days after the event giving rise to such notification, provide another bond and surety, both of which shall comply with the requirements of Paragraphs 5.01.B and 5.02.

5.02   *Licensed Sureties and Insurers*

A. All bonds and insurance required by the Contract Documents to be purchased and maintained by Owner or Contractor shall be obtained from surety or insurance companies that are duly licensed or authorized in the jurisdiction in which the Project is located to issue bonds or insurance policies for the limits and coverages so required. Such surety and insurance companies shall also meet such additional requirements and qualifications as may be provided in the Supplementary Conditions.

5.03   *Certificates of Insurance*

A. Contractor shall deliver to Owner, with copies to each additional insured identified in the Supplementary Conditions, certificates of insurance (and other evidence of insurance requested by Owner or any other additional insured) which Contractor is required to purchase and maintain.

B. Owner shall deliver to Contractor, with copies to each additional insured identified in the Supplementary Conditions, certificates of insurance (and other evidence of insurance requested by Contractor or any other additional insured) which Owner is required to purchase and maintain.

5.04   *Contractor's Liability Insurance*

A. Contractor shall purchase and maintain such liability and other insurance as is appropriate for the Work being performed and as will provide protection from claims set forth below which may arise out of or result from Contractor's performance of the Work and Contractor's other obligations under the Contract Documents, whether it is to be performed by Contractor, any Subcontractor or Supplier, or by anyone directly or indirectly employed by any of them to perform any of the Work, or by anyone for whose acts any of them may be liable:

1. claims under workers' compensation, disability benefits, and other similar employee benefit acts;

2. claims for damages because of bodily injury, occupational sickness or disease, or death of Contractor's employees;

3. claims for damages because of bodily injury, sickness or disease, or death of any person other than Contractor's employees;

4. claims for damages insured by reasonably available personal injury liability coverage which are sustained:

a. by any person as a result of an offense directly or indirectly related to the employment of such person by Contractor, or

b. by any other person for any other reason;

5. claims for damages, other than to the Work itself, because of injury to or destruction of tangible property wherever located, including loss of use resulting therefrom; and

6. claims for damages because of bodily injury or death of any person or property damage arising out of the ownership, maintenance or use of any motor vehicle.

B. The policies of insurance required by this Paragraph 5.04 shall:

1. with respect to insurance required by Paragraphs 5.04.A.3 through 5.04.A.6 inclusive, include as additional insured (subject to any customary exclusion regarding professional liability) Owner and Engineer, and any other individuals or entities identified in the Supplementary Conditions, all of whom shall be listed as additional insureds, and include coverage for the respective officers, directors, partners, employees, agents, consultants and subcontractors of each and any of all such additional insureds, and the insurance afforded to these additional insureds shall provide primary coverage for all claims covered thereby;

2. include at least the specific coverages and be written for not less than the limits of liability provided in the Supplementary Conditions or required by Laws or Regulations, whichever is greater;

3. include completed operations insurance;

4. include contractual liability insurance covering Contractor's indemnity obligations under Paragraphs 6.11 and 6.20;

5. contain a provision or endorsement that the coverage afforded will not be canceled, materially changed or renewal refused until at least 30 days prior written notice has been given to Owner and Contractor and to each other additional insured identified in the Supplementary Conditions to whom a certificate of insurance has been issued (and the certificates of insurance furnished by the Contractor pursuant to Paragraph 5.03 will so provide);

6. remain in effect at least until final payment and at all times thereafter when Contractor may be correcting, removing, or replacing defective Work in accordance with Paragraph 13.07; and

7. with respect to completed operations insurance, and any insurance coverage written on a claims-made basis, remain in effect for at least two years after final payment.

a. Contractor shall furnish Owner and each other additional insured identified in the Supplementary Conditions, to whom a certificate of insurance has been issued, evidence satisfactory to Owner and any such additional insured of continuation of such insurance at final payment and one year thereafter.

5.05   *Owner's Liability Insurance*

A. In addition to the insurance required to be provided by Contractor under Paragraph 5.04, Owner, at Owner's option, may purchase and maintain at Owner's expense Owner's own liability insurance as will protect Owner against claims which may arise from operations under the Contract Documents.

5.06   *Property Insurance*

A. Unless otherwise provided in the Supplementary Conditions, Owner shall purchase and maintain property insurance upon the Work at the Site in the amount of the full replacement cost thereof (subject to such deductible amounts as may be provided in the Supplementary Conditions or required by Laws and Regulations). This insurance shall:

1. include the interests of Owner, Contractor, Subcontractors, and Engineer, and any other individuals or entities identified in the Supplementary Conditions, and the officers, directors, partners, employees, agents, consultants and subcontractors of each and any of them, each of whom is deemed to have an insurable interest and shall be listed as an insured or additional insured;

2. be written on a Builder's Risk "all-risk" or open peril or special causes of loss policy form that shall at least include insurance for physical loss or damage to the Work, temporary buildings, false work, and materials and equipment in transit, and shall insure against at least the following perils or causes of loss: fire, lightning, extended coverage, theft, vandalism and malicious mischief, earthquake, collapse, debris removal, demolition occasioned by enforcement of Laws and Regulations, water damage (other than caused by flood) and such other perils or causes of loss as may be specifically required by the Supplementary Conditions;

3. include expenses incurred in the repair or replacement of any insured property (including but not limited to fees and charges of engineers and architects);

4. cover materials and equipment stored at the Site or at another location that was agreed to in writing by Owner prior to being incorporated in the Work, provided that such materials and equipment have been included in an Application for Payment recommended by Engineer;

5. allow for partial utilization of the Work by Owner;

6. include testing and startup; and

7. be maintained in effect until final payment is made unless otherwise agreed to in writing by Owner, Contractor, and Engineer with 30 days written notice to each other additional insured to whom a certificate of insurance has been issued.

B. Owner shall purchase and maintain such boiler and machinery insurance or additional property insurance as may be required by the Supplementary Conditions or Laws and Regulations which will include the interests of Owner, Contractor, Subcontractors, and Engineer, and any other individuals or entities identified in the Supplementary Conditions, and the officers, directors, partners, employees, agents, consultants and subcontractors of each and any of them, each of whom is deemed to have an insurable interest and shall be listed as an insured or additional insured.

C. All the policies of insurance (and the certificates or other evidence thereof) required to be purchased and maintained in accordance with Paragraph 5.06 will contain a provision or endorsement that the coverage afforded will not be canceled or materially changed or renewal refused until at least 30 days prior written notice has been given to Owner and Contractor and to each other additional insured to whom a certificate of insurance has been issued and will contain waiver provisions in accordance with Paragraph 5.07.

D. Owner shall not be responsible for purchasing and maintaining any property insurance specified in this Paragraph 5.06 to protect the interests of Contractor, Subcontractors, or others in the Work to the extent of any deductible amounts that are identified in the Supplementary Conditions. The risk of loss within such identified deductible amount will be borne by Contractor, Subcontractors, or others suffering any such loss, and if any of them wishes property insurance coverage within the limits of such amounts, each may purchase and maintain it at the purchaser's own expense.

E. If Contractor requests in writing that other special insurance be included in the property insurance policies provided under Paragraph 5.06, Owner shall, if possible, include such insurance, and the cost thereof will be charged to Contractor by appropriate Change Order. Prior to commencement of the Work at the Site, Owner shall in writing advise Contractor whether or not such other insurance has been procured by Owner.

5.07   *Waiver of Rights*

A. Owner and Contractor intend that all policies purchased in accordance with Paragraph 5.06 will protect Owner, Contractor, Subcontractors, and Engineer, and all other individuals or entities identified in the Supplementary Conditions to be listed as insureds or additional insureds (and the officers, directors, partners, employees, agents, consultants and subcontractors of each and any of them) in such policies and will provide primary coverage for all losses and damages caused by the perils or causes of loss covered thereby. All such policies shall contain provisions to the effect that in the event of payment of any loss or damage the insurers will have no rights of recovery against any of the insureds or additional insureds thereunder. Owner and Contractor waive all rights against each other and their respective officers, directors, partners, employees, agents, consultants and subcontractors of each and any of them for all losses and damages caused by, arising out of or resulting from any of the perils or causes of loss covered by such policies and any other property insurance applicable to the Work; and, in addition, waive all such rights against Subcontractors, and Engineer, and all other individuals or entities identified in the Supplementary Conditions to be listed as insured or additional insured (and the officers, directors, partners, employees, agents, consultants and subcontractors of each and any of them) under such policies for losses and damages so caused. None of the above waivers shall extend to the rights that any party making such waiver may have to the proceeds of insurance held by Owner as trustee or otherwise payable under any policy so issued.

B. Owner waives all rights against Contractor, Subcontractors, and Engineer, and the officers, directors, partners, employees, agents, consultants and subcontractors of each and any of them for:

1. loss due to business interruption, loss of use, or other consequential loss extending beyond direct physical loss or damage to Owner's property or the Work caused by, arising out of, or resulting from fire or other perils whether or not insured by Owner; and

2. loss or damage to the completed Project or part thereof caused by, arising out of, or resulting from fire or other insured peril or cause of loss covered by any property insurance maintained on the completed Project or part thereof by Owner during partial utilization pursuant to Paragraph 14.05, after Substantial Completion pursuant to Paragraph 14.04, or after final payment pursuant to Paragraph 14.07.

C. Any insurance policy maintained by Owner covering any loss, damage or consequential loss referred to in Paragraph 5.07.B shall contain provisions to the effect that in the event of payment of any such loss, damage, or consequential loss, the insurers will have no rights of recovery against Contractor, Subcontractors, or Engineer, and the officers, directors, partners, employees, agents, consultants and subcontractors of each and any of them.

5.08   *Receipt and Application of Insurance Proceeds*

A. Any insured loss under the policies of insurance required by Paragraph 5.06 will be adjusted with Owner and made payable to Owner as fiduciary for the insureds, as their interests may appear, subject to the requirements of any applicable mortgage clause and of Paragraph 5.08.B. Owner shall deposit in a separate account any money so received and shall distribute it in accordance with such agreement as the parties in interest may reach. If no other special agreement is reached, the damaged Work shall be repaired or replaced, the moneys so received applied on account thereof, and the Work and the cost thereof covered by an appropriate Change Order .

B. Owner as fiduciary shall have power to adjust and settle any loss with the insurers unless one of the parties in interest shall object in writing within 15 days after the occurrence of loss to Owner's exercise of this power. If such objection be made, Owner as fiduciary shall make settlement with the insurers in accordance with such agreement as the parties in interest may reach. If no such agreement among the parties in interest is reached, Owner as fiduciary shall adjust and settle the loss with the insurers and, if required in writing by any party in interest, Owner as fiduciary shall give bond for the proper performance of such duties.

5.09   *Acceptance of Bonds and Insurance; Option to Replace*

A. If either Owner or Contractor has any objection to the coverage afforded by or other provisions of the bonds or insurance required to be purchased and maintained by the other party in accordance with Article 5 on the basis of non-conformance with the Contract

Documents, the objecting party shall so notify the other party in writing within 10 days after receipt of the certificates (or other evidence requested) required by Paragraph 2.01.B. Owner and Contractor shall each provide to the other such additional information in respect of insurance provided as the other may reasonably request. If either party does not purchase or maintain all of the bonds and insurance required of such party by the Contract Documents, such party shall notify the other party in writing of such failure to purchase prior to the start of the Work, or of such failure to maintain prior to any change in the required coverage. Without prejudice to any other right or remedy, the other party may elect to obtain equivalent bonds or insurance to protect such other party's interests at the expense of the party who was required to provide such coverage, and a Change Order shall be issued to adjust the Contract Price accordingly.

5.10   *Partial Utilization. Acknowledgment of Property Insurer*

A. If Owner finds it necessary to occupy or use a portion or portions of the Work prior to Substantial Completion of all the Work as provided in Paragraph 14.05, no such use or occupancy shall commence before the insurers providing the property insurance pursuant to Paragraph 5.06 have acknowledged notice thereof and in writing effected any changes in coverage necessitated thereby. The insurers providing the property insurance shall consent by endorsement on the policy or policies, but the property insurance shall not be canceled or permitted to lapse on account of any such partial use or occupancy.

ARTICLE 6 - CONTRACTOR'S RESPONSIBILITIES

6.01   *Supervision and Superintendence*

A. Contractor shall supervise, inspect, and direct the Work competently and efficiently, devoting such attention thereto and applying such skills and expertise as may be necessary to perform the Work in accordance with the Contract Documents. Contractor shall be solely responsible for the means, methods, techniques, sequences, and procedures of construction. Contractor shall not be responsible for the negligence of Owner or Engineer in the design or specification of a specific means, method, technique, sequence, or procedure of construction which is shown or indicated in and expressly required by the Contract Documents.

B. At all times during the progress of the Work, Contractor shall assign a competent resident superintendent who shall not be replaced without written notice to Owner and Engineer except under extraordinary circumstances. The superintendent will be Contractor's representative at the Site and shall have authority to act on behalf of Contractor. All communications given to or

received from the superintendent shall be binding on Contractor.

6.02   *Labor: Working Hours*

A. Contractor shall provide competent, suitably qualified personnel to survey and lay out the Work and perform construction as required by the Contract Documents. Contractor shall at all times maintain good discipline and order at the Site.

B. Except as otherwise required for the safety or protection of persons or the Work or property at the Site or adjacent thereto, and except as otherwise stated in the Contract Documents, all Work at the Site shall be performed during regular working hours. Contractor will not permit the performance of Work on a Saturday, Sunday, or any legal holiday without Owner's written consent (which will not be unreasonably withheld) given after prior written notice to Engineer.

6.03   *Services, Materials, and Equipment*

A. Unless otherwise specified in the Contract Documents, Contractor shall provide and assume full responsibility for all services, materials, equipment, labor, transportation, construction equipment and machinery, tools, appliances, fuel, power, light, heat, telephone, water, sanitary facilities, temporary facilities, and all other facilities and incidentals necessary for the performance, testing, start-up, and completion of the Work.

B. All materials and equipment incorporated into the Work shall be as specified or, if not specified, shall be of good quality and new, except as otherwise provided in the Contract Documents. All special warranties and guarantees required by the Specifications shall expressly run to the benefit of Owner. If required by Engineer, Contractor shall furnish satisfactory evidence (including reports of required tests) as to the source, kind, and quality of materials and equipment.

C. All materials and equipment shall be stored, applied, installed, connected, erected, protected, used, cleaned, and conditioned in accordance with instructions of the applicable Supplier, except as otherwise may be provided in the Contract Documents.

6.04   *Progress Schedule*

A. Contractor shall adhere to the Progress Schedule established in accordance with Paragraph 2.07 as it may be adjusted from time to time as provided below.

1. Contractor shall submit to Engineer for acceptance (to the extent indicated in Paragraph 2.07) proposed adjustments in the Progress Schedule that will not result in changing the Contract Times. Such adjustments will comply with any provisions of the General Requirements applicable thereto.

2. Proposed adjustments in the Progress Schedule that will change the Contract Times shall be submitted in accordance with the requirements of Article 12. Adjustments in Contract Times may only be made by a Change Order.

6.05   *Substitutes and "Or-Equals"*

A. Whenever an item of material or equipment is specified or described in the Contract Documents by using the name of a proprietary item or the name of a particular Supplier, the specification or description is intended to establish the type, function, appearance, and quality required. Unless the specification or description contains or is followed by words reading that no like, equivalent, or "or-equal" item or no substitution is permitted, other items of material or equipment or material or equipment of other Suppliers may be submitted to Engineer for review under the circumstances described below.

1. *"Or-Equal" Items:* If in Engineer's sole discretion an item of material or equipment proposed by Contractor is functionally equal to that named and sufficiently similar so that no change in related Work will be required, it may be considered by Engineer as an "or-equal" item, in which case review and approval of the proposed item may, in Engineer's sole discretion, be accomplished without compliance with some or all of the requirements for approval of proposed substitute items. For the purposes of this Paragraph 6.05.A.1, a proposed item of material or equipment will be considered functionally equal to an item so named if:

a. in the exercise of reasonable judgment Engineer determines that:

1) it is at least equal in materials of construction, quality, durability, appearance, strength, and design characteristics;

2) it will reliably perform at least equally well the function and achieve the results imposed by the design concept of the completed Project as a functioning whole;

3) it has a proven record of performance and availability of responsive service; and

b. Contractor certifies that, if approved and incorporated into the Work:

1) there will be no increase in cost to the Owner or increase in Contract Times, and

2) it will conform substantially to the detailed requirements of the item named in the Contract Documents.

2. Substitute Items

a. If in Engineer's sole discretion an item of material or equipment proposed by Contractor does not qualify as an "or-equal" item under Paragraph 6.05.A.1, it will be considered a proposed substitute item.

b. Contractor shall submit sufficient information as provided below to allow Engineer to determine that the item of material or equipment proposed is essentially equivalent to that named and an acceptable substitute therefor. Requests for review of proposed substitute items of material or equipment will not be accepted by Engineer from anyone other than Contractor.

c. The requirements for review by Engineer will be as set forth in Paragraph 6.05.A.2.d, as supplemented in the General Requirements and as Engineer may decide is appropriate under the circumstances.

d. Contractor shall make written application to Engineer for review of a proposed substitute item of material or equipment that Contractor seeks to furnish or use. The application:

1) shall certify that the proposed substitute item will:

a) perform adequately the functions and achieve the results called for by the general design,

b) be similar in substance to that specified, and

c) be suited to the same use as that specified;

2) will state:

a) the extent, if any, to which the use of the proposed substitute item will prejudice Contractor's achievement of Substantial Completion on time;

b) whether or not use of the proposed substitute item in the Work will require a change in any of the Contract Documents (or in the provisions of any other direct contract with Owner for other work on the Project) to adapt the design to the proposed substitute item; and

c) whether or not incorporation or use of the proposed substitute item in connection with the Work is subject to payment of any license fee or royalty;

3) will identify:

a) all variations of the proposed substitute item from that specified , and

b) available engineering, sales, maintenance, repair, and replacement services;

4) and shall contain an itemized estimate of all costs or credits that will result directly or indirectly from use of such substitute item, including costs of redesign and claims of other contractors affected by any resulting change,

B. *Substitute Construction Methods or Procedures:* If a specific means, method, technique, sequence, or procedure of construction is expressly required by the Contract Documents, Contractor may furnish or utilize a substitute means, method, technique, sequence, or procedure of construction approved by Engineer. Contractor shall submit sufficient information to allow Engineer, in Engineer's sole discretion, to determine that the substitute proposed is equivalent to that expressly called for by the Contract Documents. The requirements for review by Engineer will be similar to those provided in Paragraph 6.05.A.2.

C. *Engineer's Evaluation:* Engineer will be allowed a reasonable time within which to evaluate each proposal or submittal made pursuant to Paragraphs 6.05.A and 6.05.B. Engineer may require Contractor to furnish additional data about the proposed substitute item. Engineer will be the sole judge of acceptability. No "or equal" or substitute will be ordered, installed or utilized until Engineer's review is complete, which will be evidenced by either a Change Order for a substitute or an approved Shop Drawing for an "or equal." Engineer will advise Contractor in writing of any negative determination.

D. *Special Guarantee:* Owner may require Contractor to furnish at Contractor's expense a special performance guarantee or other surety with respect to any substitute.

E. *Engineer's Cost Reimbursement:* Engineer will record Engineer's costs in evaluating a substitute proposed or submitted by Contractor pursuant to Paragraphs 6.05.A.2 and 6.05.B Whether or not Engineer approves a substitute item so proposed or submitted by Contractor, Contractor shall reimburse Owner for the charges of Engineer for evaluating each such proposed substitute. Contractor shall also reimburse Owner for the charges of Engineer for making changes in the Contract Documents (or in the provisions of any other direct contract with Owner) resulting from the acceptance of each proposed substitute.

F. *Contractor's Expense:* Contractor shall provide all data in support of any proposed substitute or "or-equal" at Contractor's expense.

6.06 *Concerning Subcontractors, Suppliers, and Others*

A. Contractor shall not employ any Subcontractor, Supplier, or other individual or entity (including those acceptable to Owner as indicated in Paragraph 6.06.B), whether initially or as a replacement, against whom Owner may have reasonable objection. Contractor shall not be required to employ any Subcontractor, Supplier, or other individual or entity to furnish or perform any of the Work against whom Contractor has reasonable objection.

B. If the Supplementary Conditions require the identity of certain Subcontractors, Suppliers, or other individuals or entities to be submitted to Owner in advance for acceptance by Owner by a specified date prior to the Effective Date of the Agreement, and if Contractor has submitted a list thereof in accordance with the Supplementary Conditions, Owner's acceptance (either in writing or by failing to make written objection thereto by the date indicated for acceptance or objection in the Bidding Documents or the Contract Documents) of any such Subcontractor, Supplier, or other individual or entity so identified may be revoked on the basis of reasonable objection after due investigation. Contractor shall submit an acceptable replacement for the rejected Subcontractor, Supplier, or other individual or entity, and the Contract Price will be adjusted by the difference in the cost occasioned by such replacement, and an appropriate Change Order will be issued . No acceptance by Owner of any such Subcontractor, Supplier, or other individual or entity, whether initially or as a replacement, shall constitute a waiver of any right of Owner or Engineer to reject defective Work.

C. Contractor shall be fully responsible to Owner and Engineer for all acts and omissions of the Subcontractors, Suppliers, and other individuals or entities performing or furnishing any of the Work just as Contractor is responsible for Contractor's own acts and omissions. Nothing in the Contract Documents:

1. shall create for the benefit of any such Subcontractor, Supplier, or other individual or entity any contractual relationship between Owner or Engineer and any such Subcontractor, Supplier or other individual or entity, nor

2. shall anything in the Contract Documents create any obligation on the part of Owner or Engineer to pay or to see to the payment of any moneys due any such Subcontractor, Supplier, or other individual

or entity except as may otherwise be required by Laws and Regulations.

D. Contractor shall be solely responsible for scheduling and coordinating the Work of Subcontractors, Suppliers, and other individuals or entities performing or furnishing any of the Work under a direct or indirect contract with Contractor.

E. Contractor shall require all Subcontractors, Suppliers, and such other individuals or entities performing or furnishing any of the Work to communicate with Engineer through Contractor.

F. The divisions and sections of the Specifications and the identifications of any Drawings shall not control Contractor in dividing the Work among Subcontractors or Suppliers or delineating the Work to be performed by any specific trade.

G. All Work performed for Contractor by a Subcontractor or Supplier will be pursuant to an appropriate agreement between Contractor and the Subcontractor or Supplier which specifically binds the Subcontractor or Supplier to the applicable terms and conditions of the Contract Documents for the benefit of Owner and Engineer. Whenever any such agreement is with a Subcontractor or Supplier who is listed as an additional insured on the property insurance provided in Paragraph 5.06, the agreement between the Contractor and the Subcontractor or Supplier will contain provisions whereby the Subcontractor or Supplier waives all rights against Owner, Contractor, and Engineer, and all other individuals or entities identified in the Supplementary Conditions to be listed as insureds or additional insureds (and the officers, directors, partners, employees, agents, consultants and subcontractors of each and any of them) for all losses and damages caused by, arising out of, relating to, or resulting from any of the perils or causes of loss covered by such policies and any other property insurance applicable to the Work. If the insurers on any such policies require separate waiver forms to be signed by any Subcontractor or Supplier, Contractor will obtain the same.

6.07   *Patent Fees and Royalties*

A. Contractor shall pay all license fees and royalties and assume all costs incident to the use in the performance of the Work or the incorporation in the Work of any invention, design, process, product, or device which is the subject of patent rights or copyrights held by others. If a particular invention, design, process, product, or device is specified in the Contract Documents for use in the performance of the Work and if to the actual knowledge of Owner or Engineer its use is subject to patent rights or copyrights calling for the payment of any license fee or royalty to others, the existence of such rights shall be disclosed by Owner in the Contract Documents.

B. To the fullest extent permitted by Laws and Regulations, Contractor shall indemnify and hold harmless Owner and Engineer, and the officers, directors, partners, employees, agents, consultants and subcontractors of each and any of them from and against all claims, costs, losses, and damages (including but not limited to all fees and charges of engineers, architects, attorneys, and other professionals and all court or arbitration or other dispute resolution costs) arising out of or relating to any infringement of patent rights or copyrights incident to the use in the performance of the Work or resulting from the incorporation in the Work of any invention, design, process, product, or device not specified in the Contract Documents.

6.08   *Permits*

A. Unless otherwise provided in the Supplementary Conditions, Contractor shall obtain and pay for all construction permits and licenses. Owner shall assist Contractor, when necessary, in obtaining such permits and licenses. Contractor shall pay all governmental charges and inspection fees necessary for the prosecution of the Work which are applicable at the time of opening of Bids, or, if there are no Bids, on the Effective Date of the Agreement. Owner shall pay all charges of utility owners for connections for providing permanent service to the Work.

6.09   *Laws and Regulations*

A. Contractor shall give all notices required by and shall comply with all Laws and Regulations applicable to the performance of the Work. Except where otherwise expressly required by applicable Laws and Regulations, neither Owner nor Engineer shall be responsible for monitoring Contractor's compliance with any Laws or Regulations.

B. If Contractor performs any Work knowing or having reason to know that it is contrary to Laws or Regulations, Contractor shall bear all claims, costs, losses, and damages (including but not limited to all fees and charges of engineers, architects, attorneys, and other professionals and all court or arbitration or other dispute resolution costs) arising out of or relating to such Work. However, it shall not be Contractor's primary responsibility to make certain that the Specifications and Drawings are in accordance with Laws and Regulations, but this shall not relieve Contractor of Contractor's obligations under Paragraph 3.03.

C. Changes in Laws or Regulations not known at the time of opening of Bids (or, on the Effective Date of the Agreement if there were no Bids) having an effect on the cost or time of performance of the Work shall be the subject of an adjustment in Contract Price or Contract Times. If Owner and Contractor are unable to agree on entitlement to or on the amount or extent, if any, of any such adjustment, a Claim may be made therefor as provided in Paragraph 10.05.

6.10 Taxes

A. Contractor shall pay all sales, consumer, use, and other similar taxes required to be paid by Contractor in accordance with the Laws and Regulations of the place of the Project which are applicable during the performance of the Work.

6.11 Use of Site and Other Areas

A. Limitation on Use of Site and Other Areas

1. Contractor shall confine construction equipment, the storage of materials and equipment, and the operations of workers to the Site and other areas permitted by Laws and Regulations, and shall not unreasonably encumber the Site and other areas with construction equipment or other materials or equipment. Contractor shall assume full responsibility for any damage to any such land or area, or to the owner or occupant thereof, or of any adjacent land or areas resulting from the performance of the Work.

2. Should any claim be made by any such owner or occupant because of the performance of the Work, Contractor shall promptly settle with such other party by negotiation or otherwise resolve the claim by arbitration or other dispute resolution proceeding or at law.

3. To the fullest extent permitted by Laws and Regulations, Contractor shall indemnify and hold harmless Owner and Engineer, and the officers, directors, partners, employees, agents, consultants and subcontractors of each and any of them from and against all claims, costs, losses, and damages (including but not limited to all fees and charges of engineers, architects, attorneys, and other professionals and all court or arbitration or other dispute resolution costs) arising out of or relating to any claim or action, legal or equitable, brought by any such owner or occupant against Owner, Engineer, or any other party indemnified hereunder to the extent caused by or based upon Contractor's performance of the Work.

B. *Removal of Debris During Performance of the Work.* During the progress of the Work Contractor shall keep the Site and other areas free from accumulations of waste materials, rubbish, and other debris. Removal and disposal of such waste materials, rubbish, and other debris shall conform to applicable Laws and Regulations.

C. *Cleaning:* Prior to Substantial Completion of the Work Contractor shall clean the Site and the Work and make it ready for utilization by Owner. At the completion of the Work Contractor shall remove from the Site all tools, appliances, construction equipment and machinery, and surplus materials and shall restore to original condition all property not designated for alteration by the Contract Documents.

D. *Loading Structures:* Contractor shall not load nor permit any part of any structure to be loaded in any manner that will endanger the structure, nor shall Contractor subject any part of the Work or adjacent property to stresses or pressures that will endanger it.

6.12 Record Documents

A. Contractor shall maintain in a safe place at the Site one record copy of all Drawings, Specifications, Addenda, Change Orders, Work Change Directives, Field Orders, and written interpretations and clarifications in good order and annotated to show changes made during construction. These record documents together with all approved Samples and a counterpart of all approved Shop Drawings will be available to Engineer for reference. Upon completion of the Work, these record documents, Samples, and Shop Drawings will be delivered to Engineer for Owner.

6.13 Safety and Protection

A. Contractor shall be solely responsible for initiating, maintaining and supervising all safety precautions and programs in connection with the Work. Contractor shall take all necessary precautions for the safety of, and shall provide the necessary protection to prevent damage, injury or loss to:

1. all persons on the Site or who may be affected by the Work;

2. all the Work and materials and equipment to be incorporated therein, whether in storage on or off the Site; and

3. other property at the Site or adjacent thereto, including trees, shrubs, lawns, walks, pavements, roadways, structures, utilities, and Underground Facilities not designated for removal, relocation, or replacement in the course of construction.

B. Contractor shall comply with all applicable Laws and Regulations relating to the safety of persons or property, or to the protection of persons or property from damage, injury, or loss; and shall erect and maintain all necessary safeguards for such safety and protection. Contractor shall notify owners of adjacent property and of Underground Facilities and other utility owners when prosecution of the Work may affect them, and shall cooperate with them in the protection, removal, relocation, and replacement of their property.

C. All damage, injury, or loss to any property referred to in Paragraph 6.13.A.2 or 6.13.A.3 caused, directly or indirectly, in whole or in part, by Contractor, any Subcontractor, Supplier, or any other individual or entity directly or indirectly employed by any of them to perform any of the Work, or anyone for whose acts any of them may be liable, shall be remedied by Contractor (except damage or loss attributable to the fault of Drawings or Specifications or to the acts or omissions of Owner or Engineer or , or anyone employed by any of them, or anyone for whose acts any of them may be liable, and not attributable, directly or indirectly, in whole or in part, to the fault or negligence of Contractor or any Subcontractor, Supplier, or other individual or entity directly or indirectly employed by any of them).

D. Contractor's duties and responsibilities for safety and for protection of the Work shall continue until such time as all the Work is completed and Engineer has issued a notice to Owner and Contractor in accordance with Paragraph 14.07.B that the Work is acceptable (except as otherwise expressly provided in connection with Substantial Completion).

6.14   *Safety Representative*

A. Contractor shall designate a qualified and experienced safety representative at the Site whose duties and responsibilities shall be the prevention of accidents and the maintaining and supervising of safety precautions and programs.

6.15   *Hazard Communication Programs*

A. Contractor shall be responsible for coordinating any exchange of material safety data sheets or other hazard communication information required to be made available to or exchanged between or among employers at the Site in accordance with Laws or Regulations.

6.16   *Emergencies*

A. In emergencies affecting the safety or protection of persons or the Work or property at the Site or adjacent thereto, Contractor is obligated to act to prevent threatened damage, injury, or loss. Contractor shall give Engineer prompt written notice if Contractor believes that any significant changes in the Work or variations from the Contract Documents have been caused thereby or are required as a result thereof. If Engineer determines that a change in the Contract Documents is required because of the action taken by Contractor in response to such an emergency, a Work Change Directive or Change Order will be issued.

6.17   *Shop Drawings and Samples*

A. Contractor shall submit Shop Drawings and Samples to Engineer for review and approval in accordance with the acceptable Schedule of Submittals (as required by Paragraph 2.07). Each submittal will be identified as Engineer may require.

1. Shop Drawings

a. Submit number of copies specified in the General Requirements.

b. Data shown on the Shop Drawings will be complete with respect to quantities, dimensions, specified performance and design criteria, materials, and similar data to show Engineer the services, materials, and equipment Contractor proposes to provide and to enable Engineer to review the information for the limited purposes required by Paragraph 6.17.D.

2. *Samples:* Contractor shall also submit Samples to Engineer for review and approval in accordance with the acceptable schedule of Shop Drawings and Sample submittals.

a. Submit number of Samples specified in the Specifications.

b. Clearly identify each Sample as to material, Supplier, pertinent data such as catalog numbers, the use for which intended and other data as Engineer may require to enable Engineer to review the submittal for the limited purposes required by Paragraph 6.17.D.

B. Where a Shop Drawing or Sample is required by the Contract Documents or the Schedule of Submittals , any related Work performed prior to Engineer's review and approval of the pertinent submittal will be at the sole expense and responsibility of Contractor.

C. Submittal Procedures

1. Before submitting each Shop Drawing or Sample, Contractor shall have determined and verified:

a. all field measurements, quantities, dimensions, specified performance and design criteria, installation requirements, materials, catalog numbers, and similar information with respect thereto;

b. the suitability of all materials with respect to intended use, fabrication, shipping, handling, storage, assembly, and installation pertaining to the performance of the Work;

c. all information relative to Contractor's responsibilities for means, methods, techniques, sequences, and procedures of construction, and safety precautions and programs incident thereto; and

d. shall also have reviewed and coordinated each Shop Drawing or Sample with other Shop Drawings and Samples and with the requirements of the Work and the Contract Documents.

2. Each submittal shall bear a stamp or specific written certification that Contractor has satisfied Contractor's obligations under the Contract Documents

with respect to Contractor's review and approval of that submittal.

3. With each submittal, Contractor shall give Engineer specific written notice of any variations, that the Shop Drawing or Sample may have from the requirements of the Contract Documents. This notice shall be both a written communication separate from the Shop Drawing's or Sample Submittal; and, in addition, by a specific notation made on each Shop Drawing or Sample submitted to Engineer for review and approval of each such variation.

D. *Engineer's Review*

1. Engineer will provide timely review of Shop Drawings and Samples in accordance with the Schedule of Submittals acceptable to Engineer. Engineer's review and approval will be only to determine if the items covered by the submittals will, after installation or incorporation in the Work, conform to the information given in the Contract Documents and be compatible with the design concept of the completed Project as a functioning whole as indicated by the Contract Documents.

2. Engineer's review and approval will not extend to means, methods, techniques, sequences, or procedures of construction (except where a particular means, method, technique, sequence, or procedure of construction is specifically and expressly called for by the Contract Documents) or to safety precautions or programs incident thereto. The review and approval of a separate item as such will not indicate approval of the assembly in which the item functions.

3. Engineer's review and approval shall not relieve Contractor from responsibility for any variation from the requirements of the Contract Documents unless Contractor has complied with the requirements of Paragraph 6.17.C.3 and Engineer has given written approval of each such variation by specific written notation thereof incorporated in or accompanying the Shop Drawing or Sample. Engineer's review and approval shall not relieve Contractor from responsibility for complying with the requirements of Paragraph 6.17.C.1.

E. *Resubmittal Procedures*

1. Contractor shall make corrections required by Engineer and shall return the required number of corrected copies of Shop Drawings and submit, as required, new Samples for review and approval. Contractor shall direct specific attention in writing to revisions other than the corrections called for by Engineer on previous submittals.

6.18   *Continuing the Work*

A. Contractor shall carry on the Work and adhere to the Progress Schedule during all disputes or disagreements with Owner. No Work shall be delayed or postponed pending resolution of any disputes or disagreements, except as permitted by Paragraph 15.04 or as Owner and Contractor may otherwise agree in writing.

6.19   *Contractor's General Warranty and Guarantee*

A. Contractor warrants and guarantees to Owner that all Work will be in accordance with the Contract Documents and will not be defective. Engineer and its Related Entities shall be entitled to rely on representation of Contractor's warranty and guarantee.

B. Contractor's warranty and guarantee hereunder excludes defects or damage caused by:

1. abuse, modification, or improper maintenance or operation by persons other than Contractor, Subcontractors, Suppliers, or any other individual or entity for whom Contractor is responsible; or

2. normal wear and tear under normal usage.

C. Contractor's obligation to perform and complete the Work in accordance with the Contract Documents shall be absolute. None of the following will constitute an acceptance of Work that is not in accordance with the Contract Documents or a release of Contractor's obligation to perform the Work in accordance with the Contract Documents:

1. observations by Engineer;

2. recommendation by Engineer or payment by Owner of any progress or final payment;

3. the issuance of a certificate of Substantial Completion by Engineer or any payment related thereto by Owner;

4. use or occupancy of the Work or any part thereof by Owner;

5. any review and approval of a Shop Drawing or Sample submittal or the issuance of a notice of acceptability by Engineer;

6. any inspection, test, or approval by others; or

7. any correction of defective Work by Owner.

6.20   *Indemnification*

A. To the fullest extent permitted by Laws and Regulations, Contractor shall indemnify and hold harmless Owner and Engineer, and the officers, directors, partners, employees, agents, consultants and subcontractors of each and any of them from and against all claims, costs, losses, and damages (including but not limited to all fees and charges of engineers, architects, attorneys, and other professionals and all court or arbitration or other dispute resolution costs) arising out of or relating to the performance of the Work, provided that any such claim, cost, loss, or damage is attributable to bodily injury, sickness, disease, or death, or to injury to or destruction of tangible property (other than the Work itself), including the loss of use resulting therefrom but only to the extent caused by any negligent act or omission of Contractor, any Subcontractor, any Supplier, or any individual or entity directly or indirectly employed by any of them to perform any of the Work or anyone for whose acts any of them may be liable .

B. In any and all claims against Owner or Engineer or any of their respective consultants, agents, officers, directors, partners, or employees by any employee (or the survivor or personal representative of such employee) of Contractor, any Subcontractor, any Supplier, or any individual or entity directly or indirectly employed by any of them to perform any of the Work, or anyone for whose acts any of them may be liable, the indemnification obligation under Paragraph 6.20.A shall not be limited in any way by any limitation on the amount or type of damages, compensation, or benefits payable by or for Contractor or any such Subcontractor, Supplier, or other individual or entity under workers' compensation acts, disability benefit acts, or other employee benefit acts.

C. The indemnification obligations of Contractor under Paragraph 6.20.A shall not extend to the liability of Engineer and Engineer's officers, directors, partners, employees, agents, consultants and subcontractors arising out of:

1. the preparation or approval of, or the failure to prepare or approve, maps, Drawings, opinions, reports, surveys, Change Orders, designs, or Specifications; or

2. giving directions or instructions, or failing to give them, if that is the primary cause of the injury or damage.

6.21   *Delegation of Professional Design Services*

A. Contractor will not be required to provide professional design services unless such services are specifically required by the Contract Documents for a portion of the Work or unless such services are required to carry out Contractor's responsibilities for construction means, methods, techniques, sequences and procedures. Contractor shall not be required to provide professional services in violation of applicable law.

B. If professional design services or certifications by a design professional related to systems, materials or equipment are specifically required of Contractor by the Contract Documents, Owner and Engineer will specify all performance and design criteria that such services must satisfy. Contractor shall cause such services or certifications to be provided by a properly licensed professional, whose signature and seal shall appear on all drawings, calculations, specifications, certifications, Shop Drawings and other submittals prepared by such professional. Shop Drawings and other submittals related to the Work designed or certified by such professional, if prepared by others, shall bear such professional's written approval when submitted to Engineer.

C. Owner and Engineer shall be entitled to rely upon the adequacy, accuracy and completeness of the services, certifications or approvals performed by such design professionals, provided Owner and Engineer have specified to Contractor all performance and design criteria that such services must satisfy.

D. Pursuant to this Paragraph 6.21, Engineer's review and approval of design calculations and design drawings will be only for the limited purpose of checking for conformance with performance and design criteria given and the design concept expressed in the Contract Documents. Engineer's review and approval of Shop Drawings and other submittals (except design calculations and design drawings) will be only for the purpose stated in Paragraph 6.17.D.1.

E. Contractor shall not be responsible for the adequacy of the performance or design criteria required by the Contract Documents.

ARTICLE 7 - OTHER WORK AT THE SITE

7.01   *Related Work at Site*

A. Owner may perform other work related to the Project at the Site with Owner's employees, or via other direct contracts therefor, or have other work performed by utility owners. If such other work is not noted in the Contract Documents, then:

1. written notice thereof will be given to Contractor prior to starting any such other work; and

2. if Owner and Contractor are unable to agree on entitlement to or on the amount or extent, if any, of any adjustment in the Contract Price or Contract Times that should be allowed as a result of such other work, a Claim may be made therefor as provided in Paragraph 10.05.

B. Contractor shall afford each other contractor who is a party to such a direct contract, each utility owner and Owner, if Owner is performing other work with Owner's employees, proper and safe access to the Site, a reasonable opportunity for the introduction and storage of materials and equipment and the execution of such other work, and shall properly coordinate the Work with theirs. Contractor shall do all cutting, fitting, and patching of the Work that may be required to properly connect or otherwise make its several parts come together and properly integrate with such other work. Contractor shall not endanger any work of others by cutting, excavating, or otherwise altering their work and will only cut or alter their work with the written consent of Engineer and the others whose work will be affected. The duties and responsibilities of Contractor under this Paragraph are for the benefit of such utility owners and other contractors to the extent that there are comparable provisions for the benefit of Contractor in said direct contracts between Owner and such utility owners and other contractors.

C. If the proper execution or results of any part of Contractor's Work depends upon work performed by others under this Article 7, Contractor shall inspect such other work and promptly report to Engineer in writing any delays, defects, or deficiencies in such other work that render it unavailable or unsuitable for the proper execution and results of Contractor's Work. Contractor's failure to so report will constitute an acceptance of such other work as fit and proper for integration with Contractor's Work except for latent defects and deficiencies in such other work.

7.02 Coordination

A. If Owner intends to contract with others for the performance of other work on the Project at the Site, the following will be set forth in Supplementary Conditions:

1. the individual or entity who will have authority and responsibility for coordination of the activities among the various contractors will be identified;

2. the specific matters to be covered by such authority and responsibility will be itemized; and

3. the extent of such authority and responsibilities will be provided.

B. Unless otherwise provided in the Supplementary Conditions, Owner shall have sole authority and responsibility for such coordination.

7.03 Legal Relationships

A. Paragraphs 7.01.A and 7.02 are not applicable for utilities not under the control of Owner.

B. Each other direct contract of Owner under Paragraph 7.01.A shall provide that the other contractor is liable to Owner and Contractor for the reasonable direct delay and disruption costs incurred by Contractor as a result of the other contractor's actions or inactions.

C. Contractor shall be liable to Owner and any other contractor for the reasonable direct delay and disruption costs incurred by such other contractor as a result of Contractor's action or inactions.

ARTICLE 8 - OWNER'S RESPONSIBILITIES

8.01 Communications to Contractor

A. Except as otherwise provided in these General Conditions, Owner shall issue all communications to Contractor through Engineer.

8.02 Replacement of Engineer

A. In case of termination of the employment of Engineer, Owner shall appoint an engineer to whom Contractor makes no reasonable objection, whose status under the Contract Documents shall be that of the former Engineer.

8.03 Furnish Data

A. Owner shall promptly furnish the data required of Owner under the Contract Documents.

8.04 Pay When Due

A. Owner shall make payments to Contractor when they are due as provided in Paragraphs 14.02.C and 14.07.C.

8.05 Lands and Easements; Reports and Tests

A. Owner's duties in respect of providing lands and easements and providing engineering surveys to establish reference points are set forth in Paragraphs 4.01 and 4.05. Paragraph 4.02 refers to Owner's identifying and making available to Contractor copies of reports of explorations and tests of subsurface conditions and drawings of physical conditions in or relating to existing surface or subsurface structures at or contiguous to the Site that have been utilized by Engineer in preparing the Contract Documents.

8.06 Insurance

A. Owner's responsibilities, if any, in respect to purchasing and maintaining liability and property insurance are set forth in Article 5.

8.07 Change Orders

A. Owner is obligated to execute Change Orders as indicated in Paragraph 10.03.

8.08 Inspections, Tests, and Approvals

A. Owner's responsibility in respect to certain inspections, tests, and approvals is set forth in Paragraph 13.03.B.

423

8.09 *Limitations on Owner's Responsibilities*

A. The Owner shall not supervise, direct, or have control or authority over, nor be responsible for, Contractor's means, methods, techniques, sequences, or procedures of construction, or the safety precautions and programs incident thereto, or for any failure of Contractor to comply with Laws and Regulations applicable to the performance of the Work. Owner will not be responsible for Contractor's failure to perform the Work in accordance with the Contract Documents.

8.10 *Undisclosed Hazardous Environmental Condition*

A. Owner's responsibility in respect to an undisclosed Hazardous Environmental Condition is set forth in Paragraph 4.06.

8.11 *Evidence of Financial Arrangements*

A. If and to the extent Owner has agreed to furnish Contractor reasonable evidence that financial arrangements have been made to satisfy Owner's obligations under the Contract Documents, Owner's responsibility in respect thereof will be as set forth in the Supplementary Conditions.

ARTICLE 9 - ENGINEER'S STATUS DURING CONSTRUCTION

9.01 *Owner's Representative*

A. Engineer will be Owner's representative during the construction period. The duties and responsibilities and the limitations of authority of Engineer as Owner's representative during construction are set forth in the Contract Documents and will not be changed without written consent of Owner and Engineer.

9.02 *Visits to Site*

A. Engineer will make visits to the Site at intervals appropriate to the various stages of construction as Engineer deems necessary in order to observe as an experienced and qualified design professional the progress that has been made and the quality of the various aspects of Contractor's executed Work. Based on information obtained during such visits and observations, Engineer, for the benefit of Owner, will determine, in general, if the Work is proceeding in accordance with the Contract Documents. Engineer will not be required to make exhaustive or continuous inspections on the Site to check the quality or quantity of the Work. Engineer's efforts will be directed toward providing for Owner a greater degree of confidence that the completed Work will conform generally to the Contract Documents. On the basis of such visits and observations, Engineer will keep Owner informed of the progress of the Work and will endeavor to guard Owner against defective Work.

B. Engineer's visits and observations are subject to all the limitations on Engineer's authority and responsibility set forth in Paragraph 9.09. Particularly, but without limitation, during or as a result of Engineer's visits or observations of Contractor's Work Engineer will not supervise, direct, control, or have authority over or be responsible for Contractor's means, methods, techniques, sequences, or procedures of construction, or the safety precautions and programs incident thereto, or for any failure of Contractor to comply with Laws and Regulations applicable to the performance of the Work.

9.03 *Project Representative*

A. If Owner and Engineer agree, Engineer will furnish a Resident Project Representative to assist Engineer in providing more extensive observation of the Work. The authority and responsibilities of any such Resident Project Representative and assistants will be as provided in the Supplementary Conditions, and limitations on the responsibilities thereof will be as provided in Paragraph 9.09. If Owner designates another representative or agent to represent Owner at the Site who is not Engineer's consultant, agent or employee, the responsibilities and authority and limitations thereon of such other individual or entity will be as provided in the Supplementary Conditions.

9.04 *Authorized Variations in Work*

A. Engineer may authorize minor variations in the Work from the requirements of the Contract Documents which do not involve an adjustment in the Contract Price or the Contract Times and are compatible with the design concept of the completed Project as a functioning whole as indicated by the Contract Documents. These may be accomplished by a Field Order and will be binding on Owner and also on Contractor, who shall perform the Work involved promptly. If Owner or Contractor believes that a Field Order justifies an adjustment in the Contract Price or Contract Times, or both, and the parties are unable to agree on entitlement to or on the amount or extent, if any, of any such adjustment , a Claim may be made therefor as provided in Paragraph 10.05.

9.05 *Rejecting Defective Work*

A. Engineer will have authority to reject Work which Engineer believes to be defective, or that Engineer believes will not produce a completed Project that conforms to the Contract Documents or that will prejudice the integrity of the design concept of the completed Project as a functioning whole as indicated by the Contract Documents. Engineer will also have authority to require special inspection or testing of the Work as provided in Paragraph 13.04, whether or not the Work is fabricated, installed, or completed.

9.06 *Shop Drawings, Change Orders and Payments*

A. In connection with Engineer's authority, and limitations thereof, as to Shop Drawings and Samples, see Paragraph 6.17.

B. In connection with Engineer's authority, and limitations thereof, as to design calculations and design drawings submitted in response to a delegation of professional design services, if any, see Paragraph 6.21.

C. In connection with Engineer's authority as to Change Orders, see Articles 10, 11, and 12.

D. In connection with Engineer's authority as to Applications for Payment, see Article 14.

9.07 *Determinations for Unit Price Work*

A. Engineer will determine the actual quantities and classifications of Unit Price Work performed by Contractor. Engineer will review with Contractor the Engineer's preliminary determinations on such matters before rendering a written decision thereon (by recommendation of an Application for Payment or otherwise). Engineer's written decision thereon will be final and binding (except as modified by Engineer to reflect changed factual conditions or more accurate data) upon Owner and Contractor, subject to the provisions of Paragraph 10.05.

9.08 *Decisions on Requirements of Contract Documents and Acceptability of Work*

A. Engineer will be the initial interpreter of the requirements of the Contract Documents and judge of the acceptability of the Work thereunder. All matters in question and other matters between Owner and Contractor arising prior to the date final payment is due relating to the acceptability of the Work, and the interpretation of the requirements of the Contract Documents pertaining to the performance of the Work, will be referred initially to Engineer in writing within 30 days of the event giving rise to the question

B. Engineer will, with reasonable promptness, render a written decision on the issue referred. If Owner or Contractor believe that any such decision entitles them to an adjustment in the Contract Price or Contract Times or both, a Claim may be made under Paragraph 10.05. The date of Engineer's decision shall be the date of the event giving rise to the issues referenced for the purposes of Paragraph 10.05.B.

C. Engineer's written decision on the issue referred will be final and binding on Owner and Contractor, subject to the provisions of Paragraph 10.05.

D. When functioning as interpreter and judge under this Paragraph 9.08, Engineer will not show partiality to Owner or Contractor and will not be liable in connection with any interpretation or decision rendered in good faith in such capacity.

9.09 *Limitations on Engineer's Authority and Responsibilities*

A. Neither Engineer's authority or responsibility under this Article 9 or under any other provision of the Contract Documents nor any decision made by Engineer in good faith either to exercise or not exercise such authority or responsibility or the undertaking, exercise, or performance of any authority or responsibility by Engineer shall create, impose, or give rise to any duty in contract, tort, or otherwise owed by Engineer to Contractor, any Subcontractor, any Supplier, any other individual or entity, or to any surety for or employee or agent of any of them.

B. Engineer will not supervise, direct, control, or have authority over or be responsible for Contractor's means, methods, techniques, sequences, or procedures of construction, or the safety precautions and programs incident thereto, or for any failure of Contractor to comply with Laws and Regulations applicable to the performance of the Work. Engineer will not be responsible for Contractor's failure to perform the Work in accordance with the Contract Documents.

C. Engineer will not be responsible for the acts or omissions of Contractor or of any Subcontractor, any Supplier, or of any other individual or entity performing any of the Work.

D. Engineer's review of the final Application for Payment and accompanying documentation and all maintenance and operating instructions, schedules, guarantees, bonds, certificates of inspection, tests and approvals, and other documentation required to be delivered by Paragraph 14.07.A will only be to determine generally that their content complies with the requirements of, and in the case of certificates of inspections, tests, and approvals that the results certified indicate compliance with the Contract Documents.

E. The limitations upon authority and responsibility set forth in this Paragraph 9.09 shall also apply to the Resident Project Representative, if any, and assistants, if any.

ARTICLE 10 - CHANGES IN THE WORK; CLAIMS

10.01 *Authorized Changes in the Work*

A. Without invalidating the Contract and without notice to any surety, Owner may, at any time or from time to time, order additions, deletions, or revisions in the Work by a Change Order, or a Work Change Directive. Upon receipt of any such document, Contractor shall promptly proceed with the Work involved which will be performed under the applicable conditions of the Contract Documents (except as otherwise specifically provided).

B. If Owner and Contractor are unable to agree on entitlement to, or on the amount or extent, if any, of an adjustment in the Contract Price or Contract Times, or both, that should be allowed as a result of a Work Change Directive, a Claim may be made therefor as provided in Paragraph 10.05.

10.02 Unauthorized Changes in the Work

A. Contractor shall not be entitled to an increase in the Contract Price or an extension of the Contract Times with respect to any work performed that is not required by the Contract Documents as amended, modified, or supplemented as provided in Paragraph 3.04, except in the case of an emergency as provided in Paragraph 6.16 or in the case of uncovering Work as provided in Paragraph 13.04.B.

10.03 Execution of Change Orders

A. Owner and Contractor shall execute appropriate Change Orders recommended by Engineer covering:

1. changes in the Work which are: (i) ordered by Owner pursuant to Paragraph 10.01.A, (ii) required because of acceptance of defective Work under Paragraph 13.08.A or Owner's correction of defective Work under Paragraph 13.09, or (iii) agreed to by the parties;

2. changes in the Contract Price or Contract Times which are agreed to by the parties, including any undisputed sum or amount of time for Work actually performed in accordance with a Work Change Directive; and

3. changes in the Contract Price or Contract Times which embody the substance of any written decision rendered by Engineer pursuant to Paragraph 10.05; provided that, in lieu of executing any such Change Order, an appeal may be taken from any such decision in accordance with the provisions of the Contract Documents and applicable Laws and Regulations, but during any such appeal, Contractor shall carry on the Work and adhere to the Progress Schedule as provided in Paragraph 6.18.A.

10.04 Notification to Surety

A. If notice of any change affecting the general scope of the Work or the provisions of the Contract Documents (including, but not limited to, Contract Price or Contract Times) is required by the provisions of any bond to be given to a surety, the giving of any such notice will be Contractor's responsibility. The amount of each applicable bond will be adjusted to reflect the effect of any such change.

10.05 Claims

A. *Engineer's Decision Required*: All Claims, except those waived pursuant to Paragraph 14.09, shall be referred to the Engineer for decision. A decision by Engineer shall be required as a condition precedent to any exercise by Owner or Contractor of any rights or remedies either may otherwise have under the Contract Documents or by Laws and Regulations in respect of such Claims.

B. *Notice*: Written notice stating the general nature of each Claim, shall be delivered by the claimant to Engineer and the other party to the Contract promptly (but in no event later than 30 days) after the start of the event giving rise thereto. The responsibility to substantiate a Claim shall rest with the party making the Claim. Notice of the amount or extent of the Claim, with supporting data shall be delivered to the Engineer and the other party to the Contract within 60 days after the start of such event (unless Engineer allows additional time for claimant to submit additional or more accurate data in support of such Claim). A Claim for an adjustment in Contract Price shall be prepared in accordance with the provisions of Paragraph 12.01.B. A Claim for an adjustment in Contract Time shall be prepared in accordance with the provisions of Paragraph 12.02.B. Each Claim shall be accompanied by claimant's written statement that the adjustment claimed is the entire adjustment to which the claimant believes it is entitled as a result of said event. The opposing party shall submit any response to Engineer and the claimant within 30 days after receipt of the claimant's last submittal (unless Engineer allows additional time).

C. *Engineer's Action*: Engineer will review each Claim and, within 30 days after receipt of the last submittal of the claimant or the last submittal of the opposing party, if any, take one of the following actions in writing:

1. deny the Claim in whole or in part,

2. approve the Claim, or

3. notify the parties that the Engineer is unable to resolve the Claim if, in the Engineer's sole discretion, it would be inappropriate for the Engineer to do so. For purposes of further resolution of the Claim, such notice shall be deemed a denial.

D. In the event that Engineer does not take action on a Claim within said 30 days, the Claim shall be deemed denied.

E. Engineer's written action under Paragraph 10.05.C or denial pursuant to Paragraphs 10.05.C.3 or 10.05.D will be final and binding upon Owner and Contractor, unless Owner or Contractor invoke the dispute resolution procedure set forth in Article 16 within 30 days of such action or denial.

F. No Claim for an adjustment in Contract Price or Contract Times will be valid if not submitted in accordance with this Paragraph 10.05.

ARTICLE 11 - COST OF THE WORK; ALLOWANCES; UNIT PRICE WORK

11.01   Cost of the Work

A. *Costs Included:* The term Cost of the Work means the sum of all costs, except those excluded in Paragraph 11.01.B, necessarily incurred and paid by Contractor in the proper performance of the Work. When the value of any Work covered by a Change Order or when a Claim for an adjustment in Contract Price is determined on the basis of Cost of the Work, the costs to be reimbursed to Contractor will be only those additional or incremental costs required because of the change in the Work or because of the event giving rise to the Claim. Except as otherwise may be agreed to in writing by Owner, such costs shall be in amounts no higher than those prevailing in the locality of the Project, shall include only the following items, and shall not include any of the costs itemized in Paragraph 11.01.B.

1. Payroll costs for employees in the direct employ of Contractor in the performance of the Work under schedules of job classifications agreed upon by Owner and Contractor. Such employees shall include, without limitation, superintendents, foremen, and other personnel employed full time at the Site. Payroll costs for employees not employed full time on the Work shall be apportioned on the basis of their time spent on the Work. Payroll costs shall include, but not be limited to, salaries and wages plus the cost of fringe benefits, which shall include social security contributions, unemployment, excise, and payroll taxes, workers' compensation, health and retirement benefits, bonuses, sick leave, vacation and holiday pay applicable thereto. The expenses of performing Work outside of regular working hours, on Saturday, Sunday, or legal holidays, shall be included in the above to the extent authorized by Owner.

2. Cost of all materials and equipment furnished and incorporated in the Work, including costs of transportation and storage thereof, and Suppliers' field services required in connection therewith. All cash discounts shall accrue to Contractor unless Owner deposits funds with Contractor with which to make payments, in which case the cash discounts shall accrue to Owner. All trade discounts, rebates and refunds and returns from sale of surplus materials and equipment shall accrue to Owner, and Contractor shall make provisions so that they may be obtained.

3. Payments made by Contractor to Subcontractors for Work performed by Subcontractors. If required by Owner, Contractor shall obtain competitive bids from subcontractors acceptable to Owner and Contractor and shall deliver such bids to Owner, who will then determine, with the advice of Engineer, which bids, if any, will be acceptable. If any subcontract provides that the Subcontractor is to be paid on the basis of Cost of the Work plus a fee, the Subcontractor's Cost of the Work and fee shall be determined in the same manner as Contractor's Cost of the Work and fee as provided in this Paragraph 11.01.

4. Costs of special consultants (including but not limited to Engineers, architects, testing laboratories, surveyors, attorneys, and accountants) employed for services specifically related to the Work.

5. Supplemental costs including the following:

a. The proportion of necessary transportation, travel, and subsistence expenses of Contractor's employees incurred in discharge of duties connected with the Work.

b. Cost, including transportation and maintenance, of all materials, supplies, equipment, machinery, appliances, office, and temporary facilities at the Site, and hand tools not owned by the workers, which are consumed in the performance of the Work, and cost, less market value, of such items used but not consumed which remain the property of Contractor.

c. Rentals of all construction equipment and machinery, and the parts thereof whether rented from Contractor or others in accordance with rental agreements approved by Owner with the advice of Engineer, and the costs of transportation, loading, unloading, assembly, dismantling, and removal thereof. All such costs shall be in accordance with the terms of said rental agreements. The rental of any such equipment, machinery, or parts shall cease when the use thereof is no longer necessary for the Work.

d. Sales, consumer, use, and other similar taxes related to the Work, and for which Contractor is liable, imposed by Laws and Regulations.

e. Deposits lost for causes other than negligence of Contractor, any Subcontractor, or anyone directly or indirectly employed by any of them or for whose acts any of them may be liable, and royalty payments and fees for permits and licenses.

f. Losses and damages (and related expenses) caused by damage to the Work, not compensated by insurance or otherwise, sustained by Contractor in connection with the performance of the Work (except losses and damages within the deductible amounts of property insurance established in accordance with Paragraph

5.06.D), provided such losses and damages have resulted from causes other than the negligence of Contractor, any Subcontractor, or anyone directly or indirectly employed by any of them or for whose acts any of them may be liable. Such losses shall include settlements made with the written consent and approval of Owner. No such losses, damages, and expenses shall be included in the Cost of the Work for the purpose of determining Contractor's fee.

g. The cost of utilities, fuel, and sanitary facilities at the Site.

h. Minor expenses such as telegrams, long distance telephone calls, telephone service at the Site, expresses, and similar petty cash items in connection with the Work.

i. The costs of premiums for all bonds and insurance Contractor is required by the Contract Documents to purchase and maintain.

B. *Costs Excluded:* The term Cost of the Work shall not include any of the following items:

1. Payroll costs and other compensation of Contractor's officers, executives, principals (of partnerships and sole proprietorships), general managers, safety managers, engineers, architects, estimators, attorneys, auditors, accountants, purchasing and contracting agents, expediters, timekeepers, clerks, and other personnel employed by Contractor, whether at the Site or in Contractor's principal or branch office for general administration of the Work and not specifically included in the agreed upon schedule of job classifications referred to in Paragraph 11.01.A.1 or specifically covered by Paragraph 11.01.A.4, all of which are to be considered administrative costs covered by the Contractor's fee.

2. Expenses of Contractor's principal and branch offices other than Contractor's office at the Site.

3. Any part of Contractor's capital expenses, including interest on Contractor's capital employed for the Work and charges against Contractor for delinquent payments.

4. Costs due to the negligence of Contractor, any Subcontractor, or anyone directly or indirectly employed by any of them or for whose acts any of them may be liable, including but not limited to, the correction of defective Work, disposal of materials or equipment wrongly supplied, and making good any damage to property.

5. Other overhead or general expense costs of any kind and the costs of any item not specifically and expressly included in Paragraphs 11.01.A and 11.01.B.

C. *Contractor's Fee:* When all the Work is performed on the basis of cost-plus, Contractor's fee shall be determined as set forth in the Agreement. When the value of any Work covered by a Change Order or when a Claim for an adjustment in Contract Price is determined on the basis of Cost of the Work, Contractor's fee shall be determined as set forth in Paragraph 12.01.C.

D. *Documentation:* Whenever the Cost of the Work for any purpose is to be determined pursuant to Paragraphs 11.01.A and 11.01.B, Contractor will establish and maintain records thereof in accordance with generally accepted accounting practices and submit in a form acceptable to Engineer an itemized cost breakdown together with supporting data.

11.02 *Allowances*

A. It is understood that Contractor has included in the Contract Price all allowances so named in the Contract Documents and shall cause the Work so covered to be performed for such sums and by such persons or entities as may be acceptable to Owner and Engineer.

B. *Cash Allowances*

1. Contractor agrees that:

a. the cash allowances include the cost to Contractor (less any applicable trade discounts) of materials and equipment required by the allowances to be delivered at the Site, and all applicable taxes; and

b. Contractor's costs for unloading and handling on the Site, labor, installation , overhead, profit, and other expenses contemplated for the cash allowances have been included in the Contract Price and not in the allowances, and no demand for additional payment on account of any of the foregoing will be valid.

C. Contingency Allowance

1. Contractor agrees that a contingency allowance, if any, is for the sole use of Owner to cover unanticipated costs.

D. Prior to final payment, an appropriate Change Order will be issued as recommended by Engineer to reflect actual amounts due Contractor on account of Work covered by allowances, and the Contract Price shall be correspondingly adjusted.

11.03 *Unit Price Work*

A. Where the Contract Documents provide that all or part of the Work is to be Unit Price Work, initially the Contract Price will be deemed to include for all Unit Price Work an amount equal to the sum of the unit price for each separately identified item of Unit Price Work times the estimated quantity of each item as indicated in the Agreement.

B. The estimated quantities of items of Unit Price Work are not guaranteed and are solely for the purpose of comparison of Bids and determining an initial Contract Price. Determinations of the actual quantities and classifications of Unit Price Work performed by Contractor will be made by Engineer subject to the provisions of Paragraph 9.07.

C. Each unit price will be deemed to include an amount considered by Contractor to be adequate to cover Contractor's overhead and profit for each separately identified item.

D. Owner or Contractor may make a Claim for an adjustment in the Contract Price in accordance with Paragraph 10.05 if:

1. the quantity of any item of Unit Price Work performed by Contractor differs materially and significantly from the estimated quantity of such item indicated in the Agreement; and

2. there is no corresponding adjustment with respect any other item of Work; and

3. Contractor believes that Contractor is entitled to an increase in Contract Price as a result of having incurred additional expense or Owner believes that Owner is entitled to a decrease in Contract Price and the parties are unable to agree as to the amount of any such increase or decrease.

ARTICLE 12 - CHANGE OF CONTRACT PRICE; CHANGE OF CONTRACT TIMES

12.01 *Change of Contract Price*

A. The Contract Price may only be changed by a Change Order. Any Claim for an adjustment in the Contract Price shall be based on written notice submitted by the party making the Claim to the Engineer and the other party to the Contract in accordance with the provisions of Paragraph 10.05.

B. The value of any Work covered by a Change Order or of any Claim for an adjustment in the Contract Price will be determined as follows:

1. where the Work involved is covered by unit prices contained in the Contract Documents, by application of such unit prices to the quantities of the items involved (subject to the provisions of Paragraph 11.03); or

2. where the Work involved is not covered by unit prices contained in the Contract Documents, by a mutually agreed lump sum (which may include an allowance for overhead and profit not necessarily in accordance with Paragraph 12.01.C.2); or

3. where the Work involved is not covered by unit prices contained in the Contract Documents and agreement to a lump sum is not reached under Paragraph 12.01.B.2, on the basis of the Cost of the Work (determined as provided in Paragraph 11.01) plus a Contractor's fee for overhead and profit (determined as provided in Paragraph 12.01.C).

C. *Contractor's Fee:* The Contractor's fee for overhead and profit shall be determined as follows:

1. a mutually acceptable fixed fee; or

2. if a fixed fee is not agreed upon, then a fee based on the following percentages of the various portions of the Cost of the Work:

a. for costs incurred under Paragraphs 11.01.A.1 and 11.01.A.2, the Contractor's fee shall be 15 percent;

b. for costs incurred under Paragraph 11.01.A.3, the Contractor's fee shall be five percent;

c. where one or more tiers of subcontracts are on the basis of Cost of the Work plus a fee and no fixed fee is agreed upon, the intent of Paragraph 12.01.C.2.a is that the Subcontractor who actually performs the Work, at whatever tier, will be paid a fee of 15 percent of the costs incurred by such Subcontractor under Paragraphs 11.01.A.1 and 11.01.A.2 and that any higher tier Subcontractor and Contractor will each be paid a fee of five percent of the amount paid to the next lower tier Subcontractor;

d. no fee shall be payable on the basis of costs itemized under Paragraphs 11.01.A.4, 11.01.A.5, and 11.01.B;

e. the amount of credit to be allowed by Contractor to Owner for any change which results in a net decrease in cost will be the amount of the actual net decrease in cost plus a deduction in Contractor's fee by an amount equal to five percent of such net decrease; and

f. when both additions and credits are involved in any one change, the adjustment in Contractor's fee shall be computed on the basis of the net change in accordance with Paragraphs 12.01.C.2.a through 12.01.C.2.e, inclusive.

12.02 *Change of Contract Times*

A. The Contract Times may only be changed by a Change Order. Any Claim for an adjustment in the Contract Times shall be based on written notice submitted by the party making the Claim to the Engineer and the other party to the Contract in accordance with the provisions of Paragraph 10.05.

B. Any adjustment of the Contract Times covered by a Change Order or any Claim for an adjustment in the Contract Times will be determined in accordance with the provisions of this Article 12.

12.03 *Delays*

A. Where Contractor is prevented from completing any part of the Work within the Contract Times due to delay beyond the control of Contractor, the Contract Times will be extended in an amount equal to the time lost due to such delay if a Claim is made therefor as provided in Paragraph 12.02.A. Delays beyond the control of Contractor shall include, but not be limited to, acts or neglect by Owner, acts or neglect of utility owners or other contractors performing other work as contemplated by Article 7, fires, floods, epidemics, abnormal weather conditions, or acts of God.

B. If Owner, Engineer, or other contractors or utility owners performing other work for Owner as contemplated by Article 7, or anyone for whom Owner is responsible, delays, disrupts, or interferes with the performance or progress of the Work, then Contractor shall be entitled to an equitable adjustment in the Contract Price or the Contract Times , or both. Contractor's entitlement to an adjustment of the Contract Times is conditioned on such adjustment being essential to Contractor's ability to complete the Work within the Contract Times.

C If Contractor is delayed in the performance or progress of the Work by fire, flood, epidemic, abnormal weather conditions, acts of God, acts or failures to act of utility owners not under the control of Owner, or other causes not the fault of and beyond control of Owner and Contractor, then Contractor shall be entitled to an equitable adjustment in Contract Times, if such adjustment is essential to Contractor's ability to complete the Work within the Contract Times. Such an adjustment shall be Contractor's sole and exclusive remedy for the delays described in this Paragraph 12.03.C.

D. Owner, Engineer and the Related Entities of each of them shall not be liable to Contractor for any claims, costs, losses, or damages (including but not limited to all fees and charges of Engineers, architects, attorneys, and other professionals and all court or arbitration or other dispute resolution costs) sustained by Contractor on or in connection with any other project or anticipated project.

E. Contractor shall not be entitled to an adjustment in Contract Price or Contract Times for delays within the control of Contractor. Delays attributable to and within the control of a Subcontractor or Supplier shall be deemed to be delays within the control of Contractor.

ARTICLE 13 - TESTS AND INSPECTIONS; CORRECTION, REMOVAL OR ACCEPTANCE OF DEFECTIVE WORK

13.01 *Notice of Defects*

A. Prompt notice of all defective Work of which Owner or Engineer has actual knowledge will be given to Contractor. All defective Work may be rejected, corrected, or accepted as provided in this Article 13.

13.02 *Access to Work*

A. Owner, Engineer, their consultants and other representatives and personnel of Owner, independent testing laboratories, and governmental agencies with jurisdictional interests will have access to the Site and the Work at reasonable times for their observation, inspecting, and testing. Contractor shall provide them proper and safe conditions for such access and advise them of Contractor's Site safety procedures and programs so that they may comply therewith as applicable.

13.03 *Tests and Inspections*

A. Contractor shall give Engineer timely notice of readiness of the Work for all required inspections, tests, or approvals and shall cooperate with inspection and testing personnel to facilitate required inspections or tests.

B. Owner shall employ and pay for the services of an independent testing laboratory to perform all inspections, tests, or approvals required by the Contract Documents except:

1. for inspections, tests, or approvals covered by Paragraphs 13.03.C and 13.03.D below;

2. that costs incurred in connection with tests or inspections conducted pursuant to Paragraph 13.04.B shall be paid as provided in said Paragraph 13.04.C; and

3. as otherwise specifically provided in the Contract Documents.

C. If Laws or Regulations of any public body having jurisdiction require any Work (or part thereof) specifically to be inspected, tested, or approved by an employee or other representative of such public body, Contractor shall assume full responsibility for arranging and obtaining such inspections, tests, or approvals, pay all costs in connection therewith, and furnish Engineer the required certificates of inspection or approval.

D. Contractor shall be responsible for arranging and obtaining and shall pay all costs in connection with any inspections, tests, or approvals required for Owner's and Engineer's acceptance of materials or equipment to be incorporated in the Work; or acceptance of materials, mix designs, or equipment submitted for approval prior to Contractor's purchase thereof for incorporation in the Work. Such inspections, tests, or approvals shall be performed by organizations acceptable to Owner and Engineer.

E. If any Work (or the work of others) that is to be inspected, tested, or approved is covered by Contractor without written concurrence of Engineer, it must, if requested by Engineer, be uncovered for observation.

F. Uncovering Work as provided in Paragraph 13.03.E shall be at Contractor's expense unless Contractor has given Engineer timely notice of Contractor's intention to cover the same and Engineer has not acted with reasonable promptness in response to such notice.

13.04   *Uncovering Work*

A. If any Work is covered contrary to the written request of Engineer, it must, if requested by Engineer, be uncovered for Engineer's observation and replaced at Contractor's expense.

B. If Engineer considers it necessary or advisable that covered Work be observed by Engineer or inspected or tested by others, Contractor, at Engineer's request, shall uncover, expose, or otherwise make available for observation, inspection, or testing as Engineer may require, that portion of the Work in question, furnishing all necessary labor, material, and equipment.

C. If it is found that the uncovered Work is defective, Contractor shall pay all claims, costs, losses, and damages (including but not limited to all fees and charges of engineers, architects, attorneys, and other professionals and all court or arbitration or other dispute resolution costs) arising out of or relating to such uncovering, exposure, observation, inspection, and testing, and of satisfactory replacement or reconstruction (including but not limited to all costs of repair or replacement of work of others); and Owner shall be entitled to an appropriate decrease in the Contract Price. If the parties are unable to agree as to the amount thereof, Owner may make a Claim therefor as provided in Paragraph 10.05.

D. If, the uncovered Work is not found to be defective, Contractor shall be allowed an increase in the Contract Price or an extension of the Contract Times, or both, directly attributable to such uncovering, exposure, observation, inspection, testing, replacement, and reconstruction. If the parties are unable to agree as to the amount or extent thereof, Contractor may make a Claim therefor as provided in Paragraph 10.05.

13.05   *Owner May Stop the Work*

A. If the Work is defective, or Contractor fails to supply sufficient skilled workers or suitable materials or equipment, or fails to perform the Work in such a way that the completed Work will conform to the Contract Documents, Owner may order Contractor to stop the Work, or any portion thereof, until the cause for such order has been eliminated; however, this right of Owner to stop the Work shall not give rise to any duty on the part of Owner to exercise this right for the benefit of Contractor, any Subcontractor, any Supplier, any other individual or entity, or any surety for, or employee or agent of any of them.

13.06   *Correction or Removal of Defective Work*

A. Promptly after receipt of notice, Contractor shall correct all defective Work, whether or not fabricated, installed, or completed, or, if the Work has been rejected by Engineer, remove it from the Project and replace it with Work that is not defective. Contractor shall pay all claims, costs, losses, and damages (including but not limited to all fees and charges of engineers, architects, attorneys, and other professionals and all court or arbitration or other dispute resolution costs) arising out of or relating to such correction or removal (including but not limited to all costs of repair or replacement of work of others).

B. When correcting defective Work under the terms of this Paragraph 13.06 or Paragraph 13.07, Contractor shall take no action that would void or otherwise impair Owner's special warranty and guarantee, if any, on said Work.

13.07   *Correction Period*

A. If within one year after the date of Substantial Completion (or such longer period of time as may be prescribed by the terms of any applicable special guarantee required by the Contract Documents) or by any specific provision of the Contract Documents, any Work is found to be defective, or if the repair of any damages to the land or areas made available for Contractor's use by Owner or permitted by Laws and Regulations as contemplated in Paragraph 6.11.A is found to be defective, Contractor shall promptly, without cost to Owner and in accordance with Owner's written instructions:

1. repair such defective land or areas; or

2. correct such defective Work; or

Standard General Conditions of the Construction Contract

3. if the defective Work has been rejected by Owner, remove it from the Project and replace it with Work that is not defective, and

4. satisfactorily correct or repair or remove and replace any damage to other Work, to the work of others or other land or areas resulting therefrom.

B. If Contractor does not promptly comply with the terms of Owner's written instructions, or in an emergency where delay would cause serious risk of loss or damage, Owner may have the defective Work corrected or repaired or may have the rejected Work removed and replaced. All claims, costs, losses, and damages (including but not limited to all fees and charges of engineers, architects, attorneys, and other professionals and all court or arbitration or other dispute resolution costs) arising out of or relating to such correction or repair or such removal and replacement (including but not limited to all costs of repair or replacement of work of others) will be paid by Contractor.

C. In special circumstances where a particular item of equipment is placed in continuous service before Substantial Completion of all the Work, the correction period for that item may start to run from an earlier date if so provided in the Specifications .

D. Where defective Work (and damage to other Work resulting therefrom) has been corrected or removed and replaced under this Paragraph 13.07, the correction period hereunder with respect to such Work will be extended for an additional period of one year after such correction or removal and replacement has been satisfactorily completed.

E. Contractor's obligations under this Paragraph 13.07 are in addition to any other obligation or warranty. The provisions of this Paragraph 13.07 shall not be construed as a substitute for or a waiver of the provisions of any applicable statute of limitation or repose.

13.08   *Acceptance of Defective Work*

A. If, instead of requiring correction or removal and replacement of defective Work, Owner (and, prior to Engineer's recommendation of final payment, Engineer) prefers to accept it, Owner may do so. Contractor shall pay all claims, costs, losses, and damages (including but not limited to all fees and charges of engineers, architects, attorneys, and other professionals and all court or arbitration or other dispute resolution costs) attributable to Owner's evaluation of and determination to accept such defective Work (such costs to be approved by Engineer as to reasonableness) and the diminished value of the Work to the extent not otherwise paid by Contractor pursuant to this sentence. If any such acceptance occurs prior to Engineer's recommendation of final payment, a Change Order will be issued incorporating the necessary revisions in the Contract Documents with respect to the Work, and Owner shall be entitled to an appropriate decrease in the

Contract Price, reflecting the diminished value of Work so accepted. If the parties are unable to agree as to the amount thereof, Owner may make a Claim therefor as provided in Paragraph 10.05. If the acceptance occurs after such recommendation, an appropriate amount will be paid by Contractor to Owner.

13.09   *Owner May Correct Defective Work*

A. If Contractor fails within a reasonable time after written notice from Engineer to correct defective Work or to remove and replace rejected Work as required by Engineer in accordance with Paragraph 13.06.A, or if Contractor fails to perform the Work in accordance with the Contract Documents, or if Contractor fails to comply with any other provision of the Contract Documents, Owner may, after seven days written notice to Contractor, correct or remedy any such deficiency.

B. In exercising the rights and remedies under this Paragraph 13.09, Owner shall proceed expeditiously. In connection with such corrective or remedial action, Owner may exclude Contractor from all or part of the Site, take possession of all or part of the Work and suspend Contractor's services related thereto, take possession of Contractor's tools, appliances, construction equipment and machinery at the Site, and incorporate in the Work all materials and equipment stored at the Site or for which Owner has paid Contractor but which are stored elsewhere. Contractor shall allow Owner, Owner's representatives, agents and employees, Owner's other contractors, and Engineer and Engineer's consultants access to the Site to enable Owner to exercise the rights and remedies under this Paragraph.

C. All claims, costs, losses, and damages (including but not limited to all fees and charges of engineers, architects, attorneys, and other professionals and all court or arbitration or other dispute resolution costs) incurred or sustained by Owner in exercising the rights and remedies under this Paragraph 13.09 will be charged against Contractor, and a Change Order will be issued incorporating the necessary revisions in the Contract Documents with respect to the Work; and Owner shall be entitled to an appropriate decrease in the Contract Price. If the parties are unable to agree as to the amount of the adjustment, Owner may make a Claim therefor as provided in Paragraph 10.05. Such claims, costs, losses and damages will include but not be limited to all costs of repair, or replacement of work of others destroyed or damaged by correction, removal, or replacement of Contractor's defective Work.

D. Contractor shall not be allowed an extension of the Contract Times because of any delay in the performance of the Work attributable to the exercise by Owner of Owner's rights and remedies under this Paragraph 13.09.

ARTICLE 14 - PAYMENTS TO CONTRACTOR AND COMPLETION

14.01 *Schedule of Values*

A. The Schedule of Values established as provided in Paragraph 2.07.A will serve as the basis for progress payments and will be incorporated into a form of Application for Payment acceptable to Engineer. Progress payments on account of Unit Price Work will be based on the number of units completed.

14.02 *Progress Payments*

A. Applications for Payments

1. At least 20 days before the date established in the Agreement for each progress payment (but not more often than once a month), Contractor shall submit to Engineer for review an Application for Payment filled out and signed by Contractor covering the Work completed as of the date of the Application and accompanied by such supporting documentation as is required by the Contract Documents. If payment is requested on the basis of materials and equipment not incorporated in the Work but delivered and suitably stored at the Site or at another location agreed to in writing, the Application for Payment shall also be accompanied by a bill of sale, invoice, or other documentation warranting that Owner has received the materials and equipment free and clear of all Liens and evidence that the materials and equipment are covered by appropriate property insurance or other arrangements to protect Owner's interest therein, all of which must be satisfactory to Owner.

2. Beginning with the second Application for Payment, each Application shall include an affidavit of Contractor stating that all previous progress payments received on account of the Work have been applied on account to discharge Contractor's legitimate obligations associated with prior Applications for Payment.

3. The amount of retainage with respect to progress payments will be as stipulated in the Agreement.

B. Review of Applications

1. Engineer will, within 10 days after receipt of each Application for Payment, either indicate in writing a recommendation of payment and present the Application to Owner or return the Application to Contractor indicating in writing Engineer's reasons for refusing to recommend payment. In the latter case, Contractor may make the necessary corrections and resubmit the Application.

2. Engineer's recommendation of any payment requested in an Application for Payment will constitute a representation by Engineer to Owner, based on Engineer's observations on the Site of the executed Work as an experienced and qualified design professional and on Engineer's review of the Application for Payment and the accompanying data and schedules, that to the best of Engineer's knowledge, information and belief:

a. the Work has progressed to the point indicated;

b. the quality of the Work is generally in accordance with the Contract Documents (subject to an evaluation of the Work as a functioning whole prior to or upon Substantial Completion, to the results of any subsequent tests called for in the Contract Documents, to a final determination of quantities and classifications for Unit Price Work under Paragraph 9.07, and to any other qualifications stated in the recommendation); and

c. the conditions precedent to Contractor's being entitled to such payment appear to have been fulfilled in so far as it is Engineer's responsibility to observe the Work.

3. By recommending any such payment Engineer will not thereby be deemed to have represented that:

a. inspections made to check the quality or the quantity of the Work as it has been performed have been exhaustive, extended to every aspect of the Work in progress, or involved detailed inspections of the Work beyond the responsibilities specifically assigned to Engineer in the Contract Documents; or

b. that there may not be other matters or issues between the parties that might entitle Contractor to be paid additionally by Owner or entitle Owner to withhold payment to Contractor.

4. Neither Engineer's review of Contractor's Work for the purposes of recommending payments nor Engineer's recommendation of any payment, including final payment, will impose responsibility on Engineer:

a. to supervise, direct, or control the Work, or

b. for the means, methods, techniques, sequences, or procedures of construction, or the safety precautions and programs incident thereto, or

c. for Contractor's failure to comply with Laws and Regulations applicable to Contractor's performance of the Work, or

d. to make any examination to ascertain how or for what purposes Contractor has used the moneys paid on account of the Contract Price, or

e. to determine that title to any of the Work, materials, or equipment has passed to Owner free and clear of any Liens.

5. Engineer may refuse to recommend the whole or any part of any payment if, in Engineer's opinion, it would be incorrect to make the representations to Owner stated in Paragraph 14.02.B.2. Engineer may also refuse to recommend any such payment or, because of subsequently discovered evidence or the results of subsequent inspections or tests, revise or revoke any such payment recommendation previously made, to such extent as may be necessary in Engineer's opinion to protect Owner from loss because:

a. the Work is defective, or completed Work has been damaged, requiring correction or replacement;

b. the Contract Price has been reduced by Change Orders;

c. Owner has been required to correct defective Work or complete Work in accordance with Paragraph 13.09; or

d. Engineer has actual knowledge of the occurrence of any of the events enumerated in Paragraph 15.02.A.

C. *Payment Becomes Due*

1. Ten days after presentation of the Application for Payment to Owner with Engineer's recommendation, the amount recommended will (subject to the provisions of Paragraph 14.02.D) become due, and when due will be paid by Owner to Contractor.

D. *Reduction in Payment*

1. Owner may refuse to make payment of the full amount recommended by Engineer because:

a. claims have been made against Owner on account of Contractor's performance or furnishing of the Work;

b. Liens have been filed in connection with the Work, except where Contractor has delivered a specific bond satisfactory to Owner to secure the satisfaction and discharge of such Liens;

c. there are other items entitling Owner to a set-off against the amount recommended; or

d. Owner has actual knowledge of the occurrence of any of the events enumerated in Paragraphs 14.02.B.5.a through 14.02.B.5.c or Paragraph 15.02.A.

2. If Owner refuses to make payment of the full amount recommended by Engineer, Owner will give Contractor immediate written notice (with a copy to Engineer) stating the reasons for such action and promptly pay Contractor any amount remaining after deduction of the amount so withheld. Owner shall promptly pay Contractor the amount so withheld, or any adjustment thereto agreed to by Owner and Contractor, when Contractor corrects to Owner's satisfaction the reasons for such action.

3. If it is subsequently determined that Owner's refusal of payment was not justified, the amount wrongfully withheld shall be treated as an amount due as determined by Paragraph 14.02.C.1.

14.03 *Contractor's Warranty of Title*

A. Contractor warrants and guarantees that title to all Work, materials, and equipment covered by any Application for Payment, whether incorporated in the Project or not, will pass to Owner no later than the time of payment free and clear of all Liens.

14.04 *Substantial Completion*

A. When Contractor considers the entire Work ready for its intended use Contractor shall notify Owner and Engineer in writing that the entire Work is substantially complete (except for items specifically listed by Contractor as incomplete) and request that Engineer issue a certificate of Substantial Completion.

B. Promptly after Contractor's notification, , Owner, Contractor, and Engineer shall make an inspection of the Work to determine the status of completion. If Engineer does not consider the Work substantially complete, Engineer will notify Contractor in writing giving the reasons therefor.

C. If Engineer considers the Work substantially complete, Engineer will deliver to Owner a tentative certificate of Substantial Completion which shall fix the date of Substantial Completion. There shall be attached to the certificate a tentative list of items to be completed or corrected before final payment. Owner shall have seven days after receipt of the tentative certificate during which to make written objection to Engineer as to any provisions of the certificate or attached list. If, after considering such objections, Engineer concludes that the Work is not substantially complete, Engineer will within 14 days after submission of the tentative certificate to Owner notify Contractor in writing, stating the reasons therefor. If, after consideration of Owner's objections, Engineer considers the Work substantially complete, Engineer will within said 14 days execute and deliver to Owner and Contractor a definitive certificate of Substantial Completion (with a revised tentative list of items to be completed or corrected) reflecting such changes from the tentative certificate as Engineer believes justified after consideration of any objections from Owner.

D. At the time of delivery of the tentative certificate of Substantial Completion, Engineer will deliver to Owner and Contractor a written recommendation as to division of responsibilities pending final payment between Owner and Contractor with respect to security, operation, safety, and protection of the Work, maintenance, heat, utilities, insurance, and warranties and guarantees. Unless Owner and Contractor agree otherwise in writing and so inform Engineer in writing prior to Engineer's issuing the definitive certificate of Substantial Completion, Engineer's aforesaid recommendation will be binding on Owner and Contractor until final payment.

E. Owner shall have the right to exclude Contractor from the Site after the date of Substantial Completion subject to allowing Contractor reasonable access to complete or correct items on the tentative list.

14.05 *Partial Utilization*

A. Prior to Substantial Completion of all the Work, Owner may use or occupy any substantially completed part of the Work which has specifically been identified in the Contract Documents, or which Owner, Engineer, and Contractor agree constitutes a separately functioning and usable part of the Work that can be used by Owner for its intended purpose without significant interference with Contractor's performance of the remainder of the Work, subject to the following conditions.

1. Owner at any time may request Contractor in writing to permit Owner to use or occupy any such part of the Work which Owner believes to be ready for its intended use and substantially complete. If and when Contractor agrees that such part of the Work is substantially complete, Contractor will certify to Owner and Engineer that such part of the Work is substantially complete and request Engineer to issue a certificate of Substantial Completion for that part of the Work.

2. Contractor at any time may notify Owner and Engineer in writing that Contractor considers any such part of the Work ready for its intended use and substantially complete and request Engineer to issue a certificate of Substantial Completion for that part of the Work.

3. Within a reasonable time after either such request, Owner, Contractor, and Engineer shall make an inspection of that part of the Work to determine its status of completion. If Engineer does not consider that part of the Work to be substantially complete, Engineer will notify Owner and Contractor in writing giving the reasons therefor. If Engineer considers that part of the Work to be substantially complete, the provisions of Paragraph 14.04 will apply with respect to certification of Substantial Completion of that part of the Work and the division of responsibility in respect thereof and access thereto.

4. No use or occupancy or separate operation of part of the Work may occur prior to compliance with the requirements of Paragraph 5.10 regarding property insurance.

14.06 *Final Inspection*

A. Upon written notice from Contractor that the entire Work or an agreed portion thereof is complete, Engineer will promptly make a final inspection with Owner and Contractor and will notify Contractor in writing of all particulars in which this inspection reveals that the Work is incomplete or defective. Contractor shall immediately take such measures as are necessary to complete such Work or remedy such deficiencies.

14.07 *Final Payment*

A. Application for Payment

1. After Contractor has, in the opinion of Engineer, satisfactorily completed all corrections identified during the final inspection and has delivered, in accordance with the Contract Documents, all maintenance and operating instructions, schedules, guarantees, bonds, certificates or other evidence of insurance certificates of inspection, marked-up record documents (as provided in Paragraph 6.12), and other documents, Contractor may make application for final payment following the procedure for progress payments.

2. The final Application for Payment shall be accompanied (except as previously delivered) by:

   a. all documentation called for in the Contract Documents, including but not limited to the evidence of insurance required by Paragraph 5.04.B.7;

   b. consent of the surety, if any, to final payment;

   c. a list of all Claims against Owner that Contractor believes are unsettled; and

   d. complete and legally effective releases or waivers (satisfactory to Owner) of all Lien rights arising out of or Liens filed in connection with the Work.

3. In lieu of the releases or waivers of Liens specified in Paragraph 14.07.A.2 and as approved by Owner, Contractor may furnish receipts or releases in full and an affidavit of Contractor that: (i) the releases and receipts include all labor, services, material, and equipment for which a Lien could be filed; and (ii) all payrolls, material and equipment bills, and other indebtedness connected with the Work for which Owner or Owner's property might in any way be responsible have been paid or otherwise satisfied. If any Subcontractor or Supplier fails to furnish such a release or receipt in full, Contractor may furnish a bond or other collateral

satisfactory to Owner to indemnify Owner against any Lien.

B. *Engineer's Review of Application and Acceptance*

1. If, on the basis of Engineer's observation of the Work during construction and final inspection, and Engineer's review of the final Application for Payment and accompanying documentation as required by the Contract Documents, Engineer is satisfied that the Work has been completed and Contractor's other obligations under the Contract Documents have been fulfilled, Engineer will, within ten days after receipt of the final Application for Payment, indicate in writing Engineer's recommendation of payment and present the Application for Payment to Owner for payment. At the same time Engineer will also give written notice to Owner and Contractor that the Work is acceptable subject to the provisions of Paragraph 14.09. Otherwise, Engineer will return the Application for Payment to Contractor, indicating in writing the reasons for refusing to recommend final payment, in which case Contractor shall make the necessary corrections and resubmit the Application for Payment.

C. Payment Becomes Due

1. Thirty days after the presentation to Owner of the Application for Payment and accompanying documentation, the amount recommended by Engineer, less any sum Owner is entitled to set off against Engineer's recommendation, including but not limited to liquidated damages, will become due and , will be paid by Owner to Contractor.

14.08   *Final Completion Delayed*

A. If, through no fault of Contractor, final completion of the Work is significantly delayed, and if Engineer so confirms, Owner shall, upon receipt of Contractor's final Application for Payment (for Work fully completed and accepted) and recommendation of Engineer, and without terminating the Contract, make payment of the balance due for that portion of the Work fully completed and accepted. If the remaining balance to be held by Owner for Work not fully completed or corrected is less than the retainage stipulated in the Agreement, and if bonds have been furnished as required in Paragraph 5.01, the written consent of the surety to the payment of the balance due for that portion of the Work fully completed and accepted shall be submitted by Contractor to Engineer with the Application for such payment. Such payment shall be made under the terms and conditions governing final payment, except that it shall not constitute a waiver of Claims.

14.09   *Waiver of Claims*

A. The making and acceptance of final payment will constitute:

1. a waiver of all Claims by Owner against Contractor, except Claims arising from unsettled Liens, from defective Work appearing after final inspection pursuant to Paragraph 14.06, from failure to comply with the Contract Documents or the terms of any special guarantees specified therein, or from Contractor's continuing obligations under the Contract Documents; and

2. a waiver of all Claims by Contractor against Owner other than those previously made in accordance with the requirements herein and expressly acknowledged by Owner in writing as still unsettled.

ARTICLE 15 - SUSPENSION OF WORK AND TERMINATION

15.01   *Owner May Suspend Work*

A. At any time and without cause, Owner may suspend the Work or any portion thereof for a period of not more than 90 consecutive days by notice in writing to Contractor and Engineer which will fix the date on which Work will be resumed. Contractor shall resume the Work on the date so fixed. Contractor shall be granted an adjustment in the Contract Price or an extension of the Contract Times, or both, directly attributable to any such suspension if Contractor makes a Claim therefor as provided in Paragraph 10.05.

15.02   *Owner May Terminate for Cause*

A. The occurrence of any one or more of the following events will justify termination for cause:

1. Contractor's persistent failure to perform the Work in accordance with the Contract Documents (including, but not limited to, failure to supply sufficient skilled workers or suitable materials or equipment or failure to adhere to the Progress Schedule established under Paragraph 2.07 as adjusted from time to time pursuant to Paragraph 6.04);

2. Contractor's disregard of Laws or Regulations of any public body having jurisdiction;

3. Contractor's disregard of the authority of Engineer; or

4. Contractor's violation in any substantial way of any provisions of the Contract Documents.

B. If one or more of the events identified in Paragraph 15.02.A occur, Owner may, after giving Contractor (and surety) seven days written notice of its intent to terminate the services of Contractor:

---

1. exclude Contractor from the Site, and take possession of the Work and of all Contractor's tools, appliances, construction equipment, and machinery at the Site, and use the same to the full extent they could be used by Contractor (without liability to Contractor for trespass or conversion),

2. incorporate in the Work all materials and equipment stored at the Site or for which Owner has paid Contractor but which are stored elsewhere, and

3. complete the Work as Owner may deem expedient.

C. If Owner proceeds as provided in Paragraph 15.02.B, Contractor shall not be entitled to receive any further payment until the Work is completed. If the unpaid balance of the Contract Price exceeds all claims, costs, losses, and damages (including but not limited to all fees and charges of engineers, architects, attorneys, and other professionals and all court or arbitration or other dispute resolution costs) sustained by Owner arising out of or relating to completing the Work, such excess will be paid to Contractor. If such claims, costs, losses, and damages exceed such unpaid balance, Contractor shall pay the difference to Owner. Such claims, costs, losses, and damages incurred by Owner will be reviewed by Engineer as to their reasonableness and, when so approved by Engineer, incorporated in a Change Order. When exercising any rights or remedies under this Paragraph Owner shall not be required to obtain the lowest price for the Work performed.

D. Notwithstanding Paragraphs 15.02.B and 15.02.C, Contractor's services will not be terminated if Contractor begins within seven days of receipt of notice of intent to terminate to correct its failure to perform and proceeds diligently to cure such failure within no more than 30 days of receipt of said notice.

E. Where Contractor's services have been so terminated by Owner, the termination will not affect any rights or remedies of Owner against Contractor then existing or which may thereafter accrue. Any retention or payment of moneys due Contractor by Owner will not release Contractor from liability.

F. If and to the extent that Contractor has provided a performance bond under the provisions of Paragraph 5.01.A, the termination procedures of that bond shall supersede the provisions of Paragraphs 15.02.B, and 15.02.C.

15.03   *Owner May Terminate For Convenience*

A. Upon seven days written notice to Contractor and Engineer, Owner may, without cause and without prejudice to any other right or remedy of Owner, terminate the Contract. In such case, Contractor shall be paid for (without duplication of any items):

1. completed and acceptable Work executed in accordance with the Contract Documents prior to the effective date of termination, including fair and reasonable sums for overhead and profit on such Work;

2. expenses sustained prior to the effective date of termination in performing services and furnishing labor, materials, or equipment as required by the Contract Documents in connection with uncompleted Work, plus fair and reasonable sums for overhead and profit on such expenses;

3. all claims, costs, losses, and damages (including but not limited to all fees and charges of engineers, architects, attorneys, and other professionals and all court or arbitration or other dispute resolution costs) incurred in settlement of terminated contracts with Subcontractors, Suppliers, and others; and

4. reasonable expenses directly attributable to termination.

B. Contractor shall not be paid on account of loss of anticipated profits or revenue or other economic loss arising out of or resulting from such termination.

15.04   *Contractor May Stop Work or Terminate*

A. If, through no act or fault of Contractor, (i) the Work is suspended for more than 90 consecutive days by Owner or under an order of court or other public authority, or (ii) Engineer fails to act on any Application for Payment within 30 days after it is submitted, or (iii) Owner fails for 30 days to pay Contractor any sum finally determined to be due, then Contractor may, upon seven days written notice to Owner and Engineer, and provided Owner or Engineer do not remedy such suspension or failure within that time, terminate the Contract and recover from Owner payment on the same terms as provided in Paragraph 15.03.

B. In lieu of terminating the Contract and without prejudice to any other right or remedy, if Engineer has failed to act on an Application for Payment within 30 days after it is submitted, or Owner has failed for 30 days to pay Contractor any sum finally determined to be due, Contractor may, seven days after written notice to Owner and Engineer, stop the Work until payment is made of all such amounts due Contractor, including interest thereon. The provisions of this Paragraph 15.04 are not intended to preclude Contractor from making a Claim under Paragraph 10.05 for an adjustment in Contract Price or Contract Times or otherwise for expenses or damage directly attributable to Contractor's stopping the Work as permitted by this Paragraph.

Standard General Conditions of the Construction Contract

ARTICLE 16 - DISPUTE RESOLUTION

16.01 Methods and Procedures

A. Either Owner or Contractor may request mediation of any Claim submitted to Engineer for a decision under Paragraph 10.05 before such decision becomes final and binding. The mediation will be governed by the Construction Industry Mediation Rules of the American Arbitration Association in effect as of the Effective Date of the Agreement. The request for mediation shall be submitted in writing to the American Arbitration Association and the other party to the Contract. Timely submission of the request shall stay the effect of Paragraph 10.05.E.

B. Owner and Contractor shall participate in the mediation process in good faith. The process shall be concluded within 60 days of filing of the request. The date of termination of the mediation shall be determined by application of the mediation rules referenced above.

C. If the Claim is not resolved by mediation, Engineer's action under Paragraph 10.05.C or a denial pursuant to Paragraphs 10.05.C.3 or 10.05.D shall become final and binding 30 days after termination of the mediation unless, within that time period, Owner or Contractor:

1. elects in writing to invoke any dispute resolution process provided for in the Supplementary Conditions, or

2. agrees with the other party to submit the Claim to another dispute resolution process, or

3. gives written notice to the other party of their intent to submit the Claim to a court of competent jurisdiction.

ARTICLE 17 - MISCELLANEOUS

17.01 Giving Notice

A. Whenever any provision of the Contract Documents requires the giving of written notice, it will be deemed to have been validly given if:

1. delivered in person to the individual or to a member of the firm or to an officer of the corporation for whom it is intended, or

2. delivered at or sent by registered or certified mail, postage prepaid, to the last business address known to the giver of the notice.

17.02 Computation of Times

A. When any period of time is referred to in the Contract Documents by days, it will be computed to exclude the first and include the last day of such period. If the last day of any such period falls on a Saturday or Sunday or on a day made a legal holiday by the law of the applicable jurisdiction, such day will be omitted from the computation.

17.03 Cumulative Remedies

A. The duties and obligations imposed by these General Conditions and the rights and remedies available hereunder to the parties hereto are in addition to, and are not to be construed in any way as a limitation of, any rights and remedies available to any or all of them which are otherwise imposed or available by Laws or Regulations, by special warranty or guarantee, or by other provisions of the Contract Documents. The provisions of this Paragraph will be as effective as if repeated specifically in the Contract Documents in connection with each particular duty, obligation, right, and remedy to which they apply.

17.04 Survival of Obligations

A. All representations, indemnifications, warranties, and guarantees made in, required by, or given in accordance with the Contract Documents, as well as all continuing obligations indicated in the Contract Documents, will survive final payment, completion, and acceptance of the Work or termination or completion of the Contract or termination of the services of Contractor.

17.05 Controlling Law

A. This Contract is to be governed by the law of the state in which the Project is located.

17.06 Headings

A. Article and paragraph headings are inserted for convenience only and do not constitute parts of these General Conditions.

APPENDIX **E**

AIA Document
A201™—1997

▓AIA® Document A201™ – 1997

General Conditions of the Contract for Construction

for the following PROJECT:
(Name and location or address)

THE OWNER:
(Name and address)

This document has important
legal consequences.
Consultation with an attorney
is encouraged with respect to
its completion or modification.

This document has been
approved and endorsed by The
Associated General Contractors
of America

THE ARCHITECT:
(Name and address)

1

AIA Document A201™—1997

5

AIA Document A201™—1997

9

ARTICLE 1 GENERAL PROVISIONS

§ 1.1 BASIC DEFINITIONS

§ 1.1.1 THE CONTRACT DOCUMENTS

The Contract Documents consist of the Agreement between Owner and Contractor (hereinafter the Agreement), Conditions of the Contract (General, Supplementary and other Conditions), Drawings, Specifications, Addenda issued prior to execution of the Contract, other documents listed in the Agreement and Modifications issued after execution of the Contract. A Modification is (1) a written amendment to the Contract signed by both parties, (2) a Change Order, (3) a Construction Change Directive or (4) a written order for a minor change in the Work issued by the Architect. Unless specifically enumerated in the Agreement, the Contract Documents do not include other documents such as bidding requirements (advertisement or invitation to bid, Instructions to Bidders, sample forms, the Contractor's bid or portions of Addenda relating to bidding requirements).

§ 1.1.2 THE CONTRACT

The Contract Documents form the Contract for Construction. The Contract represents the entire and integrated agreement between the parties hereto and supersedes prior negotiations, representations or agreements, either written or oral. The Contract may be amended or modified only by a Modification. The Contract Documents shall not be construed to create a contractual relationship of any kind (1) between the Architect and Contractor, (2) between the Owner and a Subcontractor or Sub-subcontractor, (3) between the Owner and Architect or (4) between any persons or entities other than the Owner and Contractor. The Architect shall, however, be entitled to performance and enforcement of obligations under the Contract intended to facilitate performance of the Architect's duties.

§ 1.1.3 THE WORK

The term "Work" means the construction and services required by the Contract Documents, whether completed or partially completed, and includes all other labor, materials, equipment and services provided or to be provided by the Contractor to fulfill the Contractor's obligations. The Work may constitute the whole or a part of the Project.

§ 1.1.4 THE PROJECT

The Project is the total construction of which the Work performed under the Contract Documents may be the whole or a part and which may include construction by the Owner or by separate contractors.

§ 1.1.5 THE DRAWINGS

The Drawings are the graphic and pictorial portions of the Contract Documents showing the design, location and dimensions of the Work, generally including plans, elevations, sections, details, schedules and diagrams.

§ 1.1.6 THE SPECIFICATIONS

The Specifications are that portion of the Contract Documents consisting of the written requirements for materials, equipment, systems, standards and workmanship for the Work, and performance of related services.

§ 1.1.7 THE PROJECT MANUAL

The Project Manual is a volume assembled for the Work which may include the bidding requirements, sample forms, Conditions of the Contract and Specifications.

§ 1.2 CORRELATION AND INTENT OF THE CONTRACT DOCUMENTS

§ 1.2.1 The intent of the Contract Documents is to include all items necessary for the proper execution and completion of the Work by the Contractor. The Contract Documents are complementary, and what is required by one shall be as binding as if required by all; performance by the Contractor shall be required only to the extent consistent with the Contract Documents and reasonably inferable from them as being necessary to produce the indicated results.

§ 1.2.2 Organization of the Specifications into divisions, sections and articles, and arrangement of Drawings shall not control the Contractor in dividing the Work among Subcontractors or in establishing the extent of Work to be performed by any trade.

§ 1.2.3 Unless otherwise stated in the Contract Documents, words which have well-known technical or construction industry meanings are used in the Contract Documents in accordance with such recognized meanings.

§ 1.3 CAPITALIZATION

§ 1.3.1 Terms capitalized in these General Conditions include those which are (1) specifically defined, (2) the titles of numbered articles or (3) the titles of other documents published by the American Institute of Architects.

AIA Document A201™—1997

§ 1.4 INTERPRETATION

§ 1.4.1 In the interest of brevity the Contract Documents frequently omit modifying words such as "all" and "any" and articles such as "the" and "an," but the fact that a modifier or an article is absent from one statement and appears in another is not intended to affect the interpretation of either statement.

§ 1.5 EXECUTION OF CONTRACT DOCUMENTS

§ 1.5.1 The Contract Documents shall be signed by the Owner and Contractor. If either the Owner or Contractor or both do not sign all the Contract Documents, the Architect shall identify such unsigned Documents upon request.

§ 1.5.2 Execution of the Contract by the Contractor is a representation that the Contractor has visited the site, become generally familiar with local conditions under which the Work is to be performed and correlated personal observations with requirements of the Contract Documents.

§ 1.6 OWNERSHIP AND USE OF DRAWINGS, SPECIFICATIONS AND OTHER INSTRUMENTS OF SERVICE

§ 1.6.1 The Drawings, Specifications and other documents, including those in electronic form, prepared by the Architect and the Architect's consultants are Instruments of Service through which the Work to be executed by the Contractor is described. The Contractor may retain one record set. Neither the Contractor nor any Subcontractor, Sub-subcontractor or material or equipment supplier shall own or claim a copyright in the Drawings, Specifications and other documents prepared by the Architect or the Architect's consultants, and unless otherwise indicated the Architect and the Architect's consultants shall be deemed the authors of them and will retain all common law, statutory and other reserved rights, in addition to the copyrights. All copies of Instruments of Service, except the Contractor's record set, shall be returned or suitably accounted for to the Architect, on request, upon completion of the Work. The Drawings, Specifications and other documents prepared by the Architect and the Architect's consultants, and copies thereof furnished to the Contractor, are for use solely with respect to this Project. They are not to be used by the Contractor or any Subcontractor, Sub-subcontractor or material or equipment supplier on other projects or for additions to this Project outside the scope of the Work without the specific written consent of the Owner, Architect and the Architect's consultants. The Contractor, Subcontractors, Sub-subcontractors and material or equipment suppliers are authorized to use and reproduce applicable portions of the Drawings, Specifications and other documents prepared by the Architect and the Architect's consultants appropriate to and for use in the execution of their Work under the Contract Documents. All copies made under this authorization shall bear the statutory copyright notice, if any, shown on the Drawings, Specifications and other documents prepared by the Architect and the Architect's consultants. Submittal or distribution to meet official regulatory requirements or for other purposes in connection with this Project is not to be construed as publication in derogation of the Architect's or Architect's consultants' copyrights or other reserved rights.

ARTICLE 2 OWNER

§ 2.1 GENERAL

§ 2.1.1 The Owner is the person or entity identified as such in the Agreement and is referred to throughout the Contract Documents as if singular in number. The Owner shall designate in writing a representative who shall have express authority to bind the Owner with respect to all matters requiring the Owner's approval or authorization. Except as otherwise provided in Section 4.2.1, the Architect does not have such authority. The term "Owner" means the Owner or the Owner's authorized representative.

§ 2.1.2 The Owner shall furnish to the Contractor within fifteen days after receipt of a written request, information necessary and relevant for the Contractor to evaluate, give notice of or enforce mechanic's lien rights. Such information shall include a correct statement of the record legal title to the property on which the Project is located, usually referred to as the site, and the Owner's interest therein.

§ 2.2 INFORMATION AND SERVICES REQUIRED OF THE OWNER

§ 2.2.1 The Owner shall, at the written request of the Contractor, prior to commencement of the Work and thereafter, furnish to the Contractor reasonable evidence that financial arrangements have been made to fulfill the Owner's obligations under the Contract. Furnishing of such evidence shall be a condition precedent to commencement or continuation of the Work. After such evidence has been furnished, the Owner shall not materially vary such financial arrangements without prior notice to the Contractor.

§ 2.2.2 Except for permits and fees, including those required under Section 3.7.1, which are the responsibility of the Contractor under the Contract Documents, the Owner shall secure and pay for necessary approvals, easements, assessments and charges required for construction, use or occupancy of permanent structures or for permanent changes in existing facilities.

11

§ 2.2.3 The Owner shall furnish surveys describing physical characteristics, legal limitations and utility locations for the site of the Project, and a legal description of the site. The Contractor shall be entitled to rely on the accuracy of information furnished by the Owner but shall exercise proper precautions relating to the safe performance of the Work.

§ 2.2.4 Information or services required of the Owner by the Contract Documents shall be furnished by the Owner with reasonable promptness. Any other information or services relevant to the Contractor's performance of the Work under the Owner's control shall be furnished by the Owner after receipt from the Contractor of a written request for such information or services.

§ 2.2.5 Unless otherwise provided in the Contract Documents, the Contractor will be furnished, free of charge, such copies of Drawings and Project Manuals as are reasonably necessary for execution of the Work.

§ 2.3 OWNER'S RIGHT TO STOP THE WORK
§ 2.3.1 If the Contractor fails to correct Work which is not in accordance with the requirements of the Contract Documents as required by Section 12.2 or persistently fails to carry out Work in accordance with the Contract Documents, the Owner may issue a written order to the Contractor to stop the Work, or any portion thereof, until the cause for such order has been eliminated; however, the right of the Owner to stop the Work shall not give rise to a duty on the part of the Owner to exercise this right for the benefit of the Contractor or any other person or entity, except to the extent required by Section 6.1.3.

§ 2.4 OWNER'S RIGHT TO CARRY OUT THE WORK
§ 2.4.1 If the Contractor defaults or neglects to carry out the Work in accordance with the Contract Documents and fails within a seven-day period after receipt of written notice from the Owner to commence and continue correction of such default or neglect with diligence and promptness, the Owner may after such seven-day period give the Contractor a second written notice to correct such deficiencies within a three-day period. If the Contractor within such three-day period after receipt of such second notice fails to commence and continue to correct any deficiencies, the Owner may, without prejudice to other remedies the Owner may have, correct such deficiencies. In such case an appropriate Change Order shall be issued deducting from payments then or thereafter due the Contractor the reasonable cost of correcting such deficiencies, including Owner's expenses and compensation for the Architect's additional services made necessary by such default, neglect or failure. Such action by the Owner and amounts charged to the Contractor are both subject to prior approval of the Architect. If payments then or thereafter due the Contractor are not sufficient to cover such amounts, the Contractor shall pay the difference to the Owner.

ARTICLE 3 CONTRACTOR
§ 3.1 GENERAL
§ 3.1.1 The Contractor is the person or entity identified as such in the Agreement and is referred to throughout the Contract Documents as if singular in number. The term "Contractor" means the Contractor or the Contractor's authorized representative.

§ 3.1.2 The Contractor shall perform the Work in accordance with the Contract Documents.

§ 3.1.3 The Contractor shall not be relieved of obligations to perform the Work in accordance with the Contract Documents either by activities or duties of the Architect in the Architect's administration of the Contract, or by tests, inspections or approvals required or performed by persons other than the Contractor.

§ 3.2 REVIEW OF CONTRACT DOCUMENTS AND FIELD CONDITIONS BY CONTRACTOR
§ 3.2.1 Since the Contract Documents are complementary, before starting each portion of the Work, the Contractor shall carefully study and compare the various Drawings and other Contract Documents relative to that portion of the Work, as well as the information furnished by the Owner pursuant to Section 2.2.3, shall take field measurements of any existing conditions related to that portion of the Work and shall observe any conditions at the site affecting it. These obligations are for the purpose of facilitating construction by the Contractor and are not for the purpose of discovering errors, omissions, or inconsistencies in the Contract Documents; however, any errors, inconsistencies or omissions discovered by the Contractor shall be reported promptly to the Architect as a request for information in such form as the Architect may require.

§ 3.2.2 Any design errors or omissions noted by the Contractor during this review shall be reported promptly to the Architect, but it is recognized that the Contractor's review is made in the Contractor's capacity as a contractor and not as a licensed design professional unless otherwise specifically provided in the Contract Documents. The

AIA Document A201™—1997

Contractor is not required to ascertain that the Contract Documents are in accordance with applicable laws, statutes, ordinances, building codes, and rules and regulations, but any nonconformity discovered by or made known to the Contractor shall be reported promptly to the Architect.

§ 3.2.3 If the Contractor believes that additional cost or time is involved because of clarifications or instructions issued by the Architect in response to the Contractor's notices or requests for information pursuant to Sections 3.2.1 and 3.2.2, the Contractor shall make Claims as provided in Sections 4.3.6 and 4.3.7. If the Contractor fails to perform the obligations of Sections 3.2.1 and 3.2.2, the Contractor shall pay such costs and damages to the Owner as would have been avoided if the Contractor had performed such obligations. The Contractor shall not be liable to the Owner or Architect for damages resulting from errors, inconsistencies or omissions in the Contract Documents or for differences between field measurements or conditions and the Contract Documents unless the Contractor recognized such error, inconsistency, omission or difference and knowingly failed to report it to the Architect.

§ 3.3 SUPERVISION AND CONSTRUCTION PROCEDURES

§ 3.3.1 The Contractor shall supervise and direct the Work, using the Contractor's best skill and attention. The Contractor shall be solely responsible for and have control over construction means, methods, techniques, sequences and procedures and for coordinating all portions of the Work under the Contract, unless the Contract Documents give other specific instructions concerning these matters. If the Contract Documents give specific instructions concerning construction means, methods, techniques, sequences or procedures, the Contractor shall evaluate the jobsite safety thereof and, except as stated below, shall be fully and solely responsible for the jobsite safety of such means, methods, techniques, sequences or procedures. If the Contractor determines that such means, methods, techniques, sequences or procedures may not be safe, the Contractor shall give timely written notice to the Owner and Architect and shall not proceed with that portion of the Work without further written instructions from the Architect. If the Contractor is then instructed to proceed with the required means, methods, techniques, sequences or procedures without acceptance of changes proposed by the Contractor, the Owner shall be solely responsible for any resulting loss or damage.

§ 3.3.2 The Contractor shall be responsible to the Owner for acts and omissions of the Contractor's employees, Subcontractors and their agents and employees, and other persons or entities performing portions of the Work for or on behalf of the Contractor or any of its Subcontractors.

§ 3.3.3 The Contractor shall be responsible for inspection of portions of Work already performed to determine that such portions are in proper condition to receive subsequent Work.

§ 3.4 LABOR AND MATERIALS

§ 3.4.1 Unless otherwise provided in the Contract Documents, the Contractor shall provide and pay for labor, materials, equipment, tools, construction equipment and machinery, water, heat, utilities, transportation, and other facilities and services necessary for proper execution and completion of the Work, whether temporary or permanent and whether or not incorporated or to be incorporated in the Work.

§ 3.4.2 The Contractor may make substitutions only with the consent of the Owner, after evaluation by the Architect and in accordance with a Change Order.

§ 3.4.3 The Contractor shall enforce strict discipline and good order among the Contractor's employees and other persons carrying out the Contract. The Contractor shall not permit employment of unfit persons or persons not skilled in tasks assigned to them.

§ 3.5 WARRANTY

§ 3.5.1 The Contractor warrants to the Owner and Architect that materials and equipment furnished under the Contract will be of good quality and new unless otherwise required or permitted by the Contract Documents, that the Work will be free from defects not inherent in the quality required or permitted, and that the Work will conform to the requirements of the Contract Documents. Work not conforming to these requirements, including substitutions not properly approved and authorized, may be considered defective. The Contractor's warranty excludes remedy for damage or defect caused by abuse, modifications not executed by the Contractor, improper or insufficient maintenance, improper operation, or normal wear and tear and normal usage. If required by the Architect, the Contractor shall furnish satisfactory evidence as to the kind and quality of materials and equipment.

13

Appendix E

§ 3.6 TAXES
§ 3.6.1 The Contractor shall pay sales, consumer, use and similar taxes for the Work provided by the Contractor which are legally enacted when bids are received or negotiations concluded, whether or not yet effective or merely scheduled to go into effect.

§ 3.7 PERMITS, FEES AND NOTICES
§ 3.7.1 Unless otherwise provided in the Contract Documents, the Contractor shall secure and pay for the building permit and other permits and governmental fees, licenses and inspections necessary for proper execution and completion of the Work which are customarily secured after execution of the Contract and which are legally required when bids are received or negotiations concluded.

§ 3.7.2 The Contractor shall comply with and give notices required by laws, ordinances, rules, regulations and lawful orders of public authorities applicable to performance of the Work.

§ 3.7.3 It is not the Contractor's responsibility to ascertain that the Contract Documents are in accordance with applicable laws, statutes, ordinances, building codes, and rules and regulations. However, if the Contractor observes that portions of the Contract Documents are at variance therewith, the Contractor shall promptly notify the Architect and Owner in writing, and necessary changes shall be accomplished by appropriate Modification.

§ 3.7.4 If the Contractor performs Work knowing it to be contrary to laws, statutes, ordinances, building codes, and rules and regulations without such notice to the Architect and Owner, the Contractor shall assume appropriate responsibility for such Work and shall bear the costs attributable to correction.

§ 3.8 ALLOWANCES
§ 3.8.1 The Contractor shall include in the Contract Sum all allowances stated in the Contract Documents. Items covered by allowances shall be supplied for such amounts and by such persons or entities as the Owner may direct, but the Contractor shall not be required to employ persons or entities to whom the Contractor has reasonable objection.

§ 3.8.2 Unless otherwise provided in the Contract Documents:
- .1 allowances shall cover the cost to the Contractor of materials and equipment delivered at the site and all required taxes, less applicable trade discounts;
- .2 Contractor's costs for unloading and handling at the site, labor, installation costs, overhead, profit and other expenses contemplated for stated allowance amounts shall be included in the Contract Sum but not in the allowances;
- .3 whenever costs are more than or less than allowances, the Contract Sum shall be adjusted accordingly by Change Order. The amount of the Change Order shall reflect (1) the difference between actual costs and the allowances under Section 3.8.2.1 and (2) changes in Contractor's costs under Section 3.8.2.2.

§ 3.8.3 Materials and equipment under an allowance shall be selected by the Owner in sufficient time to avoid delay in the Work.

§ 3.9 SUPERINTENDENT
§ 3.9.1 The Contractor shall employ a competent superintendent and necessary assistants who shall be in attendance at the Project site during performance of the Work. The superintendent shall represent the Contractor, and communications given to the superintendent shall be as binding as if given to the Contractor. Important communications shall be confirmed in writing. Other communications shall be similarly confirmed on written request in each case.

§ 3.10 CONTRACTOR'S CONSTRUCTION SCHEDULES
§ 3.10.1 The Contractor, promptly after being awarded the Contract, shall prepare and submit for the Owner's and Architect's information a Contractor's construction schedule for the Work. The schedule shall not exceed time limits current under the Contract Documents, shall be revised at appropriate intervals as required by the conditions of the Work and Project, shall be related to the entire Project to the extent required by the Contract Documents, and shall provide for expeditious and practicable execution of the Work.

AIA Document A201™—1997

§ 3.10.2 The Contractor shall prepare and keep current, for the Architect's approval, a schedule of submittals which is coordinated with the Contractor's construction schedule and allows the Architect reasonable time to review submittals.

§ 3.10.3 The Contractor shall perform the Work in general accordance with the most recent schedules submitted to the Owner and Architect.

§ 3.11 DOCUMENTS AND SAMPLES AT THE SITE

§ 3.11.1 The Contractor shall maintain at the site for the Owner one record copy of the Drawings, Specifications, Addenda, Change Orders and other Modifications, in good order and marked currently to record field changes and selections made during construction, and one record copy of approved Shop Drawings, Product Data, Samples and similar required submittals. These shall be available to the Architect and shall be delivered to the Architect for submittal to the Owner upon completion of the Work.

§ 3.12 SHOP DRAWINGS, PRODUCT DATA AND SAMPLES

§ 3.12.1 Shop Drawings are drawings, diagrams, schedules and other data specially prepared for the Work by the Contractor or a Subcontractor, Sub-subcontractor, manufacturer, supplier or distributor to illustrate some portion of the Work.

§ 3.12.2 Product Data are illustrations, standard schedules, performance charts, instructions, brochures, diagrams and other information furnished by the Contractor to illustrate materials or equipment for some portion of the Work.

§ 3.12.3 Samples are physical examples which illustrate materials, equipment or workmanship and establish standards by which the Work will be judged.

§ 3.12.4 Shop Drawings, Product Data, Samples and similar submittals are not Contract Documents. The purpose of their submittal is to demonstrate for those portions of the Work for which submittals are required by the Contract Documents the way by which the Contractor proposes to conform to the information given and the design concept expressed in the Contract Documents. Review by the Architect is subject to the limitations of Section 4.2.7. Informational submittals upon which the Architect is not expected to take responsive action may be so identified in the Contract Documents. Submittals which are not required by the Contract Documents may be returned by the Architect without action.

§ 3.12.5 The Contractor shall review for compliance with the Contract Documents, approve and submit to the Architect Shop Drawings, Product Data, Samples and similar submittals required by the Contract Documents with reasonable promptness and in such sequence as to cause no delay in the Work or in the activities of the Owner or of separate contractors. Submittals which are not marked as reviewed for compliance with the Contract Documents and approved by the Contractor may be returned by the Architect without action.

§ 3.12.6 By approving and submitting Shop Drawings, Product Data, Samples and similar submittals, the Contractor represents that the Contractor has determined and verified materials, field measurements and field construction criteria related thereto, or will do so, and has checked and coordinated the information contained within such submittals with the requirements of the Work and of the Contract Documents.

§ 3.12.7 The Contractor shall perform no portion of the Work for which the Contract Documents require submittal and review of Shop Drawings, Product Data, Samples or similar submittals until the respective submittal has been approved by the Architect.

§ 3.12.8 The Work shall be in accordance with approved submittals except that the Contractor shall not be relieved of responsibility for deviations from requirements of the Contract Documents by the Architect's approval of Shop Drawings, Product Data, Samples or similar submittals unless the Contractor has specifically informed the Architect in writing of such deviation at the time of submittal and (1) the Architect has given written approval to the specific deviation as a minor change in the Work, or (2) a Change Order or Construction Change Directive has been issued authorizing the deviation. The Contractor shall not be relieved of responsibility for errors or omissions in Shop Drawings, Product Data, Samples or similar submittals by the Architect's approval thereof.

§ 3.12.9 The Contractor shall direct specific attention, in writing or on resubmitted Shop Drawings, Product Data, Samples or similar submittals, to revisions other than those requested by the Architect on previous submittals. In the absence of such written notice the Architect's approval of a resubmission shall not apply to such revisions.

15

§ 3.12.10 The Contractor shall not be required to provide professional services which constitute the practice of architecture or engineering unless such services are specifically required by the Contract Documents for a portion of the Work or unless the Contractor needs to provide such services in order to carry out the Contractor's responsibilities for construction means, methods, techniques, sequences and procedures. The Contractor shall not be required to provide professional services in violation of applicable law. If professional design services or certifications by a design professional related to systems, materials or equipment are specifically required of the Contractor by the Contract Documents, the Owner and the Architect will specify all performance and design criteria that such services must satisfy. The Contractor shall cause such services or certifications to be provided by a properly licensed design professional, whose signature and seal shall appear on all drawings, calculations, specifications, certifications, Shop Drawings and other submittals prepared by such professional. Shop Drawings and other submittals related to the Work designed or certified by such professional, if prepared by others, shall bear such professional's written approval when submitted to the Architect. The Owner and the Architect shall be entitled to rely upon the adequacy, accuracy and completeness of the services, certifications or approvals performed by such design professionals, provided the Owner and Architect have specified to the Contractor all performance and design criteria that such services must satisfy. Pursuant to this Section 3.12.10, the Architect will review, approve or take other appropriate action on submittals only for the limited purpose of checking for conformance with information given and the design concept expressed in the Contract Documents. The Contractor shall not be responsible for the adequacy of the performance or design criteria required by the Contract Documents.

§ 3.13 USE OF SITE
§ 3.13.1 The Contractor shall confine operations at the site to areas permitted by law, ordinances, permits and the Contract Documents and shall not unreasonably encumber the site with materials or equipment.

§ 3.14 CUTTING AND PATCHING
§ 3.14.1 The Contractor shall be responsible for cutting, fitting or patching required to complete the Work or to make its parts fit together properly.

§ 3.14.2 The Contractor shall not damage or endanger a portion of the Work or fully or partially completed construction of the Owner or separate contractors by cutting, patching or otherwise altering such construction, or by excavation. The Contractor shall not cut or otherwise alter such construction by the Owner or a separate contractor except with written consent of the Owner and of such separate contractor; such consent shall not be unreasonably withheld. The Contractor shall not unreasonably withhold from the Owner or a separate contractor the Contractor's consent to cutting or otherwise altering the Work.

§ 3.15 CLEANING UP
§ 3.15.1 The Contractor shall keep the premises and surrounding area free from accumulation of waste materials or rubbish caused by operations under the Contract. At completion of the Work, the Contractor shall remove from and about the Project waste materials, rubbish, the Contractor's tools, construction equipment, machinery and surplus materials.

§ 3.15.2 If the Contractor fails to clean up as provided in the Contract Documents, the Owner may do so and the cost thereof shall be charged to the Contractor.

§ 3.16 ACCESS TO WORK
§ 3.16.1 The Contractor shall provide the Owner and Architect access to the Work in preparation and progress wherever located.

§ 3.17 ROYALTIES, PATENTS AND COPYRIGHTS
§ 3.17.1 The Contractor shall pay all royalties and license fees. The Contractor shall defend suits or claims for infringement of copyrights and patent rights and shall hold the Owner and Architect harmless from loss on account thereof, but shall not be responsible for such defense or loss when a particular design, process or product of a particular manufacturer or manufacturers is required by the Contract Documents or where the copyright violations are contained in Drawings, Specifications or other documents prepared by the Owner or Architect. However, if the Contractor has reason to believe that the required design, process or product is an infringement of a copyright or a patent, the Contractor shall be responsible for such loss unless such information is promptly furnished to the Architect.

AIA Document A201™—1997

§ 3.18 INDEMNIFICATION

§ 3.18.1 To the fullest extent permitted by law and to the extent claims, damages, losses or expenses are not covered by Project Management Protective Liability insurance purchased by the Contractor in accordance with Section 11.3, the Contractor shall indemnify and hold harmless the Owner, Architect, Architect's consultants, and agents and employees of any of them from and against claims, damages, losses and expenses, including but not limited to attorneys' fees, arising out of or resulting from performance of the Work, provided that such claim, damage, loss or expense is attributable to bodily injury, sickness, disease or death, or to injury to or destruction of tangible property (other than the Work itself), but only to the extent caused by the negligent acts or omissions of the Contractor, a Subcontractor, anyone directly or indirectly employed by them or anyone for whose acts they may be liable, regardless of whether or not such claim, damage, loss or expense is caused in part by a party indemnified hereunder. Such obligation shall not be construed to negate, abridge, or reduce other rights or obligations of indemnity which would otherwise exist as to a party or person described in this Section 3.18.

§ 3.18.2 In claims against any person or entity indemnified under this Section 3.18 by an employee of the Contractor, a Subcontractor, anyone directly or indirectly employed by them or anyone for whose acts they may be liable, the indemnification obligation under Section 3.18.1 shall not be limited by a limitation on amount or type of damages, compensation or benefits payable by or for the Contractor or a Subcontractor under workers' compensation acts, disability benefit acts or other employee benefit acts.

ARTICLE 4 ADMINISTRATION OF THE CONTRACT
§ 4.1 ARCHITECT

§ 4.1.1 The Architect is the person lawfully licensed to practice architecture or an entity lawfully practicing architecture identified as such in the Agreement and is referred to throughout the Contract Documents as if singular in number. The term "Architect" means the Architect or the Architect's authorized representative.

§ 4.1.2 Duties, responsibilities and limitations of authority of the Architect as set forth in the Contract Documents shall not be restricted, modified or extended without written consent of the Owner, Contractor and Architect. Consent shall not be unreasonably withheld.

§ 4.1.3 If the employment of the Architect is terminated, the Owner shall employ a new Architect against whom the Contractor has no reasonable objection and whose status under the Contract Documents shall be that of the former Architect.

§ 4.2 ARCHITECT'S ADMINISTRATION OF THE CONTRACT

§ 4.2.1 The Architect will provide administration of the Contract as described in the Contract Documents, and will be an Owner's representative (1) during construction, (2) until final payment is due and (3) with the Owner's concurrence, from time to time during the one-year period for correction of Work described in Section 12.2. The Architect will have authority to act on behalf of the Owner only to the extent provided in the Contract Documents, unless otherwise modified in writing in accordance with other provisions of the Contract.

§ 4.2.2 The Architect, as a representative of the Owner, will visit the site at intervals appropriate to the stage of the Contractor's operations (1) to become generally familiar with and to keep the Owner informed about the progress and quality of the portion of the Work completed, (2) to endeavor to guard the Owner against defects and deficiencies in the Work, and (3) to determine in general if the Work is being performed in a manner indicating that the Work, when fully completed, will be in accordance with the Contract Documents. However, the Architect will not be required to make exhaustive or continuous on-site inspections to check the quality or quantity of the Work. The Architect will neither have control over or charge of, nor be responsible for the construction means, methods, techniques, sequences or procedures, or for the safety precautions and programs in connection with the Work, since these are solely the Contractor's rights and responsibilities under the Contract Documents, except as provided in Section 3.3.1.

§ 4.2.3 The Architect will not be responsible for the Contractor's failure to perform the Work in accordance with the requirements of the Contract Documents. The Architect will not have control over or charge of and will not be responsible for acts or omissions of the Contractor, Subcontractors, or their agents or employees, or any other persons or entities performing portions of the Work.

§ 4.2.4 Communications Facilitating Contract Administration. Except as otherwise provided in the Contract Documents or when direct communications have been specially authorized, the Owner and Contractor shall endeavor to communicate with each other through the Architect about matters arising out of or relating to the

Contract. Communications by and with the Architect's consultants shall be through the Architect. Communications by and with Subcontractors and material suppliers shall be through the Contractor. Communications by and with separate contractors shall be through the Owner.

§ 4.2.5 Based on the Architect's evaluations of the Contractor's Applications for Payment, the Architect will review and certify the amounts due the Contractor and will issue Certificates for Payment in such amounts.

§ 4.2.6 The Architect will have authority to reject Work that does not conform to the Contract Documents. Whenever the Architect considers it necessary or advisable, the Architect will have authority to require inspection or testing of the Work in accordance with Sections 13.5.2 and 13.5.3, whether or not such Work is fabricated, installed or completed. However, neither this authority of the Architect nor a decision made in good faith either to exercise or not to exercise such authority shall give rise to a duty or responsibility of the Architect to the Contractor, Subcontractors, material and equipment suppliers, their agents or employees, or other persons or entities performing portions of the Work.

§ 4.2.7 The Architect will review and approve or take other appropriate action upon the Contractor's submittals such as Shop Drawings, Product Data and Samples, but only for the limited purpose of checking for conformance with information given and the design concept expressed in the Contract Documents. The Architect's action will be taken with such reasonable promptness as to cause no delay in the Work or in the activities of the Owner, Contractor or separate contractors, while allowing sufficient time in the Architect's professional judgment to permit adequate review. Review of such submittals is not conducted for the purpose of determining the accuracy and completeness of other details such as dimensions and quantities, or for substantiating instructions for installation or performance of equipment or systems, all of which remain the responsibility of the Contractor as required by the Contract Documents. The Architect's review of the Contractor's submittals shall not relieve the Contractor of the obligations under Sections 3.3, 3.5 and 3.12. The Architect's review shall not constitute approval of safety precautions or, unless otherwise specifically stated by the Architect, of any construction means, methods, techniques, sequences or procedures. The Architect's approval of a specific item shall not indicate approval of an assembly of which the item is a component.

§ 4.2.8 The Architect will prepare Change Orders and Construction Change Directives, and may authorize minor changes in the Work as provided in Section 7.4.

§ 4.2.9 The Architect will conduct inspections to determine the date or dates of Substantial Completion and the date of final completion, will receive and forward to the Owner, for the Owner's review and records, written warranties and related documents required by the Contract and assembled by the Contractor, and will issue a final Certificate for Payment upon compliance with the requirements of the Contract Documents.

§ 4.2.10 If the Owner and Architect agree, the Architect will provide one or more project representatives to assist in carrying out the Architect's responsibilities at the site. The duties, responsibilities and limitations of authority of such project representatives shall be as set forth in an exhibit to be incorporated in the Contract Documents.

§ 4.2.11 The Architect will interpret and decide matters concerning performance under and requirements of, the Contract Documents on written request of either the Owner or Contractor. The Architect's response to such requests will be made in writing within any time limits agreed upon or otherwise with reasonable promptness. If no agreement is made concerning the time within which interpretations required of the Architect shall be furnished in compliance with this Section 4.2, then delay shall not be recognized on account of failure by the Architect to furnish such interpretations until 15 days after written request is made for them.

§ 4.2.12 Interpretations and decisions of the Architect will be consistent with the intent of and reasonably inferable from the Contract Documents and will be in writing or in the form of drawings. When making such interpretations and initial decisions, the Architect will endeavor to secure faithful performance by both Owner and Contractor, will not show partiality to either and will not be liable for results of interpretations or decisions so rendered in good faith.

§ 4.2.13 The Architect's decisions on matters relating to aesthetic effect will be final if consistent with the intent expressed in the Contract Documents.

§ 4.3 CLAIMS AND DISPUTES

§ 4.3.1 Definition. A Claim is a demand or assertion by one of the parties seeking, as a matter of right, adjustment or interpretation of Contract terms, payment of money, extension of time or other relief with respect to the terms of

18

AIA Document A201™—1997

the Contract. The term "Claim" also includes other disputes and matters in question between the Owner and Contractor arising out of or relating to the Contract. Claims must be initiated by written notice. The responsibility to substantiate Claims shall rest with the party making the Claim.

§ 4.3.2 Time Limits on Claims. Claims by either party must be initiated within 21 days after occurrence of the event giving rise to such Claim or within 21 days after the claimant first recognizes the condition giving rise to the Claim, whichever is later. Claims must be initiated by written notice to the Architect and the other party.

§ 4.3.3 Continuing Contract Performance. Pending final resolution of a Claim except as otherwise agreed in writing or as provided in Section 9.7.1 and Article 14, the Contractor shall proceed diligently with performance of the Contract and the Owner shall continue to make payments in accordance with the Contract Documents.

§ 4.3.4 Claims for Concealed or Unknown Conditions. If conditions are encountered at the site which are (1) subsurface or otherwise concealed physical conditions which differ materially from those indicated in the Contract Documents or (2) unknown physical conditions of an unusual nature, which differ materially from those ordinarily found to exist and generally recognized as inherent in construction activities of the character provided for in the Contract Documents, then notice by the observing party shall be given to the other party promptly before conditions are disturbed and in no event later than 21 days after first observance of the conditions. The Architect will promptly investigate such conditions and, if they differ materially and cause an increase or decrease in the Contractor's cost of, or time required for, performance of any part of the Work, will recommend an equitable adjustment in the Contract Sum or Contract Time, or both. If the Architect determines that the conditions at the site are not materially different from those indicated in the Contract Documents and that no change in the terms of the Contract is justified, the Architect shall so notify the Owner and Contractor in writing, stating the reasons. Claims by either party in opposition to such determination must be made within 21 days after the Architect has given notice of the decision. If the conditions encountered are materially different, the Contract Sum and Contract Time shall be equitably adjusted, but if the Owner and Contractor cannot agree on an adjustment in the Contract Sum or Contract Time, the adjustment shall be referred to the Architect for initial determination, subject to further proceedings pursuant to Section 4.4.

§ 4.3.5 Claims for Additional Cost. If the Contractor wishes to make Claim for an increase in the Contract Sum, written notice as provided herein shall be given before proceeding to execute the Work. Prior notice is not required for Claims relating to an emergency endangering life or property arising under Section 10.6.

§ 4.3.6 If the Contractor believes additional cost is involved for reasons including but not limited to (1) a written interpretation from the Architect, (2) an order by the Owner to stop the Work where the Contractor was not at fault, (3) a written order for a minor change in the Work issued by the Architect, (4) failure of payment by the Owner, (5) termination of the Contract by the Owner, (6) Owner's suspension or (7) other reasonable grounds, Claim shall be filed in accordance with this Section 4.3.

§ 4.3.7 Claims for Additional Time
§ 4.3.7.1 If the Contractor wishes to make Claim for an increase in the Contract Time, written notice as provided herein shall be given. The Contractor's Claim shall include an estimate of cost and of probable effect of delay on progress of the Work. In the case of a continuing delay only one Claim is necessary.

§ 4.3.7.2 If adverse weather conditions are the basis for a Claim for additional time, such Claim shall be documented by data substantiating that weather conditions were abnormal for the period of time, could not have been reasonably anticipated and had an adverse effect on the scheduled construction.

§ 4.3.8 Injury or Damage to Person or Property. If either party to the Contract suffers injury or damage to person or property because of an act or omission of the other party, or of others for whose acts such party is legally responsible, written notice of such injury or damage, whether or not insured, shall be given to the other party within a reasonable time not exceeding 21 days after discovery. The notice shall provide sufficient detail to enable the other party to investigate the matter.

§ 4.3.9 If unit prices are stated in the Contract Documents or subsequently agreed upon, and if quantities originally contemplated are materially changed in a proposed Change Order or Construction Change Directive so that application of such unit prices to quantities of Work proposed will cause substantial inequity to the Owner or Contractor, the applicable unit prices shall be equitably adjusted.

19

Appendix E

§ 4.3.10 Claims for Consequential Damages. The Contractor and Owner waive Claims against each other for consequential damages arising out of or relating to this Contract. This mutual waiver includes:

.1   damages incurred by the Owner for rental expenses, for losses of use, income, profit, financing, business and reputation, and for loss of management or employee productivity or of the services of such persons; and

.2   damages incurred by the Contractor for principal office expenses including the compensation of personnel stationed there, for losses of financing, business and reputation, and for loss of profit except anticipated profit arising directly from the Work.

This mutual waiver is applicable, without limitation, to all consequential damages due to either party's termination in accordance with Article 14. Nothing contained in this Section 4.3.10 shall be deemed to preclude an award of liquidated direct damages, when applicable, in accordance with the requirements of the Contract Documents.

§ 4.4 RESOLUTION OF CLAIMS AND DISPUTES

§ 4.4.1 Decision of Architect. Claims, including those alleging an error or omission by the Architect but excluding those arising under Sections 10.3 through 10.5, shall be referred initially to the Architect for decision. An initial decision by the Architect shall be required as a condition precedent to mediation, arbitration or litigation of all Claims between the Contractor and Owner arising prior to the date final payment is due, unless 30 days have passed after the Claim has been referred to the Architect with no decision having been rendered by the Architect. The Architect will not decide disputes between the Contractor and persons or entities other than the Owner.

§ 4.4.2 The Architect will review Claims and within ten days of the receipt of the Claim take one or more of the following actions: (1) request additional supporting data from the claimant or a response with supporting data from the other party, (2) reject the Claim in whole or in part, (3) approve the Claim, (4) suggest a compromise, or (5) advise the parties that the Architect is unable to resolve the Claim if the Architect lacks sufficient information to evaluate the merits of the Claim or if the Architect concludes that, in the Architect's sole discretion, it would be inappropriate for the Architect to resolve the Claim.

§ 4.4.3 In evaluating Claims, the Architect may, but shall not be obligated to, consult with or seek information from either party or from persons with special knowledge or expertise who may assist the Architect in rendering a decision. The Architect may request the Owner to authorize retention of such persons at the Owner's expense.

§ 4.4.4 If the Architect requests a party to provide a response to a Claim or to furnish additional supporting data, such party shall respond, within ten days after receipt of such request, and shall either provide a response on the requested supporting data, advise the Architect when the response or supporting data will be furnished or advise the Architect that no supporting data will be furnished. Upon receipt of the response or supporting data, if any, the Architect will either reject or approve the Claim in whole or in part.

§ 4.4.5 The Architect will approve or reject Claims by written decision, which shall state the reasons therefor and which shall notify the parties of any change in the Contract Sum or Contract Time or both. The approval or rejection of a Claim by the Architect shall be final and binding on the parties but subject to mediation and arbitration.

§ 4.4.6 When a written decision of the Architect states that (1) the decision is final but subject to mediation and arbitration and (2) a demand for arbitration of a Claim covered by such decision must be made within 30 days after the date on which the party making the demand receives the final written decision, then failure to demand arbitration within said 30 days' period shall result in the Architect's decision becoming final and binding upon the Owner and Contractor. If the Architect renders a decision after arbitration proceedings have been initiated, such decision may be entered as evidence, but shall not supersede arbitration proceedings unless the decision is acceptable to all parties concerned.

§ 4.4.7 Upon receipt of a Claim against the Contractor or at any time thereafter, the Architect or the Owner may, but is not obligated to, notify the surety, if any, of the nature and amount of the Claim. If the Claim relates to a possibility of a Contractor's default, the Architect or the Owner may, but is not obligated to, notify the surety and request the surety's assistance in resolving the controversy.

§ 4.4.8 If a Claim relates to or is the subject of a mechanic's lien, the party asserting such Claim may proceed in accordance with applicable law to comply with the lien notice or filing deadlines prior to resolution of the Claim by the Architect, by mediation or by arbitration.

AIA Document A201™—1997

§ 4.5 MEDIATION

§ 4.5.1 Any Claim arising out of or related to the Contract, except Claims relating to aesthetic effect and except those waived as provided for in Sections 4.3.10, 9.10.4 and 9.10.5 shall, after initial decision by the Architect or 30 days after submission of the Claim to the Architect, be subject to mediation as a condition precedent to arbitration or the institution of legal or equitable proceedings by either party.

§ 4.5.2 The parties shall endeavor to resolve their Claims by mediation which, unless the parties mutually agree otherwise, shall be in accordance with the Construction Industry Mediation Rules of the American Arbitration Association currently in effect. Request for mediation shall be filed in writing with the other party to the Contract and with the American Arbitration Association. The request may be made concurrently with the filing of a demand for arbitration but, in such event, mediation shall proceed in advance of arbitration or legal or equitable proceedings, which shall be stayed pending mediation for a period of 60 days from the date of filing, unless stayed for a longer period by agreement of the parties or court order.

§ 4.5.3 The parties shall share the mediator's fee and any filing fees equally. The mediation shall be held in the place where the Project is located, unless another location is mutually agreed upon. Agreements reached in mediation shall be enforceable as settlement agreements in any court having jurisdiction thereof.

§ 4.6 ARBITRATION

§ 4.6.1 Any Claim arising out of or related to the Contract, except Claims relating to aesthetic effect and except those waived as provided for in Sections 4.3.10, 9.10.4 and 9.10.5, shall, after decision by the Architect or 30 days after submission of the Claim to the Architect, be subject to arbitration. Prior to arbitration, the parties shall endeavor to resolve disputes by mediation in accordance with the provisions of Section 4.5.

§ 4.6.2 Claims not resolved by mediation shall be decided by arbitration which, unless the parties mutually agree otherwise, shall be in accordance with the Construction Industry Arbitration Rules of the American Arbitration Association currently in effect. The demand for arbitration shall be filed in writing with the other party to the Contract and with the American Arbitration Association, and a copy shall be filed with the Architect.

§ 4.6.3 A demand for arbitration shall be made within the time limits specified in Sections 4.4.6 and 4.6.1 as applicable, and in other cases within a reasonable time after the Claim has arisen, and in no event shall it be made after the date when institution of legal or equitable proceedings based on such Claim would be barred by the applicable statute of limitations as determined pursuant to Section 13.7.

§ 4.6.4 **Limitation on Consolidation or Joinder.** No arbitration arising out of or relating to the Contract shall include, by consolidation or joinder or in any other manner, the Architect, the Architect's employees or consultants, except by written consent containing specific reference to the Agreement and signed by the Architect, Owner, Contractor and any other person or entity sought to be joined. No arbitration shall include, by consolidation or joinder or in any other manner, parties other than the Owner, Contractor, a separate contractor as described in Article 6 and other persons substantially involved in a common question of fact or law whose presence is required if complete relief is to be accorded in arbitration. No person or entity other than the Owner, Contractor or a separate contractor as described in Article 6 shall be included as an original third party or additional third party to an arbitration whose interest or responsibility is insubstantial. Consent to arbitration involving an additional person or entity shall not constitute consent to arbitration of a Claim not described therein or with a person or entity not named or described therein. The foregoing agreement to arbitrate and other agreements to arbitrate with an additional person or entity duly consented to by parties to the Agreement shall be specifically enforceable under applicable law in any court having jurisdiction thereof.

§ 4.6.5 **Claims and Timely Assertion of Claims.** The party filing a notice of demand for arbitration must assert in the demand all Claims then known to that party on which arbitration is permitted to be demanded.

§ 4.6.6 **Judgment on Final Award.** The award rendered by the arbitrator or arbitrators shall be final, and judgment may be entered upon it in accordance with applicable law in any court having jurisdiction thereof.

ARTICLE 5 SUBCONTRACTORS

§ 5.1 DEFINITIONS

§ 5.1.1 A Subcontractor is a person or entity who has a direct contract with the Contractor to perform a portion of the Work at the site. The term "Subcontractor" is referred to throughout the Contract Documents as if singular in

21

number and means a Subcontractor or an authorized representative of the Subcontractor. The term "Subcontractor" does not include a separate contractor or subcontractors of a separate contractor.

§ 5.1.2 A Sub-subcontractor is a person or entity who has a direct or indirect contract with a Subcontractor to perform a portion of the Work at the site. The term "Sub-subcontractor" is referred to throughout the Contract Documents as if singular in number and means a Sub-subcontractor or an authorized representative of the Sub-subcontractor.

§ 5.2 AWARD OF SUBCONTRACTS AND OTHER CONTRACTS FOR PORTIONS OF THE WORK

§ 5.2.1 Unless otherwise stated in the Contract Documents or the bidding requirements, the Contractor, as soon as practicable after award of the Contract, shall furnish in writing to the Owner through the Architect the names of persons or entities (including those who are to furnish materials or equipment fabricated to a special design) proposed for each principal portion of the Work. The Architect will promptly reply to the Contractor in writing stating whether or not the Owner or the Architect, after due investigation, has reasonable objection to any such proposed person or entity. Failure of the Owner or Architect to reply promptly shall constitute notice of no reasonable objection.

§ 5.2.2 The Contractor shall not contract with a proposed person or entity to whom the Owner or Architect has made reasonable and timely objection. The Contractor shall not be required to contract with anyone to whom the Contractor has made reasonable objection.

§ 5.2.3 If the Owner or Architect has reasonable objection to a person or entity proposed by the Contractor, the Contractor shall propose another to whom the Owner or Architect has no reasonable objection. If the proposed but rejected Subcontractor was reasonably capable of performing the Work, the Contract Sum and Contract Time shall be increased or decreased by the difference, if any, occasioned by such change, and an appropriate Change Order shall be issued before commencement of the substitute Subcontractor's Work. However, no increase in the Contract Sum or Contract Time shall be allowed for such change unless the Contractor has acted promptly and responsively in submitting names as required.

§ 5.2.4 The Contractor shall not change a Subcontractor, person or entity previously selected if the Owner or Architect makes reasonable objection to such substitute.

§ 5.3 SUBCONTRACTUAL RELATIONS

§ 5.3.1 By appropriate agreement, written where legally required for validity, the Contractor shall require each Subcontractor, to the extent of the Work to be performed by the Subcontractor, to be bound to the Contractor by terms of the Contract Documents, and to assume toward the Contractor all the obligations and responsibilities, including the responsibility for safety of the Subcontractor's Work, which the Contractor, by these Documents, assumes toward the Owner and Architect. Each subcontract agreement shall preserve and protect the rights of the Owner and Architect under the Contract Documents with respect to the Work to be performed by the Subcontractor so that subcontracting thereof will not prejudice such rights, and shall allow to the Subcontractor, unless specifically provided otherwise in the subcontract agreement, the benefit of all rights, remedies and redress against the Contractor that the Contractor, by the Contract Documents, has against the Owner. Where appropriate, the Contractor shall require each Subcontractor to enter into similar agreements with Sub-subcontractors. The Contractor shall make available to each proposed Subcontractor, prior to the execution of the subcontract agreement, copies of the Contract Documents to which the Subcontractor will be bound, and, upon written request of the Subcontractor, identify to the Subcontractor terms and conditions of the proposed subcontract agreement which may be at variance with the Contract Documents. Subcontractors will similarly make copies of applicable portions of such documents available to their respective proposed Sub-subcontractors.

§ 5.4 CONTINGENT ASSIGNMENT OF SUBCONTRACTS

§ 5.4.1 Each subcontract agreement for a portion of the Work is assigned by the Contractor to the Owner provided that:

    .1   assignment is effective only after termination of the Contract by the Owner for cause pursuant to Section 14.2 and only for those subcontract agreements which the Owner accepts by notifying the Subcontractor and Contractor in writing; and

    .2   assignment is subject to the prior rights of the surety, if any, obligated under bond relating to the Contract.

AIA Document A201™—1997

§ 5.4.2 Upon such assignment, if the Work has been suspended for more than 30 days, the Subcontractor's compensation shall be equitably adjusted for increases in cost resulting from the suspension.

ARTICLE 6 CONSTRUCTION BY OWNER OR BY SEPARATE CONTRACTORS
§ 6.1 OWNER'S RIGHT TO PERFORM CONSTRUCTION AND TO AWARD SEPARATE CONTRACTS
§ 6.1.1 The Owner reserves the right to perform construction or operations related to the Project with the Owner's own forces, and to award separate contracts in connection with other portions of the Project or other construction or operations on the site under Conditions of the Contract identical or substantially similar to these including those portions related to insurance and waiver of subrogation. If the Contractor claims that delay or additional cost is involved because of such action by the Owner, the Contractor shall make such Claim as provided in Section 4.3.

§ 6.1.2 When separate contracts are awarded for different portions of the Project or other construction or operations on the site, the term "Contractor" in the Contract Documents in each case shall mean the Contractor who executes each separate Owner-Contractor Agreement.

§ 6.1.3 The Owner shall provide for coordination of the activities of the Owner's own forces and of each separate contractor with the Work of the Contractor, who shall cooperate with them. The Contractor shall participate with other separate contractors and the Owner in reviewing their construction schedules when directed to do so. The Contractor shall make any revisions to the construction schedule deemed necessary after a joint review and mutual agreement. The construction schedules shall then constitute the schedules to be used by the Contractor, separate contractors and the Owner until subsequently revised.

§ 6.1.4 Unless otherwise provided in the Contract Documents, when the Owner performs construction or operations related to the Project with the Owner's own forces, the Owner shall be deemed to be subject to the same obligations and to have the same rights which apply to the Contractor under the Conditions of the Contract, including, without excluding others, those stated in Article 3, this Article 6 and Articles 10, 11 and 12.

§ 6.2 MUTUAL RESPONSIBILITY
§ 6.2.1 The Contractor shall afford the Owner and separate contractors reasonable opportunity for introduction and storage of their materials and equipment and performance of their activities, and shall connect and coordinate the Contractor's construction and operations with theirs as required by the Contract Documents.

§ 6.2.2 If part of the Contractor's Work depends for proper execution or results upon construction or operations by the Owner or a separate contractor, the Contractor shall, prior to proceeding with that portion of the Work, promptly report to the Architect apparent discrepancies or defects in such other construction that would render it unsuitable for such proper execution and results. Failure of the Contractor so to report shall constitute an acknowledgment that the Owner's or separate contractor's completed or partially completed construction is fit and proper to receive the Contractor's Work, except as to defects not then reasonably discoverable.

§ 6.2.3 The Owner shall be reimbursed by the Contractor for costs incurred by the Owner which are payable to a separate contractor because of delays, improperly timed activities or defective construction of the Contractor. The Owner shall be responsible to the Contractor for costs incurred by the Contractor because of delays, improperly timed activities, damage to the Work or defective construction of a separate contractor.

§ 6.2.4 The Contractor shall promptly remedy damage wrongfully caused by the Contractor to completed or partially completed construction or to property of the Owner or separate contractors as provided in Section 10.2.5.

§ 6.2.5 The Owner and each separate contractor shall have the same responsibilities for cutting and patching as are described for the Contractor in Section 3.14.

§ 6.3 OWNER'S RIGHT TO CLEAN UP
§ 6.3.1 If a dispute arises among the Contractor, separate contractors and the Owner as to the responsibility under their respective contracts for maintaining the premises and surrounding area free from waste materials and rubbish, the Owner may clean up and the Architect will allocate the cost among those responsible.

23

ARTICLE 7   CHANGES IN THE WORK

§ 7.1 GENERAL

§ 7.1.1 Changes in the Work may be accomplished after execution of the Contract, and without invalidating the Contract, by Change Order, Construction Change Directive or order for a minor change in the Work, subject to the limitations stated in this Article 7 and elsewhere in the Contract Documents.

§ 7.1.2 A Change Order shall be based upon agreement among the Owner, Contractor and Architect; a Construction Change Directive requires agreement by the Owner and Architect and may or may not be agreed to by the Contractor; an order for a minor change in the Work may be issued by the Architect alone.

§ 7.1.3 Changes in the Work shall be performed under applicable provisions of the Contract Documents, and the Contractor shall proceed promptly, unless otherwise provided in the Change Order, Construction Change Directive or order for a minor change in the Work.

§ 7.2 CHANGE ORDERS

§ 7.2.1 A Change Order is a written instrument prepared by the Architect and signed by the Owner, Contractor and Architect, stating their agreement upon all of the following:

   .1   change in the Work;
   .2   the amount of the adjustment, if any, in the Contract Sum; and
   .3   the extent of the adjustment, if any, in the Contract Time.

§ 7.2.2 Methods used in determining adjustments to the Contract Sum may include those listed in Section 7.3.3.

§ 7.3 CONSTRUCTION CHANGE DIRECTIVES

§ 7.3.1 A Construction Change Directive is a written order prepared by the Architect and signed by the Owner and Architect, directing a change in the Work prior to agreement on adjustment, if any, in the Contract Sum or Contract Time, or both. The Owner may by Construction Change Directive, without invalidating the Contract, order changes in the Work within the general scope of the Contract consisting of additions, deletions or other revisions, the Contract Sum and Contract Time being adjusted accordingly.

§ 7.3.2 A Construction Change Directive shall be used in the absence of total agreement on the terms of a Change Order.

§ 7.3.3 If the Construction Change Directive provides for an adjustment to the Contract Sum, the adjustment shall be based on one of the following methods:

   .1   mutual acceptance of a lump sum properly itemized and supported by sufficient substantiating data to permit evaluation;
   .2   unit prices stated in the Contract Documents or subsequently agreed upon;
   .3   cost to be determined in a manner agreed upon by the parties and a mutually acceptable fixed or percentage fee; or
   .4   as provided in Section 7.3.6.

§ 7.3.4 Upon receipt of a Construction Change Directive, the Contractor shall promptly proceed with the change in the Work involved and advise the Architect of the Contractor's agreement or disagreement with the method, if any, provided in the Construction Change Directive for determining the proposed adjustment in the Contract Sum or Contract Time.

§ 7.3.5 A Construction Change Directive signed by the Contractor indicates the agreement of the Contractor therewith, including adjustment in Contract Sum and Contract Time or the method for determining them. Such agreement shall be effective immediately and shall be recorded as a Change Order.

§ 7.3.6 If the Contractor does not respond promptly or disagrees with the method for adjustment in the Contract Sum, the method and the adjustment shall be determined by the Architect on the basis of reasonable expenditures and savings of those performing the Work attributable to the change, including, in case of an increase in the Contract Sum, a reasonable allowance for overhead and profit. In such case, and also under Section 7.3.3.3, the Contractor

AIA Document A201™—1997

shall keep and present, in such form as the Architect may prescribe, an itemized accounting together with appropriate supporting data. Unless otherwise provided in the Contract Documents, costs for the purposes of this Section 7.3.6 shall be limited to the following:

 .1 costs of labor, including social security, old age and unemployment insurance, fringe benefits required by agreement or custom, and workers' compensation insurance;

 .2 costs of materials, supplies and equipment, including cost of transportation, whether incorporated or consumed;

 .3 rental costs of machinery and equipment, exclusive of hand tools, whether rented from the Contractor or others;

 .4 costs of premiums for all bonds and insurance, permit fees, and sales, use or similar taxes related to the Work; and

 .5 additional costs of supervision and field office personnel directly attributable to the change.

§ 7.3.7 The amount of credit to be allowed by the Contractor to the Owner for a deletion or change which results in a net decrease in the Contract Sum shall be actual net cost as confirmed by the Architect. When both additions and credits covering related Work or substitutions are involved in a change, the allowance for overhead and profit shall be figured on the basis of net increase, if any, with respect to that change.

§ 7.3.8 Pending final determination of the total cost of a Construction Change Directive to the Owner, amounts not in dispute for such changes in the Work shall be included in Applications for Payment accompanied by a Change Order indicating the parties' agreement with part or all of such costs. For any portion of such cost that remains in dispute, the Architect will make an interim determination for purposes of monthly certification for payment for those costs. That determination of cost shall adjust the Contract Sum on the same basis as a Change Order, subject to the right of either party to disagree and assert a claim in accordance with Article 4.

§ 7.3.9 When the Owner and Contractor agree with the determination made by the Architect concerning the adjustments in the Contract Sum and Contract Time, or otherwise reach agreement upon the adjustments, such agreement shall be effective immediately and shall be recorded by preparation and execution of an appropriate Change Order.

§ 7.4 MINOR CHANGES IN THE WORK
§ 7.4.1 The Architect will have authority to order minor changes in the Work not involving adjustment in the Contract Sum or extension of the Contract Time and not inconsistent with the intent of the Contract Documents. Such changes shall be effected by written order and shall be binding on the Owner and Contractor. The Contractor shall carry out such written orders promptly.

ARTICLE 8 TIME
§ 8.1 DEFINITIONS
§ 8.1.1 Unless otherwise provided, Contract Time is the period of time, including authorized adjustments, allotted in the Contract Documents for Substantial Completion of the Work.

§ 8.1.2 The date of commencement of the Work is the date established in the Agreement.

§ 8.1.3 The date of Substantial Completion is the date certified by the Architect in accordance with Section 9.8.

§ 8.1.4 The term "day" as used in the Contract Documents shall mean calendar day unless otherwise specifically defined.

§ 8.2 PROGRESS AND COMPLETION
§ 8.2.1 Time limits stated in the Contract Documents are of the essence of the Contract. By executing the Agreement the Contractor confirms that the Contract Time is a reasonable period for performing the Work.

§ 8.2.2 The Contractor shall not knowingly, except by agreement or instruction of the Owner in writing, prematurely commence operations on the site or elsewhere prior to the effective date of insurance required by Article 11 to be furnished by the Contractor and Owner. The date of commencement of the Work shall not be changed by the effective date of such insurance. Unless the date of commencement is established by the Contract Documents or a notice to proceed given by the Owner, the Contractor shall notify the Owner in writing not less than five days or other agreed period before commencing the Work to permit the timely filing of mortgages, mechanic's liens and other security interests.

Appendix E

§ 8.2.3 The Contractor shall proceed expeditiously with adequate forces and shall achieve Substantial Completion within the Contract Time.

§ 8.3 DELAYS AND EXTENSIONS OF TIME

§ 8.3.1 If the Contractor is delayed at any time in the commencement or progress of the Work by an act or neglect of the Owner or Architect, or of an employee of either, or of a separate contractor employed by the Owner, or by changes ordered in the Work, or by labor disputes, fire, unusual delay in deliveries, unavoidable casualties or other causes beyond the Contractor's control, or by delay authorized by the Owner pending mediation and arbitration, or by other causes which the Architect determines may justify delay, then the Contract Time shall be extended by Change Order for such reasonable time as the Architect may determine.

§ 8.3.2 Claims relating to time shall be made in accordance with applicable provisions of Section 4.3.

§ 8.3.3 This Section 8.3 does not preclude recovery of damages for delay by either party under other provisions of the Contract Documents.

ARTICLE 9 PAYMENTS AND COMPLETION
§ 9.1 CONTRACT SUM

§ 9.1.1 The Contract Sum is stated in the Agreement and, including authorized adjustments, is the total amount payable by the Owner to the Contractor for performance of the Work under the Contract Documents.

§ 9.2 SCHEDULE OF VALUES

§ 9.2.1 Before the first Application for Payment, the Contractor shall submit to the Architect a schedule of values allocated to various portions of the Work, prepared in such form and supported by such data to substantiate its accuracy as the Architect may require. This schedule, unless objected to by the Architect, shall be used as a basis for reviewing the Contractor's Applications for Payment.

§ 9.3 APPLICATIONS FOR PAYMENT

§ 9.3.1 At least ten days before the date established for each progress payment, the Contractor shall submit to the Architect an itemized Application for Payment for operations completed in accordance with the schedule of values. Such application shall be notarized, if required, and supported by such data substantiating the Contractor's right to payment as the Owner or Architect may require, such as copies of requisitions from Subcontractors and material suppliers, and reflecting retainage if provided for in the Contract Documents.

§ 9.3.1.1 As provided in Section 7.3.8, such applications may include requests for payment on account of changes in the Work which have been properly authorized by Construction Change Directives, or by interim determinations of the Architect, but not yet included in Change Orders.

§ 9.3.1.2 Such applications may not include requests for payment for portions of the Work for which the Contractor does not intend to pay to a Subcontractor or material supplier, unless such Work has been performed by others whom the Contractor intends to pay.

§ 9.3.2 Unless otherwise provided in the Contract Documents, payments shall be made on account of materials and equipment delivered and suitably stored at the site for subsequent incorporation in the Work. If approved in advance by the Owner, payment may similarly be made for materials and equipment suitably stored off the site at a location agreed upon in writing. Payment for materials and equipment stored on or off the site shall be conditioned upon compliance by the Contractor with procedures satisfactory to the Owner to establish the Owner's title to such materials and equipment or otherwise protect the Owner's interest, and shall include the costs of applicable insurance, storage and transportation to the site for such materials and equipment stored off the site.

§ 9.3.3 The Contractor warrants that title to all Work covered by an Application for Payment will pass to the Owner no later than the time of payment. The Contractor further warrants that upon submittal of an Application for Payment all Work for which Certificates for Payment have been previously issued and payments received from the Owner shall, to the best of the Contractor's knowledge, information and belief, be free and clear of liens, claims, security interests or encumbrances in favor of the Contractor, Subcontractors, material suppliers, or other persons or entities making a claim by reason of having provided labor, materials and equipment relating to the Work.

AIA Document A201™—1997

§ 9.4 CERTIFICATES FOR PAYMENT

§ 9.4.1 The Architect will, within seven days after receipt of the Contractor's Application for Payment, either issue to the Owner a Certificate for Payment, with a copy to the Contractor, for such amount as the Architect determines is properly due, or notify the Contractor and Owner in writing of the Architect's reasons for withholding certification in whole or in part as provided in Section 9.5.1.

§ 9.4.2 The issuance of a Certificate for Payment will constitute a representation by the Architect to the Owner, based on the Architect's evaluation of the Work and the data comprising the Application for Payment, that the Work has progressed to the point indicated and that, to the best of the Architect's knowledge, information and belief, the quality of the Work is in accordance with the Contract Documents. The foregoing representations are subject to an evaluation of the Work for conformance with the Contract Documents upon Substantial Completion, to results of subsequent tests and inspections, to correction of minor deviations from the Contract Documents prior to completion and to specific qualifications expressed by the Architect. The issuance of a Certificate for Payment will further constitute a representation that the Contractor is entitled to payment in the amount certified. However, the issuance of a Certificate for Payment will not be a representation that the Architect has (1) made exhaustive or continuous on-site inspections to check the quality or quantity of the Work, (2) reviewed construction means, methods, techniques, sequences or procedures, (3) reviewed copies of requisitions received from Subcontractors and material suppliers and other data requested by the Owner to substantiate the Contractor's right to payment, or (4) made examination to ascertain how or for what purpose the Contractor has used money previously paid on account of the Contract Sum.

§ 9.5 DECISIONS TO WITHHOLD CERTIFICATION

§ 9.5.1 The Architect may withhold a Certificate for Payment in whole or in part, to the extent reasonably necessary to protect the Owner, if in the Architect's opinion the representations to the Owner required by Section 9.4.2 cannot be made. If the Architect is unable to certify payment in the amount of the Application, the Architect will notify the Contractor and Owner as provided in Section 9.4.1. If the Contractor and Architect cannot agree on a revised amount, the Architect will promptly issue a Certificate for Payment for the amount for which the Architect is able to make such representations to the Owner. The Architect may also withhold a Certificate for Payment or, because of subsequently discovered evidence, may nullify the whole or a part of a Certificate for Payment previously issued, to such extent as may be necessary in the Architect's opinion to protect the Owner from loss for which the Contractor is responsible, including loss resulting from acts and omissions described in Section 3.3.2, because of:

    .1   defective Work not remedied;
    .2   third party claims filed or reasonable evidence indicating probable filing of such claims unless security acceptable to the Owner is provided by the Contractor;
    .3   failure of the Contractor to make payments properly to Subcontractors or for labor, materials or equipment;
    .4   reasonable evidence that the Work cannot be completed for the unpaid balance of the Contract Sum;
    .5   damage to the Owner or another contractor;
    .6   reasonable evidence that the Work will not be completed within the Contract Time, and that the unpaid balance would not be adequate to cover actual or liquidated damages for the anticipated delay; or
    .7   persistent failure to carry out the Work in accordance with the Contract Documents.

§ 9.5.2 When the above reasons for withholding certification are removed, certification will be made for amounts previously withheld.

§ 9.6 PROGRESS PAYMENTS

§ 9.6.1 After the Architect has issued a Certificate for Payment, the Owner shall make payment in the manner and within the time provided in the Contract Documents, and shall so notify the Architect.

§ 9.6.2 The Contractor shall promptly pay each Subcontractor, upon receipt of payment from the Owner, out of the amount paid to the Contractor on account of such Subcontractor's portion of the Work, the amount to which said Subcontractor is entitled, reflecting percentages actually retained from payments to the Contractor on account of such Subcontractor's portion of the Work. The Contractor shall, by appropriate agreement with each Subcontractor, require each Subcontractor to make payments to Sub-subcontractors in a similar manner.

§ 9.6.3 The Architect will, on request, furnish to a Subcontractor, if practicable, information regarding percentages of completion or amounts applied for by the Contractor and action taken thereon by the Architect and Owner on account of portions of the Work done by such Subcontractor.

§ 9.6.4 Neither the Owner nor Architect shall have an obligation to pay or to see to the payment of money to a Subcontractor except as may otherwise be required by law.

§ 9.6.5 Payment to material suppliers shall be treated in a manner similar to that provided in Sections 9.6.2, 9.6.3 and 9.6.4.

§ 9.6.6 A Certificate for Payment, a progress payment, or partial or entire use or occupancy of the Project by the Owner shall not constitute acceptance of Work not in accordance with the Contract Documents.

§ 9.6.7 Unless the Contractor provides the Owner with a payment bond in the full penal sum of the Contract Sum, payments received by the Contractor for Work properly performed by Subcontractors and suppliers shall be held by the Contractor for those Subcontractors or suppliers who performed Work or furnished materials, or both, under contract with the Contractor for which payment was made by the Owner. Nothing contained herein shall require money to be placed in a separate account and not commingled with money of the Contractor, shall create any fiduciary liability or tort liability on the part of the Contractor for breach of trust or shall entitle any person or entity to an award of punitive damages against the Contractor for breach of the requirements of this provision.

§ 9.7 FAILURE OF PAYMENT
§ 9.7.1 If the Architect does not issue a Certificate for Payment, through no fault of the Contractor, within seven days after receipt of the Contractor's Application for Payment, or if the Owner does not pay the Contractor within seven days after the date established in the Contract Documents the amount certified by the Architect or awarded by arbitration, then the Contractor may, upon seven additional days' written notice to the Owner and Architect, stop the Work until payment of the amount owing has been received. The Contract Time shall be extended appropriately and the Contract Sum shall be increased by the amount of the Contractor's reasonable costs of shut-down, delay and start-up, plus interest as provided for in the Contract Documents.

§ 9.8 SUBSTANTIAL COMPLETION
§ 9.8.1 Substantial Completion is the stage in the progress of the Work when the Work or designated portion thereof is sufficiently complete in accordance with the Contract Documents so that the Owner can occupy or utilize the Work for its intended use.

§ 9.8.2 When the Contractor considers that the Work, or a portion thereof which the Owner agrees to accept separately, is substantially complete, the Contractor shall prepare and submit to the Architect a comprehensive list of items to be completed or corrected prior to final payment. Failure to include an item on such list does not alter the responsibility of the Contractor to complete all Work in accordance with the Contract Documents.

§ 9.8.3 Upon receipt of the Contractor's list, the Architect will make an inspection to determine whether the Work or designated portion thereof is substantially complete. If the Architect's inspection discloses any item, whether or not included on the Contractor's list, which is not sufficiently complete in accordance with the Contract Documents so that the Owner can occupy or utilize the Work or designated portion thereof for its intended use, the Contractor shall, before issuance of the Certificate of Substantial Completion, complete or correct such item upon notification by the Architect. In such case, the Contractor shall then submit a request for another inspection by the Architect to determine Substantial Completion.

§ 9.8.4 When the Work or designated portion thereof is substantially complete, the Architect will prepare a Certificate of Substantial Completion which shall establish the date of Substantial Completion, shall establish responsibilities of the Owner and Contractor for security, maintenance, heat, utilities, damage to the Work and insurance, and shall fix the time within which the Contractor shall finish all items on the list accompanying the Certificate. Warranties required by the Contract Documents shall commence on the date of Substantial Completion of the Work or designated portion thereof unless otherwise provided in the Certificate of Substantial Completion.

§ 9.8.5 The Certificate of Substantial Completion shall be submitted to the Owner and Contractor for their written acceptance of responsibilities assigned to them in such Certificate. Upon such acceptance and consent of surety, if any, the Owner shall make payment of retainage applying to such Work or designated portion thereof. Such payment shall be adjusted for Work that is incomplete or not in accordance with the requirements of the Contract Documents.

§ 9.9 PARTIAL OCCUPANCY OR USE
§ 9.9.1 The Owner may occupy or use any completed or partially completed portion of the Work at any stage when such portion is designated by separate agreement with the Contractor, provided such occupancy or use is consented

AIA Document A201™—1997

to by the insurer as required under Section 11.4.1.5 and authorized by public authorities having jurisdiction over the Work. Such partial occupancy or use may commence whether or not the portion is substantially complete, provided the Owner and Contractor have accepted in writing the responsibilities assigned to each of them for payments, retainage, if any, security, maintenance, heat, utilities, damage to the Work and insurance, and have agreed in writing concerning the period for correction of the Work and commencement of warranties required by the Contract Documents. When the Contractor considers a portion substantially complete, the Contractor shall prepare and submit a list to the Architect as provided under Section 9.8.2. Consent of the Contractor to partial occupancy or use shall not be unreasonably withheld. The stage of the progress of the Work shall be determined by written agreement between the Owner and Contractor or, if no agreement is reached, by decision of the Architect.

§ 9.9.2 Immediately prior to such partial occupancy or use, the Owner, Contractor and Architect shall jointly inspect the area to be occupied or portion of the Work to be used in order to determine and record the condition of the Work.

§ 9.9.3 Unless otherwise agreed upon, partial occupancy or use of a portion or portions of the Work shall not constitute acceptance of Work not complying with the requirements of the Contract Documents.

§ 9.10 FINAL COMPLETION AND FINAL PAYMENT
§ 9.10.1 Upon receipt of written notice that the Work is ready for final inspection and acceptance and upon receipt of a final Application for Payment, the Architect will promptly make such inspection and, when the Architect finds the Work acceptable under the Contract Documents and the Contract fully performed, the Architect will promptly issue a final Certificate for Payment stating that to the best of the Architect's knowledge, information and belief, and on the basis of the Architect's on-site visits and inspections, the Work has been completed in accordance with terms and conditions of the Contract Documents and that the entire balance found to be due the Contractor and noted in the final Certificate is due and payable. The Architect's final Certificate for Payment will constitute a further representation that conditions listed in Section 9.10.2 as precedent to the Contractor's being entitled to final payment have been fulfilled.

§ 9.10.2 Neither final payment nor any remaining retained percentage shall become due until the Contractor submits to the Architect (1) an affidavit that payrolls, bills for materials and equipment, and other indebtedness connected with the Work for which the Owner or the Owner's property might be responsible or encumbered (less amounts withheld by Owner) have been paid or otherwise satisfied, (2) a certificate evidencing that insurance required by the Contract Documents to remain in force after final payment is currently in effect and will not be canceled or allowed to expire until at least 30 days' prior written notice has been given to the Owner, (3) a written statement that the Contractor knows of no substantial reason that the insurance will not be renewable to cover the period required by the Contract Documents, (4) consent of surety, if any, to final payment and (5), if required by the Owner, other data establishing payment or satisfaction of obligations, such as receipts, releases and waivers of liens, claims, security interests or encumbrances arising out of the Contract, to the extent and in such form as may be designated by the Owner. If a Subcontractor refuses to furnish a release or waiver required by the Owner, the Contractor may furnish a bond satisfactory to the Owner to indemnify the Owner against such lien. If such lien remains unsatisfied after payments are made, the Contractor shall refund to the Owner all money that the Owner may be compelled to pay in discharging such lien, including all costs and reasonable attorneys' fees.

§ 9.10.3 If, after Substantial Completion of the Work, final completion thereof is materially delayed through no fault of the Contractor or by issuance of Change Orders affecting final completion, and the Architect so confirms, the Owner shall, upon application by the Contractor and certification by the Architect, and without terminating the Contract, make payment of the balance due for that portion of the Work fully completed and accepted. If the remaining balance for Work not fully completed or corrected is less than retainage stipulated in the Contract Documents, and if bonds have been furnished, the written consent of surety to payment of the balance due for that portion of the Work fully completed and accepted shall be submitted by the Contractor to the Architect prior to certification of such payment. Such payment shall be made under terms and conditions governing final payment, except that it shall not constitute a waiver of claims.

§ 9.10.4 The making of final payment shall constitute a waiver of Claims by the Owner except those arising from:
 .1 liens, Claims, security interests or encumbrances arising out of the Contract and unsettled;
 .2 failure of the Work to comply with the requirements of the Contract Documents; or
 .3 terms of special warranties required by the Contract Documents.

Appendix E

§ 9.10.5 Acceptance of final payment by the Contractor, a Subcontractor or material supplier shall constitute a waiver of claims by that payee except those previously made in writing and identified by that payee as unsettled at the time of final Application for Payment.

ARTICLE 10 PROTECTION OF PERSONS AND PROPERTY
§ 10.1 SAFETY PRECAUTIONS AND PROGRAMS
§ 10.1.1 The Contractor shall be responsible for initiating, maintaining and supervising all safety precautions and programs in connection with the performance of the Contract.

§ 10.2 SAFETY OF PERSONS AND PROPERTY
§ 10.2.1 The Contractor shall take reasonable precautions for safety of, and shall provide reasonable protection to prevent damage, injury or loss to:

.1 employees on the Work and other persons who may be affected thereby;

.2 the Work and materials and equipment to be incorporated therein, whether in storage on or off the site, under care, custody or control of the Contractor or the Contractor's Subcontractors or Sub-subcontractors; and

.3 other property at the site or adjacent thereto, such as trees, shrubs, lawns, walks, pavements, roadways, structures and utilities not designated for removal, relocation or replacement in the course of construction.

§ 10.2.2 The Contractor shall give notices and comply with applicable laws, ordinances, rules, regulations and lawful orders of public authorities bearing on safety of persons or property or their protection from damage, injury or loss.

§ 10.2.3 The Contractor shall erect and maintain, as required by existing conditions and performance of the Contract, reasonable safeguards for safety and protection, including posting danger signs and other warnings against hazards, promulgating safety regulations and notifying owners and users of adjacent sites and utilities.

§ 10.2.4 When use or storage of explosives or other hazardous materials or equipment or unusual methods are necessary for execution of the Work, the Contractor shall exercise utmost care and carry on such activities under supervision of properly qualified personnel.

§ 10.2.5 The Contractor shall promptly remedy damage and loss (other than damage or loss insured under property insurance required by the Contract Documents) to property referred to in Sections 10.2.1.2 and 10.2.1.3 caused in whole or in part by the Contractor, a Subcontractor, a Sub-subcontractor, or anyone directly or indirectly employed by any of them, or by anyone for whose acts they may be liable and for which the Contractor is responsible under Sections 10.2.1.2 and 10.2.1.3, except damage or loss attributable to acts or omissions of the Owner or Architect or anyone directly or indirectly employed by either of them, or by anyone for whose acts either of them may be liable, and not attributable to the fault or negligence of the Contractor. The foregoing obligations of the Contractor are in addition to the Contractor's obligations under Section 3.18.

§ 10.2.6 The Contractor shall designate a responsible member of the Contractor's organization at the site whose duty shall be the prevention of accidents. This person shall be the Contractor's superintendent unless otherwise designated by the Contractor in writing to the Owner and Architect.

§ 10.2.7 The Contractor shall not load or permit any part of the construction or site to be loaded so as to endanger its safety.

§ 10.3 HAZARDOUS MATERIALS
§ 10.3.1 If reasonable precautions will be inadequate to prevent foreseeable bodily injury or death to persons resulting from a material or substance, including but not limited to asbestos or polychlorinated biphenyl (PCB), encountered on the site by the Contractor, the Contractor shall, upon recognizing the condition, immediately stop Work in the affected area and report the condition to the Owner and Architect in writing.

§ 10.3.2 The Owner shall obtain the services of a licensed laboratory to verify the presence or absence of the material or substance reported by the Contractor and, in the event such material or substance is found to be present, to verify that it has been rendered harmless. Unless otherwise required by the Contract Documents, the Owner shall furnish in writing to the Contractor and Architect the names and qualifications of persons or entities who are to perform tests verifying the presence or absence of such material or substance or who are to perform the task of removal or safe containment of such material or substance. The Contractor and the Architect will promptly reply to the Owner in

30

AIA Document A201™—1997

writing stating whether or not either has reasonable objection to the persons or entities proposed by the Owner. If either the Contractor or Architect has an objection to a person or entity proposed by the Owner, the Owner shall propose another to whom the Contractor and the Architect have no reasonable objection. When the material or substance has been rendered harmless, Work in the affected area shall resume upon written agreement of the Owner and Contractor. The Contract Time shall be extended appropriately and the Contract Sum shall be increased in the amount of the Contractor's reasonable additional costs of shut-down, delay and start-up, which adjustments shall be accomplished as provided in Article 7.

§ 10.3.3 To the fullest extent permitted by law, the Owner shall indemnify and hold harmless the Contractor, Subcontractors, Architect, Architect's consultants and agents and employees of any of them from and against claims, damages, losses and expenses, including but not limited to attorneys' fees, arising out of or resulting from performance of the Work in the affected area if in fact the material or substance presents the risk of bodily injury or death as described in Section 10.3.1 and has not been rendered harmless, provided that such claim, damage, loss or expense is attributable to bodily injury, sickness, disease or death, or to injury to or destruction of tangible property (other than the Work itself) and provided that such damage, loss or expense is not due to the sole negligence of a party seeking indemnity.

§ 10.4 The Owner shall not be responsible under Section 10.3 for materials and substances brought to the site by the Contractor unless such materials or substances were required by the Contract Documents.

§ 10.5 If, without negligence on the part of the Contractor, the Contractor is held liable for the cost of remediation of a hazardous material or substance solely by reason of performing Work as required by the Contract Documents, the Owner shall indemnify the Contractor for all cost and expense thereby incurred.

§ 10.6 EMERGENCIES
§ 10.6.1 In an emergency affecting safety of persons or property, the Contractor shall act, at the Contractor's discretion, to prevent threatened damage, injury or loss. Additional compensation or extension of time claimed by the Contractor on account of an emergency shall be determined as provided in Section 4.3 and Article 7.

ARTICLE 11 INSURANCE AND BONDS
§ 11.1 CONTRACTOR'S LIABILITY INSURANCE
§ 11.1.1 The Contractor shall purchase from and maintain in a company or companies lawfully authorized to do business in the jurisdiction in which the Project is located such insurance as will protect the Contractor from claims set forth below which may arise out of or result from the Contractor's operations under the Contract and for which the Contractor may be legally liable, whether such operations be by the Contractor or by a Subcontractor or by anyone directly or indirectly employed by any of them, or by anyone for whose acts any of them may be liable:

.1 claims under workers' compensation, disability benefit and other similar employee benefit acts which are applicable to the Work to be performed;
.2 claims for damages because of bodily injury, occupational sickness or disease, or death of the Contractor's employees;
.3 claims for damages because of bodily injury, sickness or disease, or death of any person other than the Contractor's employees;
.4 claims for damages insured by usual personal injury liability coverage;
.5 claims for damages, other than to the Work itself, because of injury to or destruction of tangible property, including loss of use resulting therefrom;
.6 claims for damages because of bodily injury, death of a person or property damage arising out of ownership, maintenance or use of a motor vehicle;
.7 claims for bodily injury or property damage arising out of completed operations; and
.8 claims involving contractual liability insurance applicable to the Contractor's obligations under Section 3.18.

§ 11.1.2 The insurance required by Section 11.1.1 shall be written for not less than limits of liability specified in the Contract Documents or required by law, whichever coverage is greater. Coverages, whether written on an occurrence or claims-made basis, shall be maintained without interruption from date of commencement of the Work until date of final payment and termination of any coverage required to be maintained after final payment.

§ 11.1.3 Certificates of insurance acceptable to the Owner shall be filed with the Owner prior to commencement of the Work. These certificates and the insurance policies required by this Section 11.1 shall contain a provision that coverages afforded under the policies will not be canceled or allowed to expire until at least 30 days' prior written notice has been given to the Owner. If any of the foregoing insurance coverages are required to remain in force after final payment and are reasonably available, an additional certificate evidencing continuation of such coverage shall be submitted with the final Application for Payment as required by Section 9.10.2. Information concerning reduction of coverage on account of revised limits or claims paid under the General Aggregate, or both, shall be furnished by the Contractor with reasonable promptness in accordance with the Contractor's information and belief.

§ 11.2 OWNER'S LIABILITY INSURANCE
§ 11.2.1 The Owner shall be responsible for purchasing and maintaining the Owner's usual liability insurance.

§ 11.3 PROJECT MANAGEMENT PROTECTIVE LIABILITY INSURANCE
§ 11.3.1 Optionally, the Owner may require the Contractor to purchase and maintain Project Management Protective Liability insurance from the Contractor's usual sources as primary coverage for the Owner's, Contractor's and Architect's vicarious liability for construction operations under the Contract. Unless otherwise required by the Contract Documents, the Owner shall reimburse the Contractor by increasing the Contract Sum to pay the cost of purchasing and maintaining such optional insurance coverage, and the Contractor shall not be responsible for purchasing any other liability insurance on behalf of the Owner. The minimum limits of liability purchased with such coverage shall be equal to the aggregate of the limits required for Contractor's Liability Insurance under Sections 11.1.1.2 through 11.1.1.5.

§ 11.3.2 To the extent damages are covered by Project Management Protective Liability insurance, the Owner, Contractor and Architect waive all rights against each other for damages, except such rights as they may have to the proceeds of such insurance. The policy shall provide for such waivers of subrogation by endorsement or otherwise.

§ 11.3.3 The Owner shall not require the Contractor to include the Owner, Architect or other persons or entities as additional insureds on the Contractor's Liability Insurance coverage under Section 11.1.

§ 11.4 PROPERTY INSURANCE
§ 11.4.1 Unless otherwise provided, the Owner shall purchase and maintain, in a company or companies lawfully authorized to do business in the jurisdiction in which the Project is located, property insurance written on a builder's risk "all-risk" or equivalent policy form in the amount of the initial Contract Sum, plus value of subsequent Contract modifications and cost of materials supplied or installed by others, comprising total value for the entire Project at the site on a replacement cost basis without optional deductibles. Such property insurance shall be maintained, unless otherwise provided in the Contract Documents or otherwise agreed in writing by all persons and entities who are beneficiaries of such insurance, until final payment has been made as provided in Section 9.10 or until no person or entity other than the Owner has an insurable interest in the property required by this Section 11.4 to be covered, whichever is later. This insurance shall include interests of the Owner, the Contractor, Subcontractors and Sub-subcontractors in the Project.

§ 11.4.1.1 Property insurance shall be on an "all-risk" or equivalent policy form and shall include, without limitation, insurance against the perils of fire (with extended coverage) and physical loss or damage including, without duplication of coverage, theft, vandalism, malicious mischief, collapse, earthquake, flood, windstorm, falsework, testing and startup, temporary buildings and debris removal including demolition occasioned by enforcement of any applicable legal requirements, and shall cover reasonable compensation for Architect's and Contractor's services and expenses required as a result of such insured loss.

§ 11.4.1.2 If the Owner does not intend to purchase such property insurance required by the Contract and with all of the coverages in the amount described above, the Owner shall so inform the Contractor in writing prior to commencement of the Work. The Contractor may then effect insurance which will protect the interests of the Contractor, Subcontractors and Sub-subcontractors in the Work, and by appropriate Change Order the cost thereof shall be charged to the Owner. If the Contractor is damaged by the failure or neglect of the Owner to purchase or maintain insurance as described above, without so notifying the Contractor in writing, then the Owner shall bear all reasonable costs properly attributable thereto.

§ 11.4.1.3 If the property insurance requires deductibles, the Owner shall pay costs not covered because of such deductibles.

§ 11.4.1.4 This property insurance shall cover portions of the Work stored off the site, and also portions of the Work in transit.

---

§ 11.4.1.5 Partial occupancy or use in accordance with Section 9.9 shall not commence until the insurance company or companies providing property insurance have consented to such partial occupancy or use by endorsement or otherwise. The Owner and the Contractor shall take reasonable steps to obtain consent of the insurance company or companies and shall, without mutual written consent, take no action with respect to partial occupancy or use that would cause cancellation, lapse or reduction of insurance.

§ 11.4.2 **Boiler and Machinery Insurance**. The Owner shall purchase and maintain boiler and machinery insurance required by the Contract Documents or by law, which shall specifically cover such insured objects during installation and until final acceptance by the Owner; this insurance shall include interests of the Owner, Contractor, Subcontractors and Sub-subcontractors in the Work, and the Owner and Contractor shall be named insureds.

§ 11.4.3 **Loss of Use Insurance**. The Owner, at the Owner's option, may purchase and maintain such insurance as will insure the Owner against loss of use of the Owner's property due to fire or other hazards, however caused. The Owner waives all rights of action against the Contractor for loss of use of the Owner's property, including consequential losses due to fire or other hazards however caused.

§ 11.4.4 If the Contractor requests in writing that insurance for risks other than those described herein or other special causes of loss be included in the property insurance policy, the Owner shall, if possible, include such insurance, and the cost thereof shall be charged to the Contractor by appropriate Change Order.

§ 11.4.5 If during the Project construction period the Owner insures properties, real or personal or both, at or adjacent to the site by property insurance under policies separate from those insuring the Project, or if after final payment property insurance is to be provided on the completed Project through a policy or policies other than those insuring the Project during the construction period, the Owner shall waive all rights in accordance with the terms of Section 11.4.7 for damages caused by fire or other causes of loss covered by this separate property insurance. All separate policies shall provide this waiver of subrogation by endorsement or otherwise.

§ 11.4.6 Before an exposure to loss may occur, the Owner shall file with the Contractor a copy of each policy that includes insurance coverages required by this Section 11.4. Each policy shall contain all generally applicable conditions, definitions, exclusions and endorsements related to this Project. Each policy shall contain a provision that the policy will not be canceled or allowed to expire, and that its limits will not be reduced, until at least 30 days' prior written notice has been given to the Contractor.

§ 11.4.7 **Waivers of Subrogation**. The Owner and Contractor waive all rights against (1) each other and any of their subcontractors, sub-subcontractors, agents and employees, each of the other, and (2) the Architect, Architect's consultants, separate contractors described in Article 6, if any, and any of their subcontractors, sub-subcontractors, agents and employees, for damages caused by fire or other causes of loss to the extent covered by property insurance obtained pursuant to this Section 11.4 or other property insurance applicable to the Work, except such rights as they have to proceeds of such insurance held by the Owner as fiduciary. The Owner or Contractor, as appropriate, shall require of the Architect, Architect's consultants, separate contractors described in Article 6, if any, and the subcontractors, sub-subcontractors, agents and employees of any of them, by appropriate agreements, written where legally required for validity, similar waivers each in favor of other parties enumerated herein. The policies shall provide such waivers of subrogation by endorsement or otherwise. A waiver of subrogation shall be effective as to a person or entity even though that person or entity would otherwise have a duty of indemnification, contractual or otherwise, did not pay the insurance premium directly or indirectly, and whether or not the person or entity had an insurable interest in the property damaged.

§ 11.4.8 A loss insured under Owner's property insurance shall be adjusted by the Owner as fiduciary and made payable to the Owner as fiduciary for the insureds, as their interests may appear, subject to requirements of any applicable mortgagee clause and of Section 11.4.10. The Contractor shall pay Subcontractors their just shares of insurance proceeds received by the Contractor, and by appropriate agreements, written where legally required for validity, shall require Subcontractors to make payments to their Sub-subcontractors in similar manner.

§ 11.4.9 If required in writing by a party in interest, the Owner as fiduciary shall, upon occurrence of an insured loss, give bond for proper performance of the Owner's duties. The cost of required bonds shall be charged against proceeds received as fiduciary. The Owner shall deposit in a separate account proceeds so received, which the Owner shall distribute in accordance with such agreement as the parties in interest may reach, or in accordance with an arbitration award in which case the procedure shall be as provided in Section 4.6. If after such loss no other

33

Appendix E

special agreement is made and unless the Owner terminates the Contract for convenience, replacement of damaged property shall be performed by the Contractor after notification of a Change in the Work in accordance with Article 7.

§ 11.4.10 The Owner as fiduciary shall have power to adjust and settle a loss with insurers unless one of the parties in interest shall object in writing within five days after occurrence of loss to the Owner's exercise of this power; if such objection is made, the dispute shall be resolved as provided in Sections 4.5 and 4.6. The Owner as fiduciary shall, in the case of arbitration, make settlement with insurers in accordance with directions of the arbitrators. If distribution of insurance proceeds by arbitration is required, the arbitrators will direct such distribution.

§ 11.5 PERFORMANCE BOND AND PAYMENT BOND
§ 11.5.1 The Owner shall have the right to require the Contractor to furnish bonds covering faithful performance of the Contract and payment of obligations arising thereunder as stipulated in bidding requirements or specifically required in the Contract Documents on the date of execution of the Contract.

§ 11.5.2 Upon the request of any person or entity appearing to be a potential beneficiary of bonds covering payment of obligations arising under the Contract, the Contractor shall promptly furnish a copy of the bonds or shall permit a copy to be made.

ARTICLE 12 UNCOVERING AND CORRECTION OF WORK
§ 12.1 UNCOVERING OF WORK
§ 12.1.1 If a portion of the Work is covered contrary to the Architect's request or to requirements specifically expressed in the Contract Documents, it must, if required in writing by the Architect, be uncovered for the Architect's examination and be replaced at the Contractor's expense without change in the Contract Time.

§ 12.1.2 If a portion of the Work has been covered which the Architect has not specifically requested to examine prior to its being covered, the Architect may request to see such Work and it shall be uncovered by the Contractor. If such Work is in accordance with the Contract Documents, costs of uncovering and replacement shall, by appropriate Change Order, be at the Owner's expense. If such Work is not in accordance with the Contract Documents, correction shall be at the Contractor's expense unless the condition was caused by the Owner or a separate contractor in which event the Owner shall be responsible for payment of such costs.

§ 12.2 CORRECTION OF WORK
§ 12.2.1 BEFORE OR AFTER SUBSTANTIAL COMPLETION
§ 12.2.1.1 The Contractor shall promptly correct Work rejected by the Architect or failing to conform to the requirements of the Contract Documents, whether discovered before or after Substantial Completion and whether or not fabricated, installed or completed. Costs of correcting such rejected Work, including additional testing and inspections and compensation for the Architect's services and expenses made necessary thereby, shall be at the Contractor's expense.

§ 12.2.2 AFTER SUBSTANTIAL COMPLETION
§ 12.2.2.1 In addition to the Contractor's obligations under Section 3.5, if, within one year after the date of Substantial Completion of the Work or designated portion thereof or after the date for commencement of warranties established under Section 9.9.1, or by terms of an applicable special warranty required by the Contract Documents, any of the Work is found to be not in accordance with the requirements of the Contract Documents, the Contractor shall correct it promptly after receipt of written notice from the Owner to do so unless the Owner has previously given the Contractor a written acceptance of such condition. The Owner shall give such notice promptly after discovery of the condition. During the one-year period for correction of Work, if the Owner fails to notify the Contractor and give the Contractor an opportunity to make the correction, the Owner waives the rights to require correction by the Contractor and to make a claim for breach of warranty. If the Contractor fails to correct nonconforming Work within a reasonable time during that period after receipt of notice from the Owner or Architect, the Owner may correct it in accordance with Section 2.4.

§ 12.2.2.2 The one-year period for correction of Work shall be extended with respect to portions of Work first performed after Substantial Completion by the period of time between Substantial Completion and the actual performance of the Work.

§ 12.2.2.3 The one-year period for correction of Work shall not be extended by corrective Work performed by the Contractor pursuant to this Section 12.2.

§ **12.2.3** The Contractor shall remove from the site portions of the Work which are not in accordance with the requirements of the Contract Documents and are neither corrected by the Contractor nor accepted by the Owner.

§ **12.2.4** The Contractor shall bear the cost of correcting destroyed or damaged construction, whether completed or partially completed, of the Owner or separate contractors caused by the Contractor's correction or removal of Work which is not in accordance with the requirements of the Contract Documents.

§ **12.2.5** Nothing contained in this Section 12.2 shall be construed to establish a period of limitation with respect to other obligations which the Contractor might have under the Contract Documents. Establishment of the one-year period for correction of Work as described in Section 12.2.2 relates only to the specific obligation of the Contractor to correct the Work, and has no relationship to the time within which the obligation to comply with the Contract Documents may be sought to be enforced, nor to the time within which proceedings may be commenced to establish the Contractor's liability with respect to the Contractor's obligations other than specifically to correct the Work.

§ **12.3 ACCEPTANCE OF NONCONFORMING WORK**
§ **12.3.1** If the Owner prefers to accept Work which is not in accordance with the requirements of the Contract Documents, the Owner may do so instead of requiring its removal and correction, in which case the Contract Sum will be reduced as appropriate and equitable. Such adjustment shall be effected whether or not final payment has been made.

ARTICLE 13 MISCELLANEOUS PROVISIONS
§ **13.1 GOVERNING LAW**
§ **13.1.1** The Contract shall be governed by the law of the place where the Project is located.

§ **13.2 SUCCESSORS AND ASSIGNS**
§ **13.2.1** The Owner and Contractor respectively bind themselves, their partners, successors, assigns and legal representatives to the other party hereto and to partners, successors, assigns and legal representatives of such other party in respect to covenants, agreements and obligations contained in the Contract Documents. Except as provided in Section 13.2.2, neither party to the Contract shall assign the Contract as a whole without written consent of the other. If either party attempts to make such an assignment without such consent, that party shall nevertheless remain legally responsible for all obligations under the Contract.

§ **13.2.2** The Owner may, without consent of the Contractor, assign the Contract to an institutional lender providing construction financing for the Project. In such event, the lender shall assume the Owner's rights and obligations under the Contract Documents. The Contractor shall execute all consents reasonably required to facilitate such assignment.

§ **13.3 WRITTEN NOTICE**
§ **13.3.1** Written notice shall be deemed to have been duly served if delivered in person to the individual or a member of the firm or entity or to an officer of the corporation for which it was intended, or if delivered at or sent by registered or certified mail to the last business address known to the party giving notice.

§ **13.4 RIGHTS AND REMEDIES**
§ **13.4.1** Duties and obligations imposed by the Contract Documents and rights and remedies available thereunder shall be in addition to and not a limitation of duties, obligations, rights and remedies otherwise imposed or available by law.

§ **13.4.2** No action or failure to act by the Owner, Architect or Contractor shall constitute a waiver of a right or duty afforded them under the Contract, nor shall such action or failure to act constitute approval of or acquiescence in a breach there under, except as may be specifically agreed in writing.

§ **13.5 TESTS AND INSPECTIONS**
§ **13.5.1** Tests, inspections and approvals of portions of the Work required by the Contract Documents or by laws, ordinances, rules, regulations or orders of public authorities having jurisdiction shall be made at an appropriate time. Unless otherwise provided, the Contractor shall make arrangements for such tests, inspections and approvals with an independent testing laboratory or entity acceptable to the Owner, or with the appropriate public authority, and shall bear all related costs of tests, inspections and approvals. The Contractor shall give the Architect timely notice of when and where tests and inspections are to be made so that the Architect may be present for such procedures. The

35

Appendix E

Owner shall bear costs of tests, inspections or approvals which do not become requirements until after bids are received or negotiations concluded.

§ 13.5.2 If the Architect, Owner or public authorities having jurisdiction determine that portions of the Work require additional testing, inspection or approval not included under Section 13.5.1, the Architect will, upon written authorization from the Owner, instruct the Contractor to make arrangements for such additional testing, inspection or approval by an entity acceptable to the Owner, and the Contractor shall give timely notice to the Architect of when and where tests and inspections are to be made so that the Architect may be present for such procedures. Such costs, except as provided in Section 13.5.3, shall be at the Owner's expense.

§ 13.5.3 If such procedures for testing, inspection or approval under Sections 13.5.1 and 13.5.2 reveal failure of the portions of the Work to comply with requirements established by the Contract Documents, all costs made necessary by such failure including those of repeated procedures and compensation for the Architect's services and expenses shall be at the Contractor's expense.

§ 13.5.4 Required certificates of testing, inspection or approval shall, unless otherwise required by the Contract Documents, be secured by the Contractor and promptly delivered to the Architect.

§ 13.5.5 If the Architect is to observe tests, inspections or approvals required by the Contract Documents, the Architect will do so promptly and, where practicable, at the normal place of testing.

§ 13.5.6 Tests or inspections conducted pursuant to the Contract Documents shall be made promptly to avoid unreasonable delay in the Work.

§ 13.6 INTEREST
§ 13.6.1 Payments due and unpaid under the Contract Documents shall bear interest from the date payment is due at such rate as the parties may agree upon in writing or, in the absence thereof, at the legal rate prevailing from time to time at the place where the Project is located.

§ 13.7 COMMENCEMENT OF STATUTORY LIMITATION PERIOD
§ 13.7.1 As between the Owner and Contractor:
.1   Before Substantial Completion. As to acts or failures to act occurring prior to the relevant date of Substantial Completion, any applicable statute of limitations shall commence to run and any alleged cause of action shall be deemed to have accrued in any and all events not later than such date of Substantial Completion;
.2   Between Substantial Completion and Final Certificate for Payment. As to acts or failures to act occurring subsequent to the relevant date of Substantial Completion and prior to issuance of the final Certificate for Payment, any applicable statute of limitations shall commence to run and any alleged cause of action shall be deemed to have accrued in any and all events not later than the date of issuance of the final Certificate for Payment; and
.3   After Final Certificate for Payment. As to acts or failures to act occurring after the relevant date of issuance of the final Certificate for Payment, any applicable statute of limitations shall commence to run and any alleged cause of action shall be deemed to have accrued in any and all events not later than the date of any act or failure to act by the Contractor pursuant to any Warranty provided under Section 3.5, the date of any correction of the Work or failure to correct the Work by the Contractor under Section 12.2, or the date of actual commission of any other act or failure to perform any duty or obligation by the Contractor or Owner, whichever occurs last.

ARTICLE 14   TERMINATION OR SUSPENSION OF THE CONTRACT
§ 14.1 TERMINATION BY THE CONTRACTOR
§ 14.1.1 The Contractor may terminate the Contract if the Work is stopped for a period of 30 consecutive days through no act or fault of the Contractor or a Subcontractor, Sub-subcontractor or their agents or employees or any other persons or entities performing portions of the Work under direct or indirect contract with the Contractor, for any of the following reasons:
.1   issuance of an order of a court or other public authority having jurisdiction which requires all Work to be stopped;
.2   an act of government, such as a declaration of national emergency which requires all Work to be stopped;

AIA Document A201™—1997

.3   because the Architect has not issued a Certificate for Payment and has not notified the Contractor of the reason for withholding certification as provided in Section 9.4.1, or because the Owner has not made payment on a Certificate for Payment within the time stated in the Contract Documents; or

.4   the Owner has failed to furnish to the Contractor promptly, upon the Contractor's request, reasonable evidence as required by Section 2.2.1.

§ 14.1.2 The Contractor may terminate the Contract if, through no act or fault of the Contractor or a Subcontractor, Sub-subcontractor or their agents or employees or any other persons or entities performing portions of the Work under direct or indirect contract with the Contractor, repeated suspensions, delays or interruptions of the entire Work by the Owner as described in Section 14.3 constitute in the aggregate more than 100 percent of the total number of days scheduled for completion, or 120 days in any 365-day period, whichever is less.

§ 14.1.3 If one of the reasons described in Section 14.1.1 or 14.1.2 exists, the Contractor may, upon seven days' written notice to the Owner and Architect, terminate the Contract and recover from the Owner payment for Work executed and for proven loss with respect to materials, equipment, tools, and construction equipment and machinery, including reasonable overhead, profit and damages.

§ 14.1.4 If the Work is stopped for a period of 60 consecutive days through no act or fault of the Contractor or a Subcontractor or their agents or employees or any other persons performing portions of the Work under contract with the Contractor because the Owner has persistently failed to fulfill the Owner's obligations under the Contract Documents with respect to matters important to the progress of the Work, the Contractor may, upon seven additional days' written notice to the Owner and the Architect, terminate the Contract and recover from the Owner as provided in Section 14.1.3.

§ 14.2 TERMINATION BY THE OWNER FOR CAUSE

§ 14.2.1 The Owner may terminate the Contract if the Contractor:

.1   persistently or repeatedly refuses or fails to supply enough properly skilled workers or proper materials;

.2   fails to make payment to Subcontractors for materials or labor in accordance with the respective agreements between the Contractor and the Subcontractors;

.3   persistently disregards laws, ordinances, or rules, regulations or orders of a public authority having jurisdiction; or

.4   otherwise is guilty of substantial breach of a provision of the Contract Documents.

§ 14.2.2 When any of the above reasons exist, the Owner, upon certification by the Architect that sufficient cause exists to justify such action, may without prejudice to any other rights or remedies of the Owner and after giving the Contractor and the Contractor's surety, if any, seven days' written notice, terminate employment of the Contractor and may, subject to any prior rights of the surety:

.1   take possession of the site and of all materials, equipment, tools, and construction equipment and machinery thereon owned by the Contractor;

.2   accept assignment of subcontracts pursuant to Section 5.4; and

.3   finish the Work by whatever reasonable method the Owner may deem expedient. Upon request of the Contractor, the Owner shall furnish to the Contractor a detailed accounting of the costs incurred by the Owner in finishing the Work.

§ 14.2.3 When the Owner terminates the Contract for one of the reasons stated in Section 14.2.1, the Contractor shall not be entitled to receive further payment until the Work is finished.

§ 14.2.4 If the unpaid balance of the Contract Sum exceeds costs of finishing the Work, including compensation for the Architect's services and expenses made necessary thereby, and other damages incurred by the Owner and not expressly waived, such excess shall be paid to the Contractor. If such costs and damages exceed the unpaid balance, the Contractor shall pay the difference to the Owner. The amount to be paid to the Contractor or Owner, as the case may be, shall be certified by the Architect, upon application, and this obligation for payment shall survive termination of the Contract.

§ 14.3 SUSPENSION BY THE OWNER FOR CONVENIENCE

§ 14.3.1 The Owner may, without cause, order the Contractor in writing to suspend, delay or interrupt the Work in whole or in part for such period of time as the Owner may determine.

§ 14.3.2 The Contract Sum and Contract Time shall be adjusted for increases in the cost and time caused by suspension, delay or interruption as described in Section 14.3.1. Adjustment of the Contract Sum shall include profit. No adjustment shall be made to the extent:

 .1 that performance is, was or would have been so suspended, delayed or interrupted by another cause for which the Contractor is responsible; or

 .2 that an equitable adjustment is made or denied under another provision of the Contract.

§ 14.4 TERMINATION BY THE OWNER FOR CONVENIENCE

§ 14.4.1 The Owner may, at any time, terminate the Contract for the Owner's convenience and without cause.

§ 14.4.2 Upon receipt of written notice from the Owner of such termination for the Owner's convenience, the Contractor shall:

 .1 cease operations as directed by the Owner in the notice;

 .2 take actions necessary, or that the Owner may direct, for the protection and preservation of the Work; and

 .3 except for Work directed to be performed prior to the effective date of termination stated in the notice, terminate all existing subcontracts and purchase orders and enter into no further subcontracts and purchase orders.

§ 14.4.3 In case of such termination for the Owner's convenience, the Contractor shall be entitled to receive payment for Work executed, and costs incurred by reason of such termination, along with reasonable overhead and profit on the Work not executed.

AIA Document A201™—1997

Acceptance The agreement to the offer and one of the elements needed to form a contract; the acceptance must mirror the offer and if it does not it is a counteroffer.

Accord and satisfaction Acceptance of something of value, usually money but that is not required, in settlement of a claim; once the accord and satisfaction is accepted no lawsuit can be prosecuted.

Act The name given to several related statutes passed at one time, for example the Social Security Act.

Actual damages All of the losses incurred by an injured party; no court awards actual damages although some courts have defined actual damages to mean all of the legally recognized losses incurred by an injured party.

Additional work Work outside the scope of the original contract and not covered by the contract.

Administrative agency A government office, bureau, or similar, created by another branch of a government to handle some particular task, such as the Department of Labor.

Adversary system of justice A legal system in which the goal of each side in a trial is to win not to find justice or truth; parties must abide by the law and ethical prohibitions in pursuit of the win.

Agency law Term given to the law covering the relationship between a principal and agent.

Agent An entity empowered to enter into contracts for another, called the principal.

Ambiguity or ambiguous A word or phrase that is capable of two or more reasonable meanings.

American rule A rule of damages that states that attorney fees are not recoverable by the winning party unless recovery is provided for in the contract or by statute.

Americans with Disabilities Act Prohibits discrimination against, and requires reasonable accommodation of, the disabled in job application, hiring, advancement, discharge, pay, training, and other terms, conditions, and privileges of employment.

Appeal The review of a lower court's decision; only issues of law are reviewable on appeal, and issues of fact are not.

Arbitration A dispute resolution process; the parties present their dispute to a person or panel, called the arbitrators, who are specifically hired by the parties to resolve the dispute; the arbitrators decide how the dispute will be resolved and force a particular decision upon the parties.

Assumption of the risk A tort defense; it prevents recovery by an injured party when the injured party knowingly assumes the risks of an injury.

Award The decision of an arbitrator; it can be enforced by filing a copy of the award in a court and following the government-provided enforcement of judgment procedures.

Back pay A sum to compensate an employee for wrongful termination; pay for work that has not actually been performed due to the wrongful termination.

Basic premise of contract law A party must perform its contract or respond in damages and parties are presumed to know the contents of their contracts.

Battle of the experts Occurs when each side hires experts whose testimony conflicts; the jury must decide which expert to believe.

Battle of the forms Occurs when the parties' forms contain additional, different, or conflicting clauses.

Bench trial A trial without a jury; the judge determines all legal and factual issues and imposes a resolution of the matter upon the parties.

Bid An offer by a contractor to an owner or a subcontractor to a general contractor to perform work or supply materials; a bid is considered an offer under the law of contract formation.

Bid bond Assures the owner that the contractor awarded the contract will sign the contract or at least begin performance on the contract; purchased from a surety that agrees to compensate the owner should the contractor fail to enter into the contract or begin performance.

Bid shopping Occurs when a contractor attempts to get a subcontractor to lower its bid after the contractor has been awarded the project.

Black letter law Older, established law that has been tested and clarified by the courts and is unlikely to change.

Boilerplate Term used for standardized specifications or language usually chosen from books and inserted into contracts, often without reconciling the boilerplate to the actual plans, specifications, or special conditions.

Bona fide occupation qualification (BFOQ) A legitimate requirement that a person applying for or holding a particular job have some characteristic that would normally be classified as illegal discrimination.

Bond or **guarantee** Similar to an insurance policy and is purchased to ensure that the contractor performs some task, such as complete a project or pay subcontractors.

British rule Attorney fees are an element of damages and payable to the winning party; compare with American Rule.

Business entity A term that may refer to any of the possible forms of doing business.

Cardinal change Occurs when the government effects an alteration in the work so drastic that it effectively requires the contactor to perform duties materially different from those originally bargained for.

Case law When appeal court judges clarify the meaning of a constitution, statute, regulation, or prior case, that clarification is a law, and the name of that law is case law, judge-made law, or common law.

Caucus Process of separating the parties during a mediation so the mediator can speak privately to each party; not allowed during trial or arbitration.

Citizen suits Many federal and state environmental acts allow individuals to enforce environmental laws.

Civil law The broad area of law dealing with all noncriminal disputes.

Close corporation A corporation whose stock is owned by a small group of people, for example members of the same family.

Code A collection of related statutes.

Combination specification A combination of both design and performance characteristics.

Commercial senselessness or **doctrine of practical impossibility** States that at some point, when the cost of performance becomes ridiculous, performance is excused and refusing to perform is not a breach of the contract.

Common law When appeal court judges clarify the meaning of a constitution, statute, regulation, or prior case, that clarification is a law, and the name of that law is case law, judge-made law, or common law.

Commonly known dangers States that manufacturers and/or sellers are not liable for injuries that the user should realize can result from the use of certain products such as knives and guns.

Compensable delay A delay caused by the owner and for which the contractor is entitled to damages.

Compensatory damages Money damages awarded to compensate an injured party for losses incurred.

Condition precedent A requirement in a contract that must be satisfied before one party is entitled to payment by another party.

Conditional waiver of lien Does not waive lien rights until payment is received by the signer; the signer maintains all of its legal rights to file a lien if it is not paid.

Consequential or special damages Damages suffered by a party in a breach of contract action due to some unique characteristic of the injured party or because of some unique situation of the injured party known to the breaching party.

Conservation easement An easement that prevents the owner of the land from certain uses, such as industrial development.

Consideration What each party to a contract receives and what each party to a contract gives up; one of the elements needed to form a contract.

Constitution A document establishing a government and outlining what it can and cannot do.

Constitutional law A constitution and the cases that have interpreted the constitution.

Construction change directive Requires the contractor to perform the work as requested and follow procedures in the contract for making a claim.

Construction manager or CM A person hired by the owner to handle the owner's duties related to a construction project; may be an employee or agent.

Constructive change Owner conduct that amounts to a change but the owner refuses to recognize it as such.

Contract damages All of the legally recognized losses incurred by a party injured by a breach of contract; designed to, as far as is practicable, put the injured party in the same position he would be in had the contract been performed.

Contract law The law that governs the enforceability of contracts.

Contributory negligence A defense stating that each entity, including the victim, is responsible for the proportion of damages attributable to their own negligent actions.

Conversion The wrongful taking of another's personal property.

Copyright Protects writings, drawings, maps, technical drawings, contracts, artwork, musical scores, and other types of materials.

Cost rule States that the damages are the costs to repair the work.

Counteroffer Terminates the offer it has countered.

Covenants or restrictive covenants Limits placed by private entities.

Criminal laws Adopted by societies to protect the society as a whole rather than any particular individual; criminal law is enforced only the government and individuals may not file criminal actions.

Damages Money paid to the injured party; also called relief.

Davis-Bacon Act Requires federally funded construction projects over $2,000 to pay the prevailing wages of the locality in which the project is located.

Defamation per se Does not require the victim to prove damages; the jury may award the victim nominal and/or punitive damages even without proof of actual or compensatory damages.

Defamation The intentional telling of a falsehood about another that causes him injury.

Defense A legally recognized excuse or a method through which a party can limit its liability.

Deponent Person giving the deposition.

Deposition A face-to-face meeting between the attorneys, a potential witness (party or not), and in the presence of a court reporter.

Design-build A form of construction project delivery where the owner retains one business entity to perform both the design and the construction of the project; compare traditional form of project delivery method.

Design specification A specification telling the contractor exactly what to do and leaving the contractor little latitude except as to means and methods.

Differing site condition or **type 1 differing site condition** A condition at the site that varies from what the plans, specifications, or other contract documents state or picture.

Direct threat A significant risk to the health or safety of others that cannot be eliminated by reasonable accommodation.

Disabled A physical or mental impairment that substantially limits one or more of the major life activities of an individual. The following are not considered disabilities under the Americans with Disabilities Act and people with these characteristics are not protected by that law: drug abuse or addiction, transvestitism, transexualism, pedophilia,

compulsive gambling, kleptomania, and pyromania.

Discrimination To make a choice between two or more alternatives based on some criteria or standard. Every law discriminates between legally acceptable and legally unacceptable behaviors. Two types of discrimination are recognized under employment law for employment-related decisions: legal discrimination and illegal discrimination.

Disparate impact discrimination Some employment-related requirement or discrimination that on its face appears to be neutral but actually affects those in a protected class differently or disproportionately.

Doctrine of objective impossibility States that if something is impossible to do, it need not be done, and the failure to do it cannot be a breach of contract.

Doctrine of practical impossibility or **commercial senselessness** States that at some point, when the cost of performance becomes ridiculous, performance is excused.

Doctrine of respondeat superior or **vicarious liability** Says employers are responsible for the torts of their employees.

Doctrine of sovereign immunity The government cannot be sued.

Doing business as certificate or a **fictitious business name statement** A government document, usually filed in the office of the county recorder where the business entity is doing business; it records the name under which the entity is doing business and the legal name of the business entity.

Double jeopardy The legal principal that a person can only be subject to criminal prosecution once by any sovereign.

Duty of good faith Exists in almost all contracts and requires the parties not to engage in acts that prevent the other party from receiving the benefit of the contract.

Easement Allows an entity to use the land of another.

Economic loss rule A party with only contract damages cannot sue for tort damages.

Economic waste Exists when the costs of repair materially exceed the value of the project, then the project should not be repaired.

Eichleay damages Recoverable damages include the amount of overhead that must be transferred to or born by other projects when a construction contract is wrongfully terminated or delayed.

Embezzlement The misappropriation of the property of a principal or employer.

Employment-at-will doctrine States that the employer may hire or fire an employee for any reason or no reason. In addition, the doctrine states the employee may quit for any reason or no reason. Many exceptions to the doctrine exist.

Employment law Law relating to the employment relationship.

Entity A general term which can refer to a person, partnership, joint venture, limited partnership, corporation, nonprofit corporation or group, church, limited liability company, professional corporation, a government or part of a government such as an administrative agency.

Exculpatory clause A clause in a contract that attempts to transfer liability from a party with power or control of a situation to one who has little or no power or control.

Exempt property Property on a list that all states have that lists properties that cannot be seized for the debts of a person.

Expert witness One who has specialized knowledge in the area, for example an engineer, architect, or contractor.

Express power Power specifically given to the agent in the agency contract.

Express warranty An assertion that a product or service will meet a certain standard and is a contract

Factual issue *See* issue of fact

Fair Labor Standards Act of 1938 (FLSA) Requires employers to pay employees a minimum wage and to pay certain employees overtime pay for work in excess of 40 hours per week.

False light Offensive publicity that attributes to the plaintiff characteristics, conduct, or beliefs that are false, such that the plaintiff is placed before the public in a false position.

Family Medical Leave Act (FMLA) Requires employers to give employees up to 12 weeks of unpaid leave per year for various family-related problems such as birth of a child or serious illness of child or other family member.

Fee simple or **fee simple absolute** The owner or owners have the most extensive interest in the currently recognized by the law.

Fiduciary duty The agent must always act in the principal's best interest and never in the agent's interest.

Field change A change order made on the site without following the procedures outlined in the contract.

Firm offer An offer by a supplier that cannot be revoked if it states it will be kept open, is in writing, and is signed by the supplier.

Fraud An intentional misstatement or failure to disclose information that leads to damage.

Friable materials Applied to concrete, masonry, wood and nonporous surfaces, including but not limited to, steel structural members (decks, beams and columns), pipes and tanks, shall be cleaned to a degree that no traces of debris or residue are visible.

Future interest or remainder interest Interest that becomes possessory at some point in the future.

General partnership A business entity operated by two or more entities for profit.

Goods All forms of tangible personal property including specifically manufactured goods, supplies, mobile homes, and materials.

Hostile work environment Exists when the work environment is characterized by severe or pervasive sexually offensive conduct.

Illegal discrimination Discrimination based on race, creed, sex, religion, national origin, citizenship, disability, pregnancy, union membership, or age.

Implied contract Formed by the acts of the parties, rather than by words.

Implied covenant to cooperate or the implied covenant not to hinder In the construction industry, the implied term of good faith.

Implied power Power to do the things necessary to carry out the express powers but not specifically outlined in the contract.

Implied warranty A warranty the law will imply into the transaction or contract whether or not the parties have so agreed.

Improvement bond Assures the buyer that the contractor has complied with wetlands improvement or other required improvements.

Incorporators Persons who file the required forms with the state to begin the formation of a corporation.

Inexcusable delay The contractor has no excuse and the contractor is responsible for paying the owner's damages.

Injunction A court order ordering one party to do or not to do something.

Insurance Provides a fund in the event of a natural disaster or fire that damages some type of property.

Interrogatories Written questions from one party to another but cannot be sent to third parties not involved in the lawsuit.

Intestate succession State law outlining who receives a person's property if they die without a valid will.

Intrusion into personal affairs or areas Arises when one party invades the personal space of another or intrudes into an area where that party has no legitimate interest.

Invasion of privacy Includes four different torts: intrusion into seclusion, false light, public disclosure of private facts, and appropriation of another's name or likeness.

Issue of fact A question dealing with what happened at some point in the past; asks "who, what, when, where, why, or how?"; decided by a jury in a jury trial or by the trial court judge in a bench trial.

Issue of law A question dealing with the meaning of a law or which law(s) apply to a situation; always decided by a judge.

Joint and severable liability Each liable entity is individually responsible for its individual proportion of the debt and also liable for all of the debt should other liable entities be unable to pay.

Joint liability Liability by two or more entities for the same injury; each party is liable for their portion of the damages.

Joint tenants or **joint tenancy** or **joint tenancy with right of survivorship** Two or more entities who own a piece of real estate as co-owners; co-owners have equal rights to the use and enjoyment of the real estate but do not have the power to transfer their share to third parties; the share of a deceased joint tenant automatically goes to the surviving joint tenant(s) and does not pass by the deceased joint tenant's will or by intestate succession.

Joint venture A temporary business relationship between two entities and having the same legal effect as a partnership.

Judge-made law When appeal court judges clarify the meaning of a constitution, statute, regulation, or prior case, that clarification is a law, and the name of that law is case law, judge-made law, or common law.

Judgment A piece of paper signed by the judge outlining who wins and what they win.

Jurisdiction A particular geographic area over which a particular government has power.

Jurors A panel or group of citizens who decide factual issues in a trial.

Jury trial A trial in which a panel of citizens, called jurors, decides the factual issues of the parties' dispute; the legal issues are decided by the judge.

Labor law Law relating specifically to unions and the right to unionize.

Larceny The wrongful taking of the property of another.

Latent ambiguity A hidden or unobvious ambiguity.

Law of sales Also called the Uniform Commercial Code or UCC; it is a set of statutes developed by experts and adopted by almost every state with minor modifications.

Legal discrimination Discrimination based on any reason that does not fit into the category of illegal discrimination; discrimination based on experience, education, family relationship, congeniality, are forms of legal discrimination.

Legal issue See **issue of law**.

Libel Any form of defamation that is printed, recorded, digitalized, or preserved in any format that allows the defamation to be copied or reproduced.

License Allows another to use the land for a limited purpose, such as temporary parking.

Licensing bond Assures the owner the contractor will comply with all licensing requirements of the applicable governments.

Lien A legally recognized obligation of the owner of a particular piece of property to pay an amount to the lien holder.

Lien laws Allow general contractors, subcontractors, material suppliers, and, in some states, laborers to file a lien on the owner's property for amounts owned to them for work done on the owner's property.

Life estate An estate that terminates upon the death of the holder of the life estate.

Limited liability business entity A business entity whose owner's have no personal liability for the business entity's debts.

Liquidated damages A specific amount of damages, usually a specific amount per day, to which the injured party is entitled; liquidated damages are in lieu of actual damages.

Material breach A very major breach.

Maxim of law Old and fundamental legal principles upon which much of the judge-made law is based.

Mediation A process used to resolve disputes; in this process a person, called a mediator, helps the parties resolve a dispute and does not force any outcome on the parties.

Meeting of the minds Requirement that both parties to a contract have the same understanding of what the contract means before a contract can be formed.

Merchant One who is normally engaged in the selling of goods of the type in the transaction.

Modern or foreseeability rule The ultimate users can sue contractors, subcontractors, or suppliers for negligence.

Motion to dismiss A motion requesting the judge to review the law and evidence and determine if the matter must go to trial.

Mutual mistake A mistake made by both parties as to some fundamental aspect of the contract; no contract is formed if a mutual mistake has been made.

Navigable waterways Broadly defined and include all waters of the United States, territorial seas. Includes all navigable waters of the United States, intrastate lakes, rivers, and streams that are utilized by interstate travelers for recreational or other purposes such as harvesting fish or shellfish sold in interstate commerce, tributaries of such waters, and dry drainages with an ordinary high-water mark that are eventually tributary to any interstate water.

Negligence Failure to act reasonably in a particular situation with resulting injury to some entity.

Negligence per se A special category of negligence and applies when a victim is injured because of the tortfeasor's failure to follow a statute or regulation.

Negotiation The attempt to resolve a dispute without outside or professional help; the parties discuss the dispute and come to a resolution.

Nominal damages Small amount of damages.

Nonconforming use A particular landowner's use that continues after zoning regulations require a different use.

Nonexempt property All of the property owned by the individual that does not appear on the exempt property list.

Nonfriable materials Applied to concrete, masonry, or wood shall be cleaned until no residue is visible other than that which is embedded in the pores, cracks, or other small voids below the surface of the material.

Nonpossessory rights Rights to use the land belonging to another.

Nuisance Arises when an owner of real property uses the property in such a way as to unreasonably interfere with neighbors' use and enjoyment of their own property.

Offer An expression of a willingness to enter into a contract; it must be definite and certain as to the major components of the proposed contract.

Onerous clause A clause in a contract that attempts to transfer liability from a party with power or control of a situation to one who has little or no power or control.

Parol evidence Evidence outside of or extraneous to the actual words of the contract.

Parole A criminal law term; used to describe a time period after a criminal has been released from jail but is still under the scrutiny of the criminal justice system.

Partnership Often used to mean a general partnership as compared to a limited partnership.

Patent ambiguity An ambiguity obvious to a reasonable contractor or owner.

Patent Open or obvious.

Patents Government protection for machines and processes.

Payment bond Assures the owner that the contractor will pay the subcontractors and suppliers.

Perfecting a lien To carefully follow the requirements of state law for enforcing the rights under the lien.

Performance or completion bond Assures the owner the contractor will complete the job.

Performance specification A more general specification detailing the desired result and leaving it to the contractor and/or specialty subcontractor to design the system to meet the desired result.

Plaintiff The party starting the lawsuit.

Plan A pictorial or visual drawing of a particular part of a construction project.

Prejudgment interest Interest on the amount of the judgment back to the date of the breach of contract.

Preliminary negotiation A statement that does not rise to the level of an offer but begins negotiation of a contract.

Premises liability If the owner fails in this duty to maintain a safe structure and a user of the structure is injured, the owner is liable.

Prevailing wages Standard wages in the area and are determined by the federal government.

Principal An entity giving power to act for the principal to another.

Procurement laws and **false claims laws** Designed to help the government entity obtain competitive rates for contracts, prevent corruption and favoritism, and prevent fraud in government contracting.

Product misuse A limited defense; if the user misuses the product and is injured, the manufacturer and/or seller may not be liable.

Professional malpractice A specialized term for negligence committed by a professional.

Profit à prendre Allows an entity to take something, such as gravel, from the land of another.

Promissory estoppel A promise which the promisor (the party making the promise, that is, the subcontractor) should reasonably expect to induce action or forbearance on the part of the promisee (the party to whom the promise is made, that is, the contractor) or a third person and that does induce such action or forbearance is binding if injustice can be avoided only by enforcement of the promise.

Protected class A term used in employment law to refer to people in the group that is protected from illegal discrimination.

Proving causation Requires the owner to prove the damage was caused by some mistake of the contractor's.

Proximate cause damages or **tort damages** All of the legally recognized losses incurred by a party injured by a tortfeasor; includes some foreseeable and some unforeseeable damages.

Public disclosure of private facts The dissemination of private or personal information about a person without their permission.

Public policy A broad term for socially but possibly not yet legally recognized, norms of a society; public policy changes over time and therefore may change the law.

Public policy exceptions Additional exceptions to the employment at will doctrine and based upon the prevailing dictates of a society.

Publicly traded corporation A corporation whose stock is bought and sold on exchanges such as the New York Stock Exchange.

Punch list A list of minor defects or problems in the construction that the contractor needs to fix or remedy prior to receiving final payment or any retainage.

Punitive damages Damages above and beyond compensatory damages and are designed to punish the tortfeasor.

Quid pro quo sexual harassment A demand for sexual favors in exchange for a job-related benefit.

Reform or reformation To change or revise.

Regulations Laws enacted by administrative agencies.

Relief and **remedy** Terms used for the list of things, such as attorney fees or punitive damages, a court can award a winning party; examples of relief or remedies include damages, injunctions, and specific performance.

Request for documents or things Sent to another party and requires that party to produce documents or other things for review or copying.

Requirements contract Obligates the buyer to purchase its requirements for a certain job from the seller.

Res ipsa loquitur Latin for "the thing speaks for itself" and, in limited cases, permits a jury to infer negligence merely from the fact the injury occurred and without proof of any unreasonable act on the part of the alleged tortfeasor.

Rescind or **rescission** Take back or nullify.

Retainage Amount due the contractor from a particular invoice but held by the owner and not paid until the end of the project.

Rule of law A method of operating a government; under this method, the law, rather than the wealth or influence of an entity, controls governmental processes; alternative methods for operating a government include theocracy, dictatorship, military government, and the outcomes of a dispute.

Severable liability Each entity is individually liable for the entire debt should any of the other liable entities not have sufficient assets to cover their portion of the debt.

Slander Spoken defamation or defamation that is not recorded in any way.

Sole proprietorship A business entity operated by one person.

Sovereign immunity An old common law doctrine that originally held no one could sue the king or sovereign.

Spearin warranty, Spearin doctrine, or the owner's warranty of the plans and specifications Liability for defective construction caused by a defective design specification rests with the owner. If the contractor has constructed the system according to the design specification, the contractor has no fault or liability for failure of the system to perform.

Special damages or **consequential damages** Damages suffered by a party in a breach of contract action due to some unique characteristic of the injured party or because of some unique situation of the injured party known to the breaching party.

Special use permit. Granted to an entity after an application is made.

Specific performance A court order to perform a contract.

Specification A written description of a particular part of a construction project.

Statute of frauds An old term used for a particular list of contracts that must be evidenced by a writing signed by the person to be charged in order to be upheld.

Statute of limitation A statute that states how long, after an event giving rise to a lawsuit occurs, the lawsuit can be filed.

Statutes Laws passed by a legislative body.

Subchapter S corporation A corporation given special tax status by the federal government; the corporation is taxed as a partnership by the federal government and not as a corporation.

Subrogation lawsuit Sureties are allowed to see reimbursement from the defaulting contractor for any payments it makes on a bond claim.

Summary adjudication A motion requesting the judge to review the law and evidence and determine if the matter must go to trial.

Summary judgment A motion requesting the judge to review the law and evidence and determine if the matter must go to trial.

Surety An entity, similar to an insurance company, that agrees to pay the owner the amount of a bond should the contractor fail to perform as required by the bond.

Tangible employment action An action taken by an employer as the result of a claim of sexual harassment; it includes firing, pay reduction, relocation to a less desirable job, or similar acts.

Tenancy by the entirety A type of co-ownership of real property similar to joint tenancy except that the co-owners must be married. Upon the death of one of the spouses, the living spouse becomes the sole owner of the real property. Limitations exist on each spouse's ability to encumber the property.

Tenants in common or **tenancy in common** Two or more entities who own a piece of real estate as co-owners; co-owners have equal rights to the use and enjoyment of the real estate and also the power to transfer their share by will or if no will is left, the property goes by intestate succession.

Theft The wrongful taking of the property of another.

Tort damages or **proximate cause damages** All of the legally recognized losses incurred by a party injured by a tortfeasor; includes foreseeable and some unforeseeable damages, although at some point unforeseeable damages become so remote as to not be recoverable.

Tort law The law dealing with noncontract injuries. The most common tort is negligence.

Tortfeasor A person who violates a legally recognized noncontract duty to another; is liable to the injured party for damages.

Trade secret A piece of confidential information owned by a business that gives it a competitive advantage over others who do not know it.

Trademark A name, picture, or drawing that distinguishes a product or service from other products or services.

Traditional form or **design-bid-build** A form of construction project delivery where the owner retains one business entity to design the project, the project is put out to bid and the owner then hires another business entity to construct the project.

Trespass to land The invasion of the land of another without permission.

Trespass to personal property The interference with the owner's right to the use and enjoyment of personal property such as cars or tools

Trial A dispute resolution process provided by a government. The parties present their dispute at trial and a resolution of the dispute is imposed upon the parties. Two types of trials exist: jury trials and bench trials.

Trial court judge The judge who presides over the jury or bench trial; only determines issues of law in a jury trial; determines issues of law and fact in a bench trial.

Type 1 differing site condition or **differing site condition** A condition at the site that varies from what the plans, specifications, or other contract documents state or picture.

Type 2 differing site condition or **unforeseen site condition** A condition at the site that was unforeseen or unexpected; this term is often specifically defined in the construction contract.

Unforeseen site condition clause A contract clause that transfers the risk of an unforeseen site condition onto the owner.

Unforeseen site condition or **type 2 differing site condition** An unexpected condition at the site; this term is often specifically defined in the construction contract.

Uniform Commercial Code (UCC) Also called the law of sales; it is a set of statutes developed by experts and adopted by almost every state with minor modifications.

Unilateral change A change made by one of the parties to contract and under traditional contract law is not allowed. Unilateral changes are allowed in construction contracts as long as the owner pays the contractor a reasonable sum for the change.

Unilateral mistake A mistake made by one party to the contract; it does not nullify the contract.

Unlimited liability business entity A business entity whose owners have unlimited personal liability for all of the business entity's debts.

Value rule A rule of damage law that states the damages recovered by the injured party are only the diminution in the value of the project due to the defects not the cost of repair; only applicable when the cost to repair is extraordinarily high as compared to the diminution of the value of the project.

Variance An exemption or change to a zoning regulation given to a particular landowner.

Voidable A contract that can be performed but one or both of the parties has the option of not performing without legal repercussions.

Waiver of lien rights A document signed by a potential lien claimant waiving the potential lien claimant's right to file a lien.

Waiver The knowing relinquishment of a known right.

Whistleblowing Informing government authorities when the employer or its agent has broken the law or intends to break the law.

Withdrawal of an offer or bid The law states that an offer or bid can be withdrawn at any time prior to acceptance.

Workers' Compensation law State government law that requires or encourages employers to obtain medical coverage for employees injured on the job; not all states require Workers' Compensation.

Wrongful discharge or **wrongful termination** Term used to cover an employee's illegal termination; the employee is entitled to damages when this occurs.

Zoning regulations Laws that limit the allowable uses of land.

INDEX

C

E

P

Q